ECONOMICS OF RESOURCES, AGRICULTURE, AND FOOD

McGraw-Hill Series in Agricultural Economics

Erickson, Akridge, Barnard, and Downey: Agribusiness Management
Ferris: Agricultural Prices and Commodity Market Analysis
Kay and Edwards: Farm Management
Schaffner, Schroeder, and Earle: Food Marketing Management: An International
  Perspective
Seitz, Nelson, and Halcrow: Economics of Resources, Agriculture, and Food

ECONOMICS OF RESOURCES, AGRICULTURE, AND FOOD

SECOND EDITION

Wesley D. Seitz
Gerald C. Nelson
Harold G. Halcrow
All University of Illinois, Urbana-Champaign

Boston   Burr Ridge, IL   Dubuque, IA   Madison, WI   New York   San Francisco   St. Louis
Bangkok   Bogotá   Caracas   Kuala Lumpur   Lisbon   London   Madrid   Mexico City
Milan   Montreal   New Delhi   Santiago   Seoul   Singapore   Sydney   Taipei   Toronto

McGraw-Hill Higher Education

*A Division of The **McGraw-Hill** Companies*

ECONOMICS OF RESOURCES, AGRICULTURE, AND FOOD
SECOND EDITION

Published by McGraw-Hill, a business unit of The McGraw-Hill Companies, Inc., 1221 Avenue of the Americas, New York, NY 10020. Copyright © 2002, 1994 by The McGraw-Hill Companies, Inc. All rights reserved. No part of this publication may be reproduced or distributed in any form or by any means, or stored in a database or retrieval system, without the prior written consent of The McGraw-Hill Companies, Inc., including, but not limited to, in any network or other electronic storage or transmissions, or broadcast for distance learning.

Some ancillaries, including electronic and print components, may not be available to customers outside the United States.

This book is printed on acid-free paper.

1 2 3 4 5 6 7 8 9 0 QPF/QPF 0 9 8 7 6 5 4 3 2 1

ISBN 0–07–025958–5

Publisher: *Edward E. Bartell*
Marketing manager: *Heather K. Wagner*
Project manager: *Richard H. Hecker*
Production supervisor: *Kara Kudronowicz*
Coordinator of freelance design: *Michelle D. Whitaker*
Freelance cover designer: *Lisa Gravunder*
Supplement producer: *Tammy Juran*
Media technology producer: *Judi David*
Compositor: *Interactive Composition Corporation*
Typeface: *10/12 Times Roman*
Printer: *Quebecor World Fairfield, PA*

Library of Congress Cataloging-in-Publication Data

Seitz, Wesley D.
    Economics of resources, agriculture, and food / Wesley D. Seitz, Gerald C. Nelson,
Harold G. Halcrow—2nd ed.
        p.   cm—(McGraw-Hill series in agricultural economics)
    Includes index.
    ISBN 0–07–025958–5
    1. Agriculture—Economic aspects.   2. Agricultural industries.   3. Agriculture and state.
I. Nelson, Gerald C. II. Halcrow, Harold G. III. Title. IV. Series.

HD1415.S395   2002
338.1—dc21

                                                        2001031266
                                                        CIP

ABOUT THE AUTHORS

WESLEY D. SEITZ is Professor and Director of Graduate Programs, Department of Agricultural and Consumer Economics at the University of Illinois, Urbana-Champaign. He was also a professor in the Institute for Environmental Studies. From 1968 to 1971 he was a member of the faculty of the College of Commerce where he taught marketing and pricing. After serving six years as head of the Department of Agricultural Economics he assumed responsibility, with Professor Nelson, for the revitalization of the introductory course in agricultural economics. Out of this effort came the award-winning educational software developed by Nelson and Seitz as well as this text. He is currently teaching the introductory course in agricultural economics using these materials. His research has focused primarily on the economics of soil conservation, with emphasis on policies to control the off-site damages of sediment. He has served on and chaired National Research Council task forces and the International Resource Policy Consortium. He served as chair of the Faculty Senate during the period of the adoption of a revision of the general education requirements at the University of Illinois at Urbana-Champaign.

GERALD C. NELSON is Associate Professor, Department of Agricultural and Consumer Economics at the University of Illinois, Urbana-Champaign. He teaches courses ranging from the introductory agricultural economics course to graduate courses in economic development. His research investigates the role of agricultural and macroeconomic policies in fostering development of third world countries. Professor Nelson has served as advisor to international aid agencies and national governments in Southeast Asia. In addition to teaching and research, Professor Nelson has been actively involved in producing educational software. He led the development of the AECONIntro software, an EDUCOM Distinguished Social Sciences Software (Economics) award winner in 1992. Professor Nelson's recent research interests include the economics of genetically modified organisms and using spatial analysis to assess environmental externalities associated with development projects.

HAROLD G. HALCROW is Professor Emeritus of Agricultural Economics at the University of Illinois, Urbana-Champaign, where he has taught and conducted research since 1957, and was head of the department for thirteen years. He has also served on the faculties of Montana State University and the University of Connecticut. He was a visiting professor at Stanford and University of California, Berkeley. He is author or coauthor of ten books, editor of two books, and has contributed a large number of research publications, journal articles, and extension reports. He served as book review editor and then editor of the *Journal of Farm Economics,* predecessor to the *American Journal of Agricultural Economics.* He is a Fellow of the American Agricultural Economics Association, and has been cited five times by the AAEA and the former American Farm Economic Association for Professional Excellence in research and communication. He served for twenty years as the AAEA representative on the Board of Directors of the National Bureau of Economic Research and has served as a consultant to the U.S. Departments of Agriculture and Commerce, and the Illinois Economic Technical Advisory Committee.

CONTENTS IN BRIEF

CONTENTS

PREFACE

TO THE INSTRUCTOR AND THE STUDENT

We live in exciting times! The information technology revolution has profoundly reshaped the international political arena, and we have only begun to see the shape of a new world order. New political realities are revolutionizing our world. These political changes have accelerated the restructuring of international economic relations that was already under way. International trade negotiators struggle to lower world trade barriers, regional trading blocs assume new importance, and many bilateral agreements are being consummated. These changes are altering the structure of international transactions in ways no one could have imagined.

Astute observers realize that these political and economic changes will recast the way the agricultural system uses resources to produce food and other products. In addition, however, we are in the throes of a biotechnology revolution that will alter the world as dramatically as did the Industrial Revolution. Although this scientific revolution holds great promise for agriculture, it also raises fears of environmental and health effects and has stimulated vigorous debate.

Today, every farmer and agribusiness competes in a world market. In this dynamic environment, agriculture must be both constant and revolutionary: constant, because of the ongoing challenge of supplying adequate food and related products to an ever-growing world population; revolutionary, because of the new research and production techniques that test the abilities of the best managers. At the same time, advanced communication systems, sophisticated financial markets, and new trading relationships enhance the interconnectedness among nations and businesses around the world.

With agricultural horizons expanding in several dimensions, agriculture can no longer be equated with farming. Yet the view of the farm as a unique economic entity remains intact. The image of the self-sufficient family farm with a red barn, a dairy herd, and a flock of chickens dies hard, even though such farms have not been typical in American agriculture for decades. A "typical" twenty-first century family farm in the United States has a million dollars in assets and a family income that is often above the national average. One must look to the developing world for "traditional" family farms, yet even there the pace of agricultural change is rapid.

While the boundaries of agricultural concerns have long included input suppliers and commodity processors, ever-increasing numbers of the decisions controlling the production of food are made in the boardrooms of large, multifaceted, multinational corporations. Agribusiness is playing an increasingly important role in agricultural production.

Perhaps a more important concern when considering the performance of agriculture is that providing adequate food is not enough. To the centuries-old concern of farmers and agricultural policymakers for protecting the productivity of the soil, environmentalists have stimulated a pervasive concern for the quality of the resource base, the human environment, and the natural habitat. One result has been the dramatic reshaping of the agricultural policy setting. What was once "farm policy" is now a comprehensive "resource, agriculture, and food policy" set in an international context.

All of these developments have made the challenge facing individuals beginning their careers as agricultural economists both more demanding and more exciting than ever before. We hope this book will introduce many young people to this fascinating arena.

An Introduction to the Contents

This book was developed to present the full scope of basic economic theory, which we illustrate with examples from resource, agricultural, and food issues of today. Students who study this book will be prepared to understand the basis for most economic and management decisions. They will have a firm foundation for more advanced work in economics, agricultural economics, and related courses in agricultural and allied disciplines. They will approach major national and international issues with greater confidence in their ability to understand them.

The first task of this text is to present the basic concepts of **microeconomics.** An understanding of how individual producers and consumers make economic decisions is fundamental. However, an understanding of the elements of **macroeconomics** is also a key to preparing new economists to deal with complex issues. With these basic tools, students will be ready to appreciate the economic aspects of the major economic issues facing agriculture. Throughout the book international examples are presented to remind students of the scope of agriculture and the pervasiveness of the economic approach to problems.

The first section of the book begins with an introduction to economics and the concepts of resources, agriculture, and food. The second section covers the microeconomic concepts of demand, production, and supply. We discuss demand first because readers can easily relate the concepts to their personal experiences as consumers. Next, the ideas of production, costs, supply, and revenue are developed to demonstrate how firms maximize profits. We then show how demand and supply interact in markets to determine equilibrium prices and quantities.

Market concepts make up the third part of this book. The main topics include the forms of market competition, international trade, the effects of trade policy, and agribusiness management and economic performance.

The fourth major section deals with macroeconomics. It covers the organization and performance of the economy, fiscal and monetary policy, and the effects of macroeconomic policies on international trade. Together, the discussions of microeconomics and macroeconomics provide a solid foundation for the fifth and final section, in which we discuss the role of agriculture in economic development, the basis for agricultural price and income policy, and public issues in resource and environmental policy.

Modifications in the Second Edition

In response to feedback from users and our experience in the classroom we have revised the first edition. The revisions include:

- Combining the discussion of the elasticity of supply and demand and the determinants of supply, demand, and elasticity with the chapters presenting demand and supply.
- Combining all of the material on resource and environmental economics into a single chapter.
- New student exercises at the end of each chapter.
- A brief section on mathematical tools and graphing.
- A section listing a number of web sites that students may find helpful in seeking additional information.
- A re-write of much of the text to improve readability.

Intended Audience

This text is intended for students who are interested in learning the principles of economics. It is particularly appropriate for students enrolled in colleges of agriculture and in some of the biological and physical sciences in liberal arts, where a familiarity with agriculture will help enliven the examples. Students from all disciplines will benefit from this text, but especially those who want to learn the principles of economics and something about agriculture. Because of the dynamic nature of agriculture and the economy, we have avoided using economic data series and currents events that quickly become dated and thus dampen the student's interest.

Alternative Uses of the Text

The introductory agricultural economics course plays different roles in different institutions. This text can serve in several of these roles:

- The entire text can be used in the typical introductory economics course to cover microeconomic principles, to present a solid introduction to macroeconomics, and to discuss important agricultural policy issues. Coverage of all this material might call for a four-hour, semester-long course.
- The macroeconomics section (Chapters 11 to 14) can be omitted if the goal is to provide a basic microeconomics course from an agricultural perspective.
- The policy section (Chapters 15, 16, and 17) can be omitted for a course in basic microeconomics, macroeconomics, agribusiness, and international trade.
- The more advanced discussions of microeconomics (Chapter 6), macroeconomics (Chapter 14), and perhaps agribusiness (Chapters 9 and 10), can be omitted to provide an introduction to microeconomics and macroeconomics and an appreciation of important agricultural issues.

Acknowledgments

The authors gratefully acknowledge the help of many people in preparing this manuscript, especially the reviewers, who provided many extremely helpful comments, but also the students who have used earlier versions of the book and made valuable suggestions. Of course any errors that remain are ours. We thank our spouses, Janice, Laurian, and Eleanor, for their patience as we worked through many revisions.

PART **ONE**

ECONOMIC SCOPE, ORGANIZATION, AND PROBLEMS OF AGRICULTURE

1

INTRODUCTION
TO THE ISSUES

CHAPTER OUTLINE

OVERVIEW

This book develops the basic principles of economics and illustrates them with examples of resource and environmental management, agricultural production and marketing, agribusiness management and consumer economics. We address three questions: What is economics? Why concentrate on resources, agriculture, and food? Why do we address economics and the economic problems of resources, agriculture, and food?

LEARNING OBJECTIVES

This chapter will help you learn:

- The basic concepts of economics
- How to visualize the economic scope and organization of agriculture
- Why we study the economic problems of resources, agriculture, and food

ECONOMICS

What is **economics,** and why is it important that we study it? The term has many definitions. Simply defined, *economics is the study of how to allocate limited resources to produce goods that help satisfy unlimited human wants.* This brief definition highlights the three important aspects of economics: limited resources, goods produced from resources, and human wants that are partly met by consuming goods. When we (and other economists) refer to goods, we generally mean both physical things as well as services such as watching a movie or financial consulting. For convenience we will refer to "goods" rather than to "goods and services." The study of economics deals with *what* will be produced from the limited resources available, *how* it will be produced, for *whom* it will be produced, and *when* it will be consumed. It is truly amazing that independent decisions of millions of individuals and organizations come together to give us a functioning economy! *Agricultural economics* focuses on the decisions relating to the production, distribution, and consumption of food, fiber, and related products. *Resource and environmental economics* focuses on the management of the land, water, and air resources from the perspectives of their use, conservation, and protection.

Resource Scarcity

The first element in the definition of economics is *scarce productive resources*. Because resources are an important part of what economics is all about, we need to have a clear understanding of what we mean by resources. A **resource** is an input provided by nature and modified by humans using technology to produce goods that satisfy human wants. Resources include land, labor, equipment, and machines (sometimes called capital goods), and also mineral and vegetable resources such as coal, iron ore, forests, and water. Combining these resources through human activity and technology produces useful outputs. Resources are sometimes called **factors of production** because they are necessary to produce goods.

One common characteristic of resources is that they have economic value. A producer generally has to pay to use them. Directly or indirectly you pay for all of the resources used to produce all of the goods you buy. You also are paying for the use of a resource when you pay for the service of having your trash landfilled or incinerated. Further, you would be imposing an environmental cost on society if you discarded your wastes by tossing them in a river or on the land and creating a pollution problem. But what about your use of the air? Although air is used in the production of many goods, the producers normally do not pay anything for its use. In this context it is a free resource. We can all use as much as we want. Of course, clean air is not always free. Industrial production, transportation, and other economic activities pollute air. Nations and individuals must choose how much they are willing to pay to keep the air clean.

Unlike air, *the supply of most resources is limited.* They are scarce, especially in useful form. And because there is a scarcity of resources, goods produced from them are also scarce. The limited supply of diamonds means that we cannot have as many large diamonds as we would like. There is a limited supply of the materials and workers needed to produce useful goods such as food, automobiles, stereos, and airplanes. Therefore the supply of these goods is inadequate to provide people with as much as they would like.

Alternative Uses

Another important characteristic of scarce resources and the goods made from them is that they have **alternative uses.** Steel can be used to make cars, planes, railroad tracks, or ships (alternative uses in production). Gasoline is used either for pleasure or for work. A hamburger can be eaten by you or by me (alternative uses in consumption). The scarcity of goods requires choices or tradeoffs by individuals and society: they choose among goods that satisfy various wants and desires. You will buy a good only if its value to you is greater than or equal to the price of that good.

Opportunity Costs

If you use a good for one purpose, you can't put it to any other alternative use. You give up the opportunity to use it elsewhere. The value of what you give up is a measure of the cost of its use. This value in an alternative use is called **opportunity cost.** For example, to go to college you must give up the opportunity to do something else. The value to you of the next best way to spend that time is your opportunity cost of going to college. For the agribusiness that chooses to introduce a new product, the opportunity cost is the return that could have been earned if a different product had been introduced. If a farmer uses a tractor for one job, the opportunity cost is the return that could be earned using that tractor in its best alternative use. If you accept an invitation to dinner with Raoul or Celeste you forgo the opportunity to have dinner with Jason or Sasha. The cost of choosing to do one thing is the value of the best forgone opportunity.

Unlimited Wants and Value

Humans seek to improve their material well being. We all have different wants; together we have unlimited wants. You will soon understand how these individual and group wants are reflected in the economy as demand for goods.

The value to an individual, to you, of a good has something to do with your desire to have it. What does it mean for "society" to value something? How does a nation decide that a good is worth so many dollars, yen, or pesos? Why does gasoline cost much less per gallon in the United States than in many other countries? In the early 1990s, the European Union (EU) faced a difficult choice as it responded to pressure from other countries to reduce its agricultural subsidies. Although EU consumers and industries outside agriculture would benefit in the long term, as would foreign countries, many EU farmers would suffer losses. How does a society weigh the income losses of EU farmers against the benefits to consumers and industries outside agriculture?

Allocation over Time

Time is another important element in economic decisions. Someone must decide whether to use a resource today or in the future. For example, once oil is pumped from the ground and burned, it is gone forever. A tree cut down for timber today cannot be logged in the future, although a sapling can be planted to replace the tree. Our ancestors did not (or could not) use all the natural resources available to them. Nor did they consume all the goods they produced. Instead, they saved and invested their savings in capital goods—buildings and equipment used to increase productive capacity in the future. You are delaying your entry into the world of employment at least in part because an education now will mean higher earnings in the future. (A wise decision even if you do not consider the fun you will have during the four years.)

Distribution

Another concern of economists is the **distribution** among various persons and groups in society of the goods for consumption. One way to think about distribution is to imagine the economy as a pie. Every person in the economy gets a slice, but some slices are larger than others. One branch of economics studies what determines the size of the slices. An example from agriculture is the distribution of food. Far too often malnutrition or famine in some part of the world is the subject of newspaper headlines. What causes hunger? Economic and political conflicts are often at the root of the problem. In some cases economists can help identify causes and suggest solutions.

Uses of Economics

The study of problems related to the production and distribution of goods involves many specialties besides economics; and economics itself has many sub-specialties. For example, some economists study a particular sector of the economy such as agriculture or mining. Others study natural resources, labor, consumers, policy formation, business decision making, finance, or institutional structures. Still others such as marketing economists and international trade economists study certain economic transactions. All economic analysis, however, falls into two broad categories: microeconomics and macroeconomics.

Microeconomics is the study of individual economic behavior. For example, a firm's decision about the quantity of a good to produce and its sale price is a microeconomic decision. Similarly, your decision on how to spend a weekly budget is a microeconomic decision. In microeconomic analysis we assume that factors outside the immediate topic of study (such as prices of other goods and income) do not change.

Macroeconomics considers the performance of a nation's economy and the international economy. At its most basic, macroeconomics is the study of determinants of the total quantity of goods produced and the distribution of income in an economy. Unemployment and inflation are key macroeconomic issues. Both are related to economic factors such as the rate of economic growth, the expansion of the nation's money and credit, and international transactions.

Positive, Normative, and Prescriptive Economics

We have talked about many different kinds of economic questions. Now we present a central classification scheme—the differences between positive, normative, and prescriptive economics. **Positive economics** deals with questions that do not involve a judgment comparing the differing values individuals place on an option (no value judgments are made). Such questions can be answered without making a judgment about what ought to be. Two plus two does equal four. For instance: Does a decline in the value of a nation's currency tend to reduce that nation's trade deficit? Does an increase in the government's price support level for an agricultural commodity cause farmers to produce more of that commodity? What will be the effect on the inputs used in production? Will a ban on the export of logs from the Philippines reduce deforestation in that country? No value judgments are needed to answer these questions, although accurate answers may be difficult to find.

Normative economics deals with value judgments. To answer normative questions, someone or some group must decide what is good or bad, fair or unfair, or develop some standard for determining what is good or fair. For instance: Should government policy guarantee that farmers get a "fair" price for their grain? What is "fair"? If food prices increase as a result, is that fair to consumers? Does it make a difference if these changes in costs and returns are large or small? Someone must decide whether the increase in value to the gainers is "worth more" than the losses in value to those made less well off. Such questions cannot be answered without some judgment concerning values.

Anyone making an economic decision must deal with both positive and normative questions. When most economists wear their professional hats, they prefer to answer positive questions, leaving normative decision making to other members of society. But economists are often asked to identify costs and returns and to prescribe the cheapest or best way to get a desired result. When economists are doing interesting work that is relevant to society, it is almost always prescriptive—it involves both positive and normative economics.

Prescriptive economics deals with the ways to achieve a desired result in the most efficient, profitable, or acceptable manner. Prescriptive economics identifies alternative ways to reach a goal and provides methods for choosing among them. Three issues are involved: (1) how to get the highest return from a given set of resources, (2) how to achieve a public policy objective at the lowest cost to society, and (3) how to structure a policy to get an acceptable allocation of costs and benefits among various individuals now and in the future. Prescriptive economics involves both positive and normative economic analyses.

Typical Economic Questions

The study of economics deals with many important questions: How can resources be made more productive? How can society be organized to use resources productively and equitably? How can resources be preserved for the future? In short, economics deals either directly or indirectly with many of the basic elements of a productive life.

To illustrate, we consider some everyday examples of questions that are important to almost everyone:

• Why isn't the price of a Big Mac twice, or half, its current price?
• Why did the price of hogs fall to record lows in 1998, and why didn't the price of pork fall significantly?
• Why can I buy a good lunch for half the price of a good dinner at the same restaurant?

Some questions that economists try to answer might not affect you directly, but they are of vital interest to the national economy and ultimately determine your job prospects:

• What explains the drastic changes in the value of the dollar in international currency markets in the last twenty years of the previous century? What should we expect as we begin the twenty-first century?
• What determines inflation rates?
• Why did the same farmland in the U.S. Corn Belt cost $4000 per acre in 1980, $1500 in 1985, $4000 in 1997, and $2500 in 1999? What will cause future price variations?
• How important is rainfall to the price of an agricultural commodity like wheat?
• Will consumers abandon shopping malls for home shopping via the Web?
• Why are most U.S. farms larger than most farms in less-developed countries (LDCs)?
• Why are crop yields on similar land higher in Europe and Japan than in the United States and Australia?

One goal of economics is to answer questions of immediate concern to the nation or to you as a consumer and future producer of goods. But economists also ask more basic questions. Three of the most important are these:

• What determines *value?* Why does a pair of designer jeans cost more than ordinary jeans? Why is the salary of a dot com employee five times that of a worker in a fast-food restaurant? Why does a football coach earn more than a professor or even the president of the university or the country?
• What determines the *distribution* of economic goods? Why are many Americans rich and overweight while many in Africa are poor and starving?
• What determines *growth* in the availability of economic goods? At any time some countries will be growing at nearly 10 percent per year while others grow slowly or even stagnate. Why?

Answering Economic Questions

Economists analyze human behavior in the context of all life's complexities and uncertainties. If economists could set up an appropriate laboratory experiment, as physical scientists do, they could answer positive economic questions with experiments. But most economists could not, in good conscience, experiment by lowering the corn price to see how many farmers go bankrupt or raise the price of milk to measure the health effects on children. Instead, economists generally observe actual conditions and infer underlying relationships from these observations. For example, the adverse weather in the major corn-producing region of the United States in the mid 1990s provided a chance to study the effect that a decline in corn production has on the corn price (corn production fell, but the corn price increased much more). This event demonstrated the effect of a drop in production on the price of corn, and it helped to confirm basic understandings about the way that the corn market operates. Another approach to economic analysis is studying milk consumption in several nations to determine the implications of differing prices on use rates.

The real world is extremely complex, and no one can deal fully with its intricacies. The "art" of economic analysis is to extract the essential elements to answer economic questions. To do this, econometricians use sophisticated statistics and mathematical models and must make many assumptions. Two of the most common and powerful assumptions are that individuals want to maximize their well-being and firms want to maximize their profits. Although these simplifying assumptions are not always realistic, they have validity in most situations and are key to the study of resource, agriculture, and food-related issues.

AGRICULTURE

In this book, we concentrate on agriculture because of its fundamental importance to society and because it provides many examples to illustrate the principles of economics. *Agriculture* refers to the complex system that begins with natural resources and involves farms, agribusinesses, and governmental organizations in providing products of the land to consumers. The term agriculture was applied originally to the growing of crops and the raising of livestock. As economic systems have become more complex, however, agriculture has taken on a much broader meaning. It also encompasses firms and industries that manufacture and market inputs for farming—farm machinery, fertilizers, farm chemicals—that is, all the commodities, services, and supplies used in modern farming. The term also includes the industries that process and market farm products to consumers—grain trading firms, meat packers, cotton mills, dairy manufacturing and marketing, fruit and vegetable dealers and processors, wholesale food firms, and finally, the retail markets. In addition, agriculture includes a public sector that devises rules and regulations; conducts scientific research; provides education, extension, market news, price forecasting services; and conducts economic analysis. These economic activities can be grouped into three subsectors that are identified primarily by function: the farm sector, agribusiness, and the public sector.

The Farm Sector

The **farm sector** includes all the firms (farms, ranches, hobby farms, ranchettes) that grow crops and raise livestock, usually for sale. Several characteristics differentiate this sector from the rest of the economy.

1 First, both crops and livestock are living organisms. This means that agricultural production is subject to the uncertainties of weather, disease, and life cycles associated with living things. Some farmers diversify their operations between crops and livestock, grow several different crops, or produce more than one kind of livestock to deal with these uncertainties. Other farmers decide that specialized operations are more profitable, and still others purchase insurance.

2 A second characteristic of the farm sector is that crops and livestock depend on land and water resources. This dependence places geographic limits on production. For example, in temperate climates in the Northern Hemisphere, farmers can use land only from spring to fall and can grow only one crop. However, winter kills most pests, so farmers start "fresh" every spring. In contrast, near the equator farmers can use their land throughout the year and can sometimes grow two, three, or even four crops a year if enough water is available. But insects and other pests thrive year-round, so these farmers must deal with their constant presence.

Crop production is more closely tied to the land resource than livestock production. Although livestock consume some of the crops used to produce products for human use, they can be grown off the land. Containment poultry and hog operations, where animals are raised entirely in buildings, are becoming increasingly common. Thus crop and livestock activities, although closely linked, have different economic problems that require different solutions.

3 A third characteristic is that the operating units, both farms and ranches, are small relative to firms in much of the rest of the economy and relative to the firms that sell inputs to and purchase the outputs from farmers and ranchers. As a result, individual producers have little power in the market to influence prices. In some countries this relative lack of power has led to the development of farmer cooperatives and also has contributed to the adoption of price and income support policies.

Most countries have a statistical definition of a farm. In the United States, for example, a farm is defined as any operation with actual or potential sales of agricultural commodities of $1000 or more in a year. Thus some very small enterprises are classified as farms. For instance, commodities worth $1000 or more can be produced by one good dairy cow or grown on as little as two acres (one hectare) of fruit trees, five acres (two hectares) of corn or soybeans, or an acre of some vegetable crops. In the United States, where the average income per person is more than $15,000, these small farms clearly cannot support a single person, much less a family. However, in some less-developed countries where the average income per person is less than $1000, a farm of one hectare can easily support a family.

4 A fourth characteristic of agriculture is that as incomes grow, consumption of most agricultural commodities, especially food, does not grow as rapidly. In a poor country, many consumers spend much of their income on food, and agricultural production makes up a large share of total output of the economy. As an economy grows, consumers spend more of their income on nonagricultural goods, and other parts of the economy assume a larger role. In a country, such as the United States, where almost everyone can afford an adequate diet, the demand for food grows more slowly than the demand for other goods. Most of the growth in demand for U.S. agricultural commodities must come from international markets.

Throughout the world, more people earn their livelihood by farming than by any other single occupation. As populations and incomes grow and markets expand, so must the output of farms. The history of agriculture in developed countries documents that farming has already undergone dramatic changes and suggests that it will continue to do so. A powerful force driving these changes is the continuing transformation of agriculture through technological advances in the physical and biological sciences. Another important force is the enhancement of human skills, or human capital, through education and training and the improvement of management through more efficient use of information technologies.

5 A fifth characteristic is that in the developed economies the demand for many agricultural commodities is relatively insensitive to retail price changes. When food prices increase, most consumers in wealthy countries reduce the amount they eat very little, if at all. And when food prices fall, consumers don't usually eat more either. This insensitivity to price changes causes instability in markets for many agricultural commodities and has been a major reason for the adoption of farm price and income support policies.

Changes in the Farm Sector   U.S. farms range from very small, mostly part-time enterprises in which the operators earn most of their income from off-farm employment, to large enterprises with annual sales in the millions of dollars. Both gross farm output and the money that farmers receive from the sale of farm commodities are highly skewed (Figure 1-1). The value of output, rather than acreage or another physical measure, is generally used as the gauge of farm size because it better reflects a farmer's ability to earn a living from the farming operation. The value of output also provides a measure for comparing different units such as a sprawling cattle ranch, an intensive vegetable crop farm, and a broiler facility housed in a single building.

The skewed distribution of U.S. farm size and income illustrated in Figure 1-1 is a relatively recent phenomenon. For more than 200 years, the goal of both colonial and national policy has been a farm sector made up of family farms. These are defined as farms large enough to support a family in comfort, but small enough for a family to do most of the work.

FIGURE 1-1   Comparisons of U.S. farm size by number and value of output.

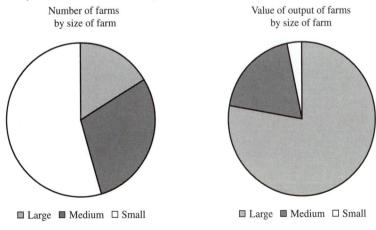

Number of farms
by size of farm

Value of output of farms
by size of farm

□ Large ■ Medium □ Small   □ Large ■ Medium □ Small

As the income needed to support a family has risen, the size of a farm required to support a family has also grown and thus the number of farms has declined. In the 1930s, the number of U.S. farms peaked at 6.8 million. Since then, the number of farms has declined to fewer than 2 million. Furthermore, about half have annual sales of $1,000 to $10,000. Families on these small farms receive the major portion of their income from non-farm sources.

The decline in the number of farms in the United States is due to technological advances in production and management, the development of markets, and government policy. Most other developed countries have experienced a similar transformation, although it differs in detail and extent. In the less-developed countries, which include more than three-fourths of the world's population, the forces that led to the transformation of agriculture are exerting the same pressures. We are already seeing similar changes in many of those countries. For example, in middle income, and even some of the poorest countries, the proportion of economic activity in the agricultural sector is declining as the other sectors grow more rapidly.

Agribusiness

When agricultural output grows, the businesses that provide farm services and supplies and that process and market farm commodities also expand. These businesses make up the agribusiness sector, which, like the farm sector, has two distinct economic subsectors. **Agribusiness** includes (1) firms and industries that produce and sell goods for use in farm production, and (2) firms and industries that buy, store, and process farm commodities and distribute them to domestic and export markets. The first group is the input subsector, and the second is the agricultural processing and marketing subsector.

Input Subsector   Firms in the **input subsector** transform minerals, raw materials, and other inputs into farm machinery, fertilizer, pesticides, and various other commodities used in growing crops and raising livestock. In less-developed countries, this subsector is small. As an economy develops, this subsector expands rapidly.

As part of the development process, new inputs sometimes complement and sometimes displace other inputs. For instance, mechanical implements displaced oxen and horses in the United States when farmers found that costs could be reduced and the work done faster and more efficiently. Inorganic fertilizer is applied when the value added to the crop is greater than the cost of fertilizer and when not enough manure is available. Even when manure is available, the cost of transporting and spreading it may be greater than its contribution to the value of the crop. Farm chemicals are used when they can control weeds, insect pests, and crop diseases at less cost than alternative methods. Firms develop to provide credit, consulting, and market advisory services, which are not direct inputs to the production process but help farmers use the physical inputs more profitably.

Processing and Marketing Subsector   The agribusiness sector buys commodities from the farm sector and transforms them into commodities more suitable for consumption:

$$\text{Corn} + \text{grass} \rightarrow \text{cattle} \rightarrow \text{processing} \rightarrow \text{hamburger}$$

The kinds and amounts of services added by these industries are reliable indicators of a country's development. In less-developed countries, a much higher proportion of the activities in the agricultural sector is devoted to farming than to agribusiness. Agricultural commodities are consumed in less-processed forms, and less variety is available. More processing is done at home. By comparison, in more developed countries, more food is available in highly processed forms, such as frozen, canned, cooked, and even ready-to-eat. In the most highly developed countries, as much as three-fourths of the total value of the final products is contributed by the **processing and marketing subsector.** Developed countries generally have an efficient, corporate-style agribusiness sector that may employ four or five people for every person working on a farm. Employment in these industries is still expanding, whereas employment in the farm sector is declining.

The Public Sector

The development and growth of agriculture is marked by important advances in an array of publicly supported services known as the **public sector.** Roads, airports, and harbors are examples of basic public sector services. Agricultural development also depends on biotechnological and economic research; primary, secondary, and higher education available to the agricultural sector; and extension and information services. Education builds human capital. People become more efficient as workers, managers, communicators, and the like. (You get a good job!) Government services such as food inspection, market news, sanitary regulations, and market supervision improve market efficiency and product quality. In most countries, governments provide or sponsor sources of financial resources such as farm credit, crop insurance, or emergency grants in addition to those provided by the private sector.

Although the farm sector is often thought of as relatively unregulated, the farming and agribusiness sectors in developed countries receive a great deal of publicly supported services. Investment in the farm sector generally becomes more productive as a result. These services stimulate economic development, change the allocation of resources within agriculture, and influence the distribution of income between agriculture and the rest of the economy. In most developed countries, the public sector also provides substantial price support for farm products, other forms of income support, and aid for foreign trade. In contrast, in many developing countries the public sector taxes agricultural production and sometimes taxes agricultural exports. Whether in developed or developing countries, the economics of the public sector is integral to the study of the economics of resources, agriculture, and food.

Agriculture has an important relationship with rural communities, most of which are affected economically by the level and distribution of farm income. The kind and quality of services available in their local communities significantly affect rural families. In many cases, as farms grow fewer and larger, rural communities see their customer base diminish and the businesses in these communities struggle to remain viable. However, in other cases, families that prefer a simpler lifestyle are moving to these communities from large cities.

RESOURCE, AGRICULTURE, AND FOOD ISSUES

There are at least four reasons for an economics text focusing on resource, agriculture, and food issues. First, agriculture controls a large portion of the nation's resources and it has a major effect on another significant portion. Second, agriculture deals with living organisms, raising many issues that are not usually adequately covered in a general economics course. Third, most agricultural issues require economic analysis; these issues include problems faced by small family farms, large corporations, government bureaucracies, and individual consumers. Fourth, much of the world's population derives its livelihood from agriculture, and all consumers depend on agricultural production for adequate and safe food. Continued population growth will magnify the importance of resource, agriculture, and food issues. (How will we feed and clothe the additional 2 billion people expected by 2025?)

Here are six agricultural economic issues. They will give you a feel of the range and scope of issues we face. Most of them are likely to remain important, and perhaps become more so in the years to come. You may spend your life grappling with one or more of them.

Issue 1: Food Availability and Safety

Providing a safe and abundant food supply is a fundamental objective of every country. Even the United States has hungry people, and malnourishment is a serious problem in many areas worldwide. A continuing challenge is to develop agricultural development policies that reduce the incidence of hunger.

Food safety is also an important issue. Contaminated food can cause illness and death. The development of improved storage, handling, and inspection techniques to ensure the cleanliness of food is an ongoing concern in all countries, but is a more substantial challenge in the less-developed countries. Over the last several decades, especially in the developed countries, people have expressed concern about the contamination of foods with agricultural chemicals, such as pesticide residues, some of which cause health problems. More recently, the safety of foods produced with genetically modified organisms has become a major food safety issue.

Issue 2: Environmental Consequences of Agricultural Production

Farmers and ranchers manage, and are therefore stewards of, approximately three-fourths of the land in the United States. They also lease several million acres of public land for grazing. Resource management and environmental issues have been on the U.S. agricultural policy agenda for more than a century. In recent years, they have assumed a more prominent role. The policy focus has shifted from—first, giving public land to private farmers, to second, encouraging soil conservation efforts to protect the land, and finally, to controlling erosion, pesticides, and other contaminants of water resources. In some states concerned citizens have been successful in reducing urban sprawl and thus maintaining green space and habitat for some wildlife, especially in filterstrips and green belts. Also efforts have been made to protect wetlands and endangered species. Attention has been devoted to natural resource policies relating to timber harvest, fishery management, and the sustainable rate of agricultural input use. Environmental problems—soil erosion, contamination of water supplies, and loss of biodiversity—have become important at national, state, and local levels. The ability of agricultural crops to store carbon may play a role in dealing with the global warming problem.

Issue 3: Managing Technological Advances in Agriculture

The transformation of agriculture has been stimulated by (1) developments in science and technology—an increasingly interconnected global economic environment, and (2) advances in management and information skills. The rate of this transformation has been accelerating. The last half of the twentieth century saw more changes in agricultural science and technology and in management and information systems than has occurred in all preceding time. The challenge is to ensure that new technologies are safe and contribute to public well-being before they are widely disseminated. At the same time, excessive regulation can hinder the implementation of new technologies that could help feed the hungry.

Issue 4: Increasing Internationalization of Agriculture

The world economy is increasingly interdependent, and agriculture is no exception. Farmers and agribusinesses face competition from other firms in both domestic and foreign markets. At the same time, increased access to foreign markets provides new market opportunities. Consumers here and abroad like having a better or cheaper product available regardless of its origin. (Do you buy the best product at the best price, whether domestic or foreign?)

Issue 5: Policy Responses to Uncertainty in Agriculture

The farm sector is at the mercy of the weather, unlike most other major industries. Drought, flood, wind, frost, plant diseases and insect infestations result in a high degree of risk and uncertainty for farmers. Because it takes only a few years of adverse conditions to force some farmers into bankruptcy, policies have been instituted to assist them through difficult times. But finding the appropriate policy responses challenges economists and fuels hot political debate. With the concern and uncertainty about global warming and the effect on agricultural production, this debate becomes more complex.

Issue 6: Decline in the Number of American Farmers

Since the mid 1930s, the number of U.S. farmers has steadily declined; now only about 2 percent of the U.S. population lives on farms. This change can be viewed from several perspectives. From the perspective of productive efficiency, it is a major achievement. The country can produce all of its own food needs and have food available for export with fewer people working on farms. People who might have been working on farms are employed in a wide range of other positions, producing other consumer goods. However, some people view the fall in the number of farmers as a significant problem, citing the Jeffersonian view that the farmer is a pillar of democracy. Others note that the reduction in the number of farmers causes rural communities to disappear, as the numerous small businesses serving farmers are replaced by fewer, but larger, firms.

While the number of farms and farmers is declining, the processing, distribution, and retailing activities continue to grow. Reasons for this growth include the increasing affluence of the society, the growing number of single-parent and dual-career families, the preference for convenience foods, and the trend to eat more meals in restaurants.

The preceding six issues are only some of the major challenges facing agriculture. Such issues can be addressed, at least in part, through an understanding of economic principles. This book is designed to serve as a foundation for anyone interested in economics and in the application of economics to resource, agriculture, and food issues. Emphasis is placed on creating a broad vision of this part of the economy and on developing the basic economic theory used in more advanced courses. In addition, this book can help to provide the foundation for students specializing in other areas of study, and to broaden their vision and increase their understanding of economics.

SUMMARY

Economics is concerned with understanding how limited resources can be allocated to satisfy unlimited wants. It deals with what will be produced, how, and for whom. Economics is concerned with how goods can be produced efficiently and how a society can distribute them equitably. Economics involves both positive and normative questions and answers. To answer economic questions, we must extract the essential features from the complexity of the real world. This involves making assumptions about human behavior and other elements of the economic world.

Agriculture includes farming and ranching; agribusinesses that produce and sell services and supplies to farmers; agribusinesses that buy, process, and distribute farm commodities; and a public sector. The transformation of agriculture generally involves the integration of farms and ranches into fewer and larger units, relatively rapid growth of agribusinesses, and the increase in output from the publicly supported sector.

A text such as this—combined with a course on the economics of resources, agriculture, and food—will help you broaden your vision and improve your ability to apply economic analysis to a broad spectrum of problems.

LOOKING AHEAD

The plan for this book is to begin with microeconomics, the decisions of the individual and the firm. These should be easier for you to understand because these are familiar decisions made in daily life. We discuss the decisions that consumers make as they purchase goods to maximize utility and the decisions that producers of goods make to maximize profits. After developing these two basic components, we show how they come together in markets to determine the prices that allocate society's resources. In exploring markets we consider the role of the government in structuring markets and address market imperfections and failures. That section is followed by a discussion of how the general economy functions and how monetary and fiscal policy might be used. The last section of the book considers the use of economics in public policy relating to agricultural development, price and income support, resource and environmental management, and consumer issues.

IMPORTANT TERMS AND CONCEPTS

agribusiness   12
agricultural economics   4
agriculture   9
alternative uses   5
distribution   6
economic growth   7
economics   4
factors of production   4
farm sector   9
input subsector   12
macroeconomics   7

microeconomics   6
normative economics   7
opportunity cost   5
positive economics   7
prescriptive economics   7
processing and marketing subsector   12
public sector   13
resource   4
resource and environmental economics   4
scarce productive resources   4

QUESTIONS AND EXERCISES

Name That Term

Read the following sentences carefully and fill in the missing term or terms.

1 _____ is the study of how to allocate limited resources to produce goods that help satisfy unlimited human wants.
2 _____ economics deals with questions that do not involve a judgment comparing the values that different individuals place on an option.
3 A _____ is an input provided by nature and modified by humans using technology to produce goods that satisfy human wants.
4 Firms in the _____ subsector transform minerals, raw materials, and other inputs into farm machinery, fertilizer, pesticides, and various other commodities used in growing crops and raising livestock.
5 The _____ of using any good or service is the value of the next best alternative use of that good or service.
6 _____ economics deals with ways to achieve a desired result in the most efficient, profitable, or acceptable way.
7 The _____ sector includes all the firms that grow crops and raise livestock, usually for sale.
8 _____ economics deals with value judgments.
9 The _____ subsector buys commodities from the farm sector and transforms them into commodities more suitable for consumption.
10 _____ refers to the complex system that begins with natural resources and involves farms, agribusiness, and government organizations in providing the products of the land to consumers.
11 _____ includes firms and industries that produce and sell goods for use in farm production, and firms and industries that buy, store, and process farm commodities and distribute them to domestic and export markets.

True/False

Read the following sentences, then decide whether each statement is true (T) or false (F) and mark it accordingly.

T F   **1** Four important economic questions are *what* will be produced from our limited resources, *how* it will be produced, *for whom* it will be produced, and *when* it will be consumed.

T F   **2** Everyone gains when correct economic policies are followed.

T F   **3** Goods not consumed today can be saved and invested in capital goods to increase productive capacities in the future.

T F   **4** Resources include land, labor, equipment, and capital goods as well as mineral and vegetable resources such as coal, iron ore, forests, and water.

T F   **5** Generally, as the economy of a less-developed country grows, its agricultural input subsector shrinks.

T F   **6** What you choose to purchase for lunch today is an example of applied microeconomics.

T F   **7** Macroeconomic performance has little effect on your everyday life.

T F   **8** Developed countries often have four or five people working in the processing and marketing subsector for every person working on a farm.

T F   **9** Firms in the farm sector are subject to the risk and uncertainties of weather, disease, and life cycles associated with living things.

T F   **10** Farms and ranches tend to be small relative to the firms that sell inputs to and purchase output from farmers and ranchers.

T F   **11** As incomes grow, consumption of most agricultural commodities, especially food, increases even more rapidly and consumers increase the share of their income that goes to food.

Multiple-Choice Questions

Circle the letter of the response that best answers the question or completes the statement.

1 Prescriptive economics involves
  a how to get the largest return from a given set of resources.
  b how to achieve a public policy objective at the lowest cost to society.
  c how to structure a policy to get a particular allocation of costs and benefits of a program or policy, now and in the future.
  d all of the above.
  e none of the above.

2 Since the 1930s the number of U.S. farms has
  a increased from about 2 million to about 6.8 million.
  b decreased from about 6.8 million to about 2 million.
  c held steady at about 6.8 million.
  d held steady at about 2 million.
  e none of the above.

3 The major source of income for families living on residential farms (with annual sales of $1000 to $10,000) in the United States is
  a corn.
  b soybeans.
  c milk.

   d tobacco.
   e nonfarm sources.
4 Presently in the United States about _____ percent of the population lives on farms.
   a 33
   b 25
   c 10
   d 5
   e 2
5 Farmers and ranchers own approximately _____ percent of the land in the United States and lease several million more acres of public land for grazing.
   a 90
   b 75
   c 50
   d 25
   e 10

Technical Training

Suppose the opportunity cost of using a good exceeds the value of using the good. For example, suppose the best alternative to answering these questions is going on a bike ride. Suppose further that going on a bike ride adds more to your long-term happiness than does the economics you would learn. What should you do?

Questions for Thought

The following questions are designed to help you apply the concepts you have learned to real-life situations. Thinking about them will help prepare you for discussion questions based on the material in this chapter.

1 Are all goods scarce? If not, name three goods or services available in unlimited quantities.
2 Do you have everything you want in this world? Do you think anyone does?
3 What is the opportunity cost of your working on this material today? What's your opportunity cost of being in school? Do all activities have opportunity costs?
4 Which is more important to life, diamonds or water? Which has a higher price? Why does the clearly more valuable commodity have a much lower price? (You might keep this "paradox of value" in mind. By the end of the course you'll have a quite sophisticated insight into this perplexing question.)
5 List three major questions studied by economists. In the study of economics, what do the terms value, distribution, and growth mean?
6 Give an example of a resource that is scarce. Is the resource a factor of production? Explain what would happen if it were no longer scarce.
7 Identify a resource or good that has alternative uses in production or consumption. What factors do you think determine how the resource or good is allocated or consumed?
8 What do you think your opportunity cost is for attending college? Why are you willing to take on this cost?
9 Give examples of positive and normative economic questions not used in the text.
10 Give a specific example of how the federal government might utilize prescriptive economics.

11 Briefly describe the three subsectors of the agricultural industry in the United States. What are the major differences you would expect in these sectors in a developing country?

12 If I offered you a choice between $100 today and $102 one year from today, which would you take? Why?

13 It has been said that "Microeconomics is a game we all play every day; macroeconomics is a spectator sport." List ten microeconomic choices you've made in the last week. How might macroeconomic forces affect you in the next few years?

14 While you and Joe are walking down the street together, you find $20. There's no way to find its rightful owner so you must decide who gets it. Consider three possible distributions:
- Joe gets $20 and you get nothing because he's poorer than you
- You get $20 and he gets nothing because you found the money and are bigger than Joe
- You "split the difference" and each gets $10.

Which distribution do you suggest? Do you think Joe agrees with you?

15 Why are there fewer farmers in the United States than there were fifty years ago? Do you think the number of farmers will continue to decrease? Why?

ANSWERS AND HINTS

Name That Term   **1.** Economics; **2.** Positive; **3.** Resource; **4.** Input; **5.** Opportunity cost; **6.** Prescriptive; **7.** Farm; **8.** Normative; **9.** Processing and marketing; **10.** Agriculture; **11.** Agribusiness

True/False   **1.** T; **2.** F; **3.** T; **4.** T; **5.** F; **6.** T; **7.** F; **8.** T; **9.** T; **10.** T; **11.** F

Multiple Choice   **1.** d; **2.** b; **3.** e; **4.** e; **5.** b

Technical Training   Get on your bike.

MICROECONOMIC CONCEPTS

2

ECONOMICS OF DEMAND

CHAPTER OUTLINE

OVERVIEW

In this chapter, we show that the way people make consumption decisions undergirds a fundamental concept in economics: the law of demand. A nearly universal characteristic of the human condition is "unlimited wants." This condition interacts with market prices and the consumer's budget to determine how much of each of the many available goods is consumed. We use this understanding to examine the effect of price changes on quantities demanded. From this analysis we can derive the "law of demand." This explains why as a product's price increases, the quantity purchased decreases, and vice versa. This inverse relationship is reflected in a downward-sloping demand curve.

Next, we turn our attention to another important economic concept—how much the quantity demanded changes when price changes. This concept is called elasticity. Elasticity measures how much the quantity demanded (or supplied) changes in response to a price change. Elasticity estimates are indispensable in understanding the organization and functioning of the economy, especially the agricultural sector. An understanding of elasticity is essential for appreciating the implications of the interaction between demand and supply, and how an economy, or sector of an economy, operates to allocate resources and products among firms and individuals.

LEARNING OBJECTIVES

This chapter will help you learn:

- About the law of demand
- How an individual's budget limits the goods that can be purchased
- What "utility" is, and how an indifference curve is derived
- How a demand curve is determined by an individual's budget and tastes and preferences
- The basic concepts of elasticity of demand, cross-price elasticity, and income elasticity
- The determinants of demand elasticity

WHAT IS THE LAW OF DEMAND?

We begin with the economic concepts that are probably the most familiar to you: *consumption* and *demand*. While we use an example that may seem trivial, it demonstrates a concept central to economics—because the choices consumers make are fundamental to the operation of an economy.

You spend at least part of every day of your life consuming things. Eating food, purchasing new clothes, going to a movie, and riding in a vehicle are all examples of consumption. The purpose of this chapter is to examine how an individual (such as you) decides what, and how much, to consume (and not to consume). As you soon will see, the decision involves an interaction among:

- How much money an individual has to spend, a budget
- The scarce goods available in the marketplace and their prices
- The individual's tastes and preferences

Once we have explained the basic process of making consumption decisions, we then can show how one of the few "laws" of economics—the law of demand—works. To start, we illustrate the law of demand with answers to some hypothetical questions:

- Would you change the number of times you go to the movies if ticket prices increase 20 percent, and all else remains the same?
- How would a doubling of gasoline prices affect the number of miles you drive during the school year?
- How many fewer times a week would you buy a hamburger if the price increased by 25 percent?

In each case, as the price goes up, the quantity you consume is likely to go down. Of course, the inverse is also true: if the price goes down, the quantity consumed is likely to go up. This inverse relationship between price and quantity consumed is called the **law of demand.** *If nothing else changes, the quantity of a good purchased varies inversely with the price of that good.* If the price of a good is high, less will be demanded than if the price is low. As the price of a good falls, more is demanded. The qualification that "nothing else changes" (such as prices of other goods or your income) is important. For example, if your weekly budget tripled while the pizza price went up by 25 percent, you might increase your pizza consumption, yet the law of demand is not violated. "All other things" were not held constant. The idea of nothing else changing is sometimes called the *ceteris paribus* assumption (Latin for "all else equal"), and we will make frequent use of this concept. In fact, anytime we state that one thing is related to another, we presume nothing else changes. In more advanced economic studies, problems with several variables are considered.

WHY IS IT CALLED THE LAW OF DEMAND?

In order to consume a good or a service, you must first purchase or "demand" the good from the marketplace.

Why does a change in the price of a good (or a service) affect the amount purchased and consumed? We begin with the observation that a consumer buys goods (such as gasoline and hamburgers) and services (such as movies and train rides) to satisfy wants and desires. The choices made depend on that person's tastes and preferences. Your friend may "hate" a good that you "love." Individuals from different age groups, social groups, countries, and cultures might have quite different tastes and preferences. But a basic characteristic of all consumers is that they have unlimited **wants.** Although consumers choose very different combinations of goods to satisfy their wants and desires, regardless of how much is now consumed, more is desirable. Don't you and your friends all want more of something, likely more of lots of things?

UNLIMITED WANTS AND ALTRUISM

Some people object to the assumption that consumers have unlimited wants. They argue that unlimited wants are not consistent with gift giving, contributions to charity, and other altruistic behaviors. In fact, altruism is not inconsistent with unlimited wants. Since individuals derive pleasure from acts of altruism, such acts also satisfy wants and desires.

Consumption and Utility

Economists use the word **utility** to describe the satisfaction derived from consuming a good or a service. Each person's utility is unique and reflects that person's *tastes and preferences.* Utility, and especially **marginal utility** (the utility provided by the last unit of a good consumed), is central to understanding consumption decisions and the law of demand.

Imagine you are 8 years old. It is a special holiday and you may eat as much candy as you want. The first piece of candy tastes wonderful. It makes a large addition to your utility. After you have eaten a few pieces, the candy does not taste quite as good, but you still enjoy each piece. The candy is contributing to your utility, but not by as much as it did with the first piece. Each additional piece provides less utility. At some point you quit eating the candy because additional pieces no longer add to your utility.

This image shows two important aspects of utility. Each piece of candy eaten increases your satisfaction, or total utility. But as you eat more and more candy, your marginal utility falls. This phenomenon is called the law of **diminishing marginal utility:** *As the amount of a good consumed increases, the addition to total utility derived from adding one more unit of that good decreases.*

NEGATIVE MARGINAL UTILITY

Does additional consumption ever result in *negative marginal utility?* The answer is yes. In the candy example, you could eat so much candy that you would become ill. Clearly the additional utility from the last piece of candy at that point would be negative. But normally you wouldn't buy a piece of candy that would make you ill. This is why economists can assume that marginal utility is positive. Consumers won't purchase and consume to the point where marginal utility falls to zero. They purchase combinations of products so that the marginal utility of each product purchased is positive.

Utility Illustrated

Early economists tried to measure utility as a means of explaining consumer behavior and the law of demand. As we show later, it is not necessary to measure utility. However, the approach will help you understand the concepts of utility, marginality, and demand. Suppose someone has developed a numerical measure of utility called **utils.** Table 2-1 and Figure 2-1 illustrate the law of diminishing marginal utility using utils derived from an individual's weekly hamburger consumption. Total utility increases with each hamburger consumed, but the marginal utility of each additional hamburger is less

TABLE 2-1   UTILITY FROM HAMBURGER CONSUMPTION

Hamburgers per week	Total utils	Utils from last hamburger
0	0	
		50
1	50	
		30
2	80	
		25
3	105	
		20
4	125	
		16
5	141	
		14
6	155	
		12
7	167	
		11
8	178	
		10
9	188	

Note: Hypothetical data.

FIGURE 2-1   Graph of Table 2-1.

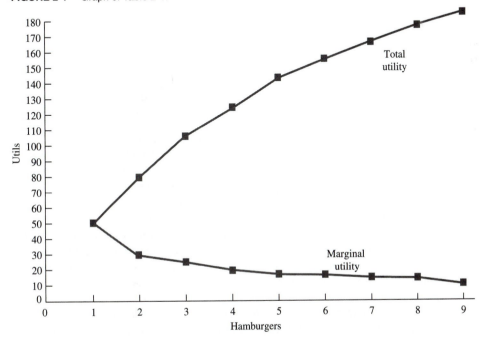

than the one that preceded it. Or, as the number of hamburgers eaten increases, the number of utils added by the last hamburger becomes smaller (diminishing marginal utility).

The Budget Constraint

Utility and diminishing marginal utility are only part of what we need to explain consumption behavior. Individuals have unlimited wants, but the goods available to meet those wants are limited. Each of us has a limited amount of money to use for consumption—a *budget.* The consumer's budget (from salary, allowance, loans, dividends, etc.) is the amount of money available for purchases in a given period. Your budget constrains or limits how much you can buy.

The size of the budget and market prices impose a constraint on consumption. You can't spend more money than you have. Your budget forces you to make choices—tradeoffs. To understand the effect of the **budget constraint** on the quantities of goods consumed, consider a simple example. Suppose a consumer has a budget of $5.00 a week to spend on only two goods—hamburgers that cost $1.25 per burger, and pizza slices that cost $0.50. What combinations of pizza slices and hamburgers can this consumer afford? What happens to the affordable combinations of pizza and hamburgers when the budget changes? What happens to the affordable combinations of pizza and hamburgers when the price of a burger or a slice of pizza changes?

CONSUMPTION AND SAVINGS

A typical individual receives income from one or more sources. Some of that is paid to the government as taxes, and part of it is saved to provide for consumption in the future. The remainder is available for current consumption. Throughout this chapter, we assume that the consumer has decided already what share of income is to be consumed at this time. The savings-consumption decision is discussed in the macroeconomics section of the book.

To answer these questions, we begin with Figure 2-2. The quantity of hamburgers consumed per week is on the horizontal (or x) axis; the quantity of pizza slices consumed per week is on the vertical (or y) axis. If all of the $5.00 budget were spent on hamburgers, hamburger consumption would be 4. If all of the budget were spent on pizza slices, 10 slices would be eaten per week. The line connecting the points 4,0 (all of the budget spent on hamburgers) and 0,10 (all of the budget spent on pizza) is called the **budget line.** It is a graphical representation of the budget constraint for this individual. Every point on the budget line is a combination of hamburgers and pizza slices that could be purchased with the $5.00 budget (if it was possible to purchase fractions of hamburgers and pizza slices). For example, at the point 2,5 (on the budget line), the cost of 2 hamburgers is $2.50 and the cost of 5 slices of pizza is $2.50. The entire $5.00, but not more, is spent.

Suppose that our hypothetical consumer tried to purchase 3 hamburgers and 5 slices, as shown by point *a* in Figure 2-2. The cost of this combination is $6.50, which is more than the budget, and thus lies to the right of the budget line. Any combination

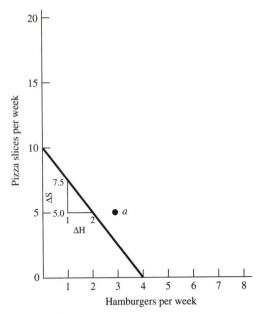

FIGURE 2-2   The budget constraint.

represented by a point to the right of the budget line costs more than the budget. All such combinations are unaffordable.

What about points to the left of the budget constraint? Any point to the left of the line represents consumption of fewer hamburgers or fewer slices (or both) than some point on the budget line. All points on and to the left of the budget line are affordable. However, buying a combination to the left of the budget line would not use all of the money designated for these goods. Only purchases on the budget constraint use all of the consumer's budget. The utility maximizing combination—where consumption is optimum—lies somewhere on the budget constraint.

As consumption changes from one point on the budget line to another, consumption of one good goes up while consumption of the other goes down. The slope of the budget line shows how much consumption of each good changes.[1] In this example, the slope of the budget line ($\Delta S \div \Delta H$) is -2.5. (The symbol Δ means change.) To consume another hamburger (costing \$1.25), consumption of pizza slices would have to drop by 2.5 (saving \$1.25). This is the same as the ratio of $-P_h \div P_s$.

The Effect of Budget Changes   What happens if the budget changes? Figure 2-3 illustrates the effect of doubling the budget. With the new budget, it is possible to buy twice as many hamburgers (the budget line meets the x axis at 8,0), twice as many slices

---

[1]The general formula for a straight line is $y = mx + b$, where m is the slope, the rise over the run, and b is the intercept of the line. The formula for the budget line is $P_xQ_x + P_yQ_y = B$ (B = the budget). The formula for the budget line can be rewritten as $Q_x = -(P_y \div P_x)Q_y + B \div P_x$. The slope is $-P_y \div P_x$, the intercept is $B \div P_x$.

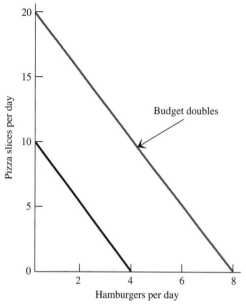

FIGURE 2-3 Effect of a doubling of the budget.

of pizza (the budget line meets the *y* axis at 0,20) or twice as large a combination of the two. A budget increase causes a *parallel shift* of the budget line to the right. A budget decrease results in a parallel shift of the budget line to the left.

The Effect of Price Changes What happens if the price of one of the goods changes? Figure 2-4 shows the effect of an increase in the hamburger price to $2.00 with the budget and pizza price unchanged. The maximum number of hamburgers drops to 2.5, but there is no change in the maximum number of pizza slices. Therefore, the budget line remains fixed on the *y* (pizza) axis and rotates to the left along the *x* (hamburger) axis. What happens if the price of hamburgers is unchanged but the price of pizza slices doubles? In this case, the maximum number of hamburgers remains constant but the maximum number of pizza slices falls from 10 to 5. The intersection of the budget line and the horizontal axis is unchanged, and the intersection with the vertical axis shifts downward (think about how this would look on a graph similar to Figure 2-4).

Here are some important points to remember about the budget constraint and the budget line.

• The end points of the budget line show how many units of a good would be bought if all of the budget were spent on that good.
• A reduction in consumption of one good allows an increase in consumption of the other good.
• Combinations to the left of the budget line are affordable but not as desirable as combinations on the budget line. Combinations to the right of the budget line are not affordable.
• A change in the budget causes a parallel shift in the budget line.

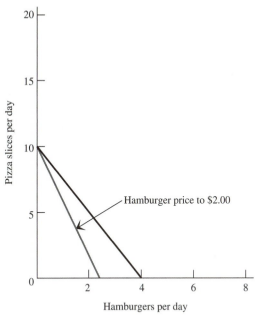

FIGURE 2-4   The hamburger price increases to $2.00.

• A change in the price of a good changes the amount of that good that is affordable but does not change the maximum amount of the second good that can be purchased. A price increase moves the point of intersection with the axis for that good toward the origin. The line rotates toward the origin.

• The slope of the budget line equals change in quantity on the y axis associated with a given change in quantity on the x axis. It is also equal to the negative inverse of the price ratio, which in Figure 2-2 is $-P_h \div P_s$ or $-1.25 \div 0.50$, or -2.5.

The Consumer Choice Problem

We now combine the behavioral concept of utility with a budget and prices to show how a consumer chooses among several goods, the *consumer choice problem. The basic problem a consumer faces is how to allocate the budget among various goods to maximize utility.* A rational consumer maximizes utility by consuming as many goods as desired, within the limits imposed by the budget. One way to think about this is that a consumer buys goods to "buy" utility. This means that the consumer buys goods that provide the most utility per dollar spent. Or, you want as much satisfaction as you can possibly get from your limited funds.

Of course, consumers choose among many goods, but the basics of the choice process can be understood by considering two goods. We continue with the example of hamburgers and pizza slices. Suppose the consumer has a budget of $18 to spend on hamburgers or pizza slices during the next week. Hamburgers cost $2, and pizza slices cost $1 each. The amount of satisfaction—utils—provided by hamburgers was given in Table 2-1. With the $18, the consumer can afford up to 9 hamburgers, which will provide up to 188 utils. Note that the first hamburger provides 50 utils and that each hamburger adds fewer utils than the previous hamburger. Remember: this is the diminishing marginal utility rule we discussed earlier.

Table 2-2 contains the pizza utility data for this consumer. Since a pizza slice costs only half as much as a hamburger, the consumer can purchase twice as many slices of pizza, up to 18 slices per week. As is true of hamburgers, the utility per slice falls as the consumer purchases and eats more pizza. The first slice provides 56 utils, the ninth slice provides 12 utils, and the eighteenth slice actually reduces utility slightly. (Don't you just "hate" being forced to eat one more piece when you really are stuffed?) Of course,

TABLE 2-2   UTILS FROM PIZZA CONSUMPTION

Pizza slices per week	Total utils	Utils from last slice
0	0	
		56
1	56	
		50
2	106	
		38
3	144	
		32
4	176	
		26
5	202	
		20
6	222	
		19
7	241	
		16
8	257	
		12
9	269	
		10
10	279	
		9
11	288	
		8
12	296	
		6
13	302	
		5
14	307	
		4
15	311	
		2
16	313	
		0
17	313	
		−1
18	312	

Note: Hypothetical data.

the consumer would not intentionally reach the point where more consumption of pizza would reduce total utility. Well before that point, the budget would be spent on other goods that have higher utilities.

The consumer's goal is to spend the $18 on hamburgers and pizza slices in the way that generates the most utils. Table 2-3 gives the various combinations on the budget line that the consumer can afford. The table also gives the number of utils per dollar spent on each good and the total utility provided by both hamburgers and pizza purchased. (Note that as you move down the table the consumer is spending more on pizza and less on hamburgers. On each line of the table the consumer is spending the budgeted $18 for pizza and hamburgers.)

This consumer will purchase 10 pizza slices and 4 hamburgers. There are two ways to see that this is the best combination. First, this combination provides more total utility, 404 utils, than any other possible combination. Second, the marginal utility per dollar spent on the two goods is equal. Therefore, the last $2 spent on 2 pizza slices and the last $2 spent to buy 1 hamburger each provide 10 utils of satisfaction. Any movement away from this combination lowers total utility. For example, giving up 1 hamburger and buying 2 more slices of pizza reduces the total utility by 3 utils (the loss in utility from 1 hamburger is slightly greater than the gain in utility from 2 pizza slices). Similarly, buying 1 more hamburger and 2 fewer pizza slices reduces total utility, in this case by 6 utils. The *decision rule that maximizes utility is to equate the cost of marginal utility of the two goods.* This is calculated as the ratios of marginal utilities divided by prices.

TABLE 2-3   FINDING THE UTILITY-MAXIMIZING COMBINATION OF HAMBURGERS AND PIZZA SLICES

Pizza slices	Slice utils	MU pizza	Burgers	Burger utils	MU burger	Total utils	MU/P burgers	MU/P slices
0	0		9	188		188		
		50			10		5	50
2	106		8	178		284		
		32			11		5.5	32
4	176		7	167		343		
		20			12		6	20
6	222		6	155		377		
		16			14		7	16
8	257		5	141		398		
		10			16		8	10
10	**279**		**4**	**125**		**404**		
		8			20		10	8
12	296		3	105		401		
		5			25		12.5	5
14	307		2	80		387		
		2			30		15	2
16	313		1	50		363		
		−1			50		25	UD
18	312		0	0		312		

Note: Hypothetical data. MU = marginal utility is the utility from the last unit consumed; UD = undefined.

COMBINATIONS THAT DO NOT MAXIMIZE UTILITY

If the consumer purchases 8 slices of pizza and 5 hamburgers, the *marginal utilities would be equal.* However, by buying 1 fewer hamburger and giving up the last 16 utils, the consumer can purchase 2 more pizza slices, which provide 22 more utils, for a net gain of 8 utils. Another strategy the consumer might have considered is to get the same amount of utility from each of the two goods. The consumer can get approximately *equal total utilities* by purchasing 4 slices of pizza and 7 hamburgers. In this case, giving up a hamburger and adding 2 slices of pizza adds 32 utils (hamburger utility falls 14, and pizza utility increases 46 utils).

We have illustrated utility maximization when the choice is between just two goods—pizza slices and hamburgers. However, the rule can be extended to cover the decision when many goods are being considered: *To maximize total utility within a budget constraint, allocate expenditures so that the cost of the marginal utilities is the same for all goods.* This means that the ratio of marginal utility to price is the same for all goods and can be stated mathematically as follows:

$$\frac{MU_1}{P_1} = \frac{MU_2}{P_2} = \frac{MU_3}{P_3} = \cdots$$

Now we can answer the question posed at the beginning of this chapter: Why does a change in price affect consumption? If the price of a good changes, the $MU \div P$ ratio also changes and is no longer equal to the ratio for the other goods consumed. The last dollar spent on that good no longer buys the same amount of utility as the last dollar spent on the other goods, and utility is no longer being maximized. Therefore, consumption of that good (and possibly of several others) must be changed to find the new utility-maximizing combination. The price of a good and the quantity purchased move in opposite directions. If price goes up, quantity purchased goes down.

THE LAW OF DEMAND USING INDIFFERENCE CURVES

If consumers were able to quantify the number of utils they receive from each product they consume, we could derive a demand curve from the utility data. But, it is impossible to measure the change in utility (satisfaction or pleasure) received from the consumption of a good. Fortunately, we don't need to measure utility to demonstrate the law of demand. We can use the consumer's ability to choose the most desirable of several alternative combinations of goods.

Again we use the hamburgers and pizza slices example. Imagine that consumers have a choice between (a) three hamburgers and two pizza slices, or (b) one hamburger and four pizza slices. One consumer might prefer option (a); another might prefer option (b). However, for a third consumer, either combination is equally desirable. For this person both combinations provide the same utility. The term used to describe this is "indifference." The third consumer is indifferent to a choice between the two alternatives.

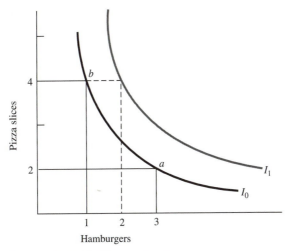

FIGURE 2-5   Two hypothetical indifference curves.

Indifference Curve

A consumer can identify many combinations of two goods that provide the same level of utility. An **indifference curve** is a graph of all these combinations. In Figure 2-5, I_0 and I_1 are examples of two such curves. An individual whose tastes and preferences are reflected in these indifference curves is indifferent to all the combinations of goods on a particular curve. Also, this consumer gets more utility from any combination on the higher curve, I_1, than any combination on the lower indifference curve, I_0. Although we show only two curves in Figure 2-5, many such curves are possible, each representing a different level of total utility. The whole set of indifference curves is called an **indifference map.**

> **WHY INDIFFERENCE CURVES NEVER INTERSECT**
>
> Since every point on an indifference curve gives the consumer the same level of satisfaction and since each curve gives a different level of satisfaction, indifference curves never intersect.

Indifference Maps and the Budget

We can use the indifference curves of an individual with that person's budget constraint to find the combination of goods that maximizes utility. To do this, remember two points. First, the highest possible consumption is found somewhere on the budget line. Second, total utility increases as a person moves to a higher indifference curve. To maximize utility, we want to find the highest indifference curve that just touches (is tangent to) the budget line. To see this, look at Figure 2-6. All the points on I_2 are to the right of the budget line. This means that they are not affordable. On I_0, two combinations, a and b, are

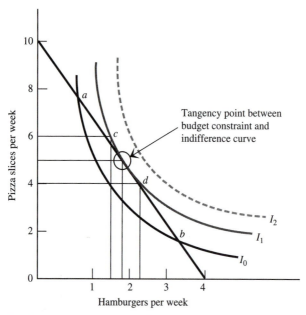

FIGURE 2-6 Finding the optimum consumption point using indifference curves.

on the budget line. Does this mean that one or both of these combinations give the maximum possible utility, given the budget? No! There are many points above I_0 that are on or below the budget line and, therefore, are both affordable and have higher utility than *a* or *b*.

There is only one point on the budget line that just touches the highest possible indifference curve. That point is found at the point of tangency of the budget line and indifference curve I_1. *The point of tangency of an indifference curve with the budget line gives the combination of goods that maximizes utility.* No matter how many indifference curves are drawn for an individual, there is only one that is tangent to the budget line.

The indifference curve can be used to find the **marginal rate of substitution.** The marginal rate of substitution is the quantity of one good a consumer is willing to give up in order to get more of the other good and get the same level of satisfaction. In Figure 2-6, the consumer would be willing to give up two slices of pizza to get 0.8 of a hamburger, as indicated by points *c* and *d*. The marginal rate of substitution is 0.8 divided by 2 or 0.4. Note that this is also the slope of the indifference curve between these two points. Therefore, at the point of tangency of an indifference curve and budget line, the marginal rate of substitution is equal to the slope of the budget line (MRS $= -P_h \div P_p$), or:

$$\text{MRS} = -\frac{P_h}{P_p}$$

This shows that the advantage of using indifference curves is that we don't need a measure of utility to derive a demand curve. We only need to determine the combination of goods that gives the highest utility. By purchasing the combination of goods that gives the most satisfaction, the consumer gets the same result as purchasing goods until the MU $\div$ P ratios for all goods are equal.

The Demand Curve Using Indifference Curves

Now we can demonstrate the law of demand. We show how the tangency between the budget constraint and an indifference curve shifts as one of the prices change. Suppose the price of pizza doubles. As we saw in Figure 2-4, a price change, *ceteris paribus,* rotates the budget constraint around its intersection with the axis of the other good. Increasing the pizza price results in a counterclockwise rotation of the budget constraint.

In Figure 2-7a, we show a consumer, Nadia, who has allocated a budget for the week of $24.00 for hamburgers and pizza. The price of pizza is $3.00. We want to know how many hamburgers Nadia will consume at several different hamburger prices. If the hamburger price is $4.00, Nadia's budget line is B_1 and she will purchase 4 hamburgers and $2\frac{2}{3}$ pizza slices. At lower hamburger prices Nadia's budget line rotates outward (to B_2, B_3, and B_4) allowing her to purchase more hamburgers. In each case she will choose a combination of hamburgers and pizza slices that allow her to reach the highest possible indifference curve on her indifference map for these two goods. Another way to say this is that the decrease in the hamburger price has increased the purchasing power (the real value) of Nadia's budget.

In Figure 2-7b we show Nadia's demand curve for hamburgers. Here we have the number of hamburgers on the horizontal axis and the price of hamburgers on the vertical axis. When we plot the price and quantity combinations, we get Nadia's demand curve. Note that for the indifference curves in Figure 2-7a, the quantities of both hamburgers and pizza are plotted. In Figure 2-7b, however, only the price and quantity of hamburgers are plotted to show the demand curve for hamburgers. In general, the **demand curve** connects all the combinations of price and quantity consumed for a particular good, given the budget constraint and the prices of other goods. *The demand curve slopes downward and to the right.* Each point on the demand curve gives a quantity of the good that a consumer will buy to maximize utility at that price. While the demand schedule and demand curve give all or many of the possible price/quantity combinations, the consumer will demand (choose) only one combination.

The same information can be presented in numerical form, called a **demand schedule** as shown in Table 2-4.

Also note that as shown in the graph drawn in Figure 2-7a, the change in the price of a hamburger causes a small change in the number of pizza slices Nadia consumes. The effect of a change in one price on the consumption of another good is called the *substitution* or *cross-price* effect. The substitution effect can be either positive or negative, depending on the shape of the indifference curve and how the price change affects the purchasing power of the budget. This is discussed later in this chapter.

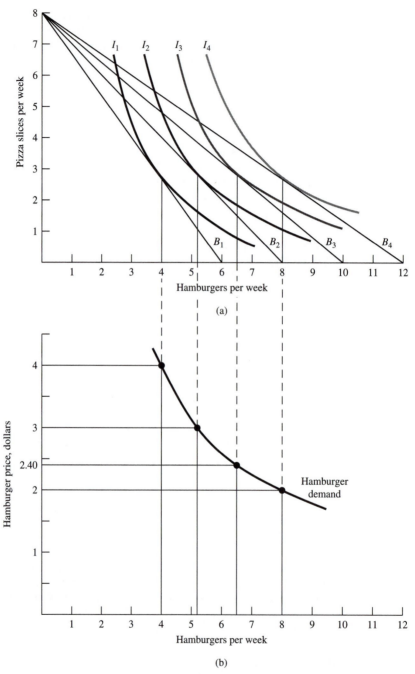

FIGURE 2-7 Derivation of a demand curve from indifference map.

TABLE 2-4 DEMAND SCHEDULE
FOR HAMBURGERS

Price	Quantity
$2.00	8.0
2.40	6.5
3.00	5.25
4.00	4.0

Note: Hypothetical data.

COMMON ERRORS

Many students confuse the budget constraint and the demand curve. While they have the same general shape, note that the axes are defined differently. Second, some students look at an indifference map such as Figure 2-7a and mistakenly indicate that the consumer would purchase 6, 8, 10, or 12 hamburgers, depending on the price of hamburgers. While the consumer can afford this number of hamburgers, these are not the points of tangency of the budget lines and the indifference curves.

This development of the concepts underlying the demand curve provides an example of the *ceteris paribus* assumption. The quantity demanded depends on a good's own price, the price of other goods, the budget, and the tastes and preferences of the consumer. If any of these variables change, the quantity demanded at a given price likewise changes. Also implicit in the discussion is the notion of a fixed period. For example, the budget might apply to a day, a week, or a month. All of these other variables—the price of other goods, the budget, tastes and preferences, and the time period—must remain constant when we construct a demand schedule and draw a demand curve. If we allow any other variable to change as we change the price, it is impossible to tell what caused the change in quantity demanded.

EFFECTIVE DEMAND

There is an important difference between desire, which is based on a person's wants, and effective demand, which combines wants with the person's budget and market prices to determine the quantity demanded. For example, a hungry person desires more food. Unfortunately, the poor who are hungry cannot translate their desires into effective demand. Their budget constraint doesn't allow them to buy enough food to satisfy their hunger.

PRICE ELASTICITY OF DEMAND

The law of demand states that consumers respond to a decline/increase in price by buying more/less of a good. **Elasticity** measures how much consumers change the amount they purchase. We show later that this is important to businesses because they can calculate the change in total revenue associated with a price change. Economists measure this responsiveness and refer to it as the price elasticity of demand. Price elasticity of demand is the percentage change in the quantity of a good demanded in response to a given percentage change in price, *ceteris paribus* (E_d = percentage change in quantity demanded ÷ percentage change in price).

The Formula for Price Elasticity of Demand

The price elasticity of demand, E_d can be calculated as:

$$E_d = \frac{\text{change in quantity demanded}}{\dfrac{\text{sum of two quantities}}{2}} \div \frac{\text{change in price}}{\dfrac{\text{sum of two prices}}{2}} = \frac{Q_2 - Q_1}{Q_2 + Q_1} \div \frac{P_2 - P_1}{P_2 + P_1}$$

This is called the arc elasticity formula because it calculates the elasticity between two points on a demand curve. Elasticity is different at each point on a linear demand curve and the closer together the points, the more accurate is the elasticity estimate. Note that the formula uses the average as the base in calculating the percent changes of both price and quantity. This gives us an elasticity estimate that is the same whether considering an increase or a decrease between two prices.

ELASTICITY OF DEMAND AS A POSITIVE VALUE

The inverse relationship between price changes and the quantity demanded produces a downward-sloping demand curve. When the price increases, quantity demanded declines and vice versa. As a result, the price elasticity of demand coefficient is a negative number. However, as a matter of convenience and to avoid possible confusion in interpretation, economists generally use the absolute value of the price elasticity coefficient, and so the values are reported as positive numbers. Some consider it confusing to say that a price elasticity of –3.0 is greater than –2.0. There is less confusion using the absolute values: price elasticity of 3.0 is greater than an elasticity of 2.0. However, this does not change the fact that E_d is negative. This problem does not arise for supply elasticity, discussed in chapter four, because calculated supply elasticity values are positive.

Own-Price Elasticity of Demand

There are three categories of **own-price elasticity of demand** illustrated in Figure 2-8. They are determined by the responsiveness of the quantity demanded to a change in the price, as follows:

• D_1 **Inelastic demand.** On a percentage basis, the change in price is greater than the change in quantity.
• D_2 **Unitary elasticity of demand.** On a percentage basis, the change in price is the same as the change in quantity.
• D_3 **Elastic demand.** On a percentage basis, the change in price is smaller than the change in quantity.

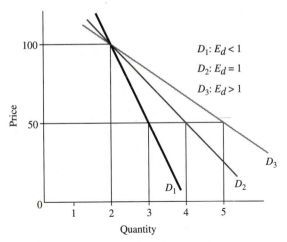

FIGURE 2-8 Inelastic, unitary elastic, and elastic demand curves.

Figure 2-8 illustrates these three cases. In each case the price is arbitrarily reduced by 50 percent. The associated percentage change in quantity is different in the three cases generating the different elasticities between the two points on each curve. The three curves in Figure 2-8 have the following arc elasticities:

Figure 2-8, D_1:

$$E_{d1} = \frac{2-3}{(3+2) \div 2} \div \frac{100-50}{(100+50) \div 2} = \frac{-1.0}{2.5} \div \frac{50}{75} = -0.6, \text{ or } 0.6 \text{ (inelastic)}$$

Figure 2-8, D_2:

$$E_{d2} = \frac{2-4}{(4+2) \div 2} \div \frac{100-50}{(100+50) \div 2} = \frac{-2.0}{2.5} \div \frac{50}{75} = -1.0, \text{ or } 1.0 \text{ (unitary elastic)}$$

Figure 2-8, D_3:

$$E_{d3} = \frac{2-5}{(5+2) \div 2} \div \frac{100-50}{(100+50) \div 2} = \frac{-3.0}{2.5} \div \frac{50}{75} = -1.3, \text{ or } 1.3 \text{ (elastic)}$$

Figure 2-9 indicates that the elasticity of demand changes along a linear demand curve. For this demand curve, the elasticity is 3.0 for the elastic section between points a and b but only 0.33 for the inelastic section between points c and d.

CHANGING ELASTICITY OF A STRAIGHT LINE

The formula for elasticity can also be written as $E = \Delta Q \div \Delta P \times P \div Q$. For a straight line, the slope $\Delta Q \div \Delta P$ is constant, but the price-quantity ratio changes as we move along the line. Therefore, the elasticity changes as we move along a straight-line demand curve.

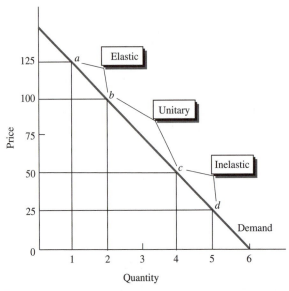

FIGURE 2-9   Elasticities on a linear demand curve.

Cross-Price Elasticity of Demand

The concept of own-price elasticity refers to the effect of a change in price of an individual good, or group of goods, on the quantity of that good demanded (or as discussed in Chapter 4, supplied). The price of one good may also affect the quantity of another good demanded, that is, **cross-price elasticity of demand.**

For goods X and Y, the cross-price elasticity of demand indicates how responsive consumer or producer purchases of X are to changes in the price of Y, and vice versa. The computation of cross-price elasticity is similar to the computation of the good's own-price elasticity of demand. It is the ratio of the percentage change in the quantity of X demanded to a percentage change in the price of Y. The coefficient of cross-price elasticity of demand (E_{dxy}) is calculated as follows:

$$E_{dxy} = \frac{\text{percentage change in quantity demanded of X}}{\text{percentage change in price of Y}}$$

The concept of cross-price elasticity of demand will help us to better understand the relationships between **substitute goods** and **complementary goods,** and to quantify some of these relationships. If X and Y are substitutes, the cross elasticity of demand (E_{dxy}) is positive. This occurs if the quantity of X demanded varies directly with a change in the price of Y. For example, an increase in the price of butter will result in consumers increasing their purchases of margarine, *ceteris paribus*. Thus the cross elasticity between butter and margarine is positive. The larger the positive coefficient, the greater is the degree of substitutability between X and Y.

When E_{dxy} is negative, goods X and Y are complementary in demand—they are used together—for example, hamburgers and hamburger buns. The larger the negative coefficient, the stronger is the complementarity between two commodities. The larger the negative value of E_{dxy}, the greater is the degree of complementarity.

An E_{dxy} coefficient of approximately zero suggests that two goods are independent in consumption. For example, a change in the price of oranges has little if any effect on the purchases of wheat or of automobiles.

DETERMINANTS OF PRICE ELASTICITY OF DEMAND

There are several **determinants of price elasticity of demand.** The law of diminishing marginal utility affects the price elasticity of demand. The marginal utility of a good determines how much consumers will increase their purchases of the good when its price decreases, *ceteris paribus.* The decrease in marginal utility depends on several factors. The first and most general factor is the extent to which one good substitutes for another without substantial loss of utility. This is called *substitutability.*

Substitutability

The ease and extent to which one good may be substituted for another is the most important determinant of the own-price elasticity of demand. Commodities that do not have close substitutes have the most inelastic demands—fluid milk, eggs, salt, potatoes, and bread, for example. The demand is very inelastic for vegetables in aggregate. However, because one vegetable may be substituted for another, the demand for each vegetable is more elastic than the elasticity of demand for the aggregate. The elasticity of demand for most livestock products is greater than the elasticity of demand for most crops, largely because of the greater ease of substitution among most individual livestock products. Thus, the elasticity of demand for any individual commodity depends on the closeness of its substitutes (which is measured by cross elasticity). A high degree of substitutability results in a high demand elasticity coefficient. This is because if the price increases for a product with many good substitutes, consumers will shift to the substitute products.

Complementarity

A high degree of complementarity among goods is associated with a lower own-price elasticity of demand. That is, complementarity tends to make the demand for a commodity more inelastic with respect to price. Because sandwich buns are complementary goods with hamburgers and other meats, the demand for buns would be inelastic. If you are going to make sandwiches for a picnic, you are quite likely to buy buns.

Besides the effects of substitutability and complementarity, there are other factors that affect price elasticity of demand. Consumer income is one of the more important determinants in markets for agricultural commodities.

Consumer Income

Consumer income affects price elasticity of demand for all consumer goods. For example, the price elasticity of demand for total food is inversely related to the level of consumer income. Rich people buy about the same quantity and quality of food regardless of food prices; they have much more price-inelastic demand for food than poor people do.

Necessities versus Luxuries

Price elasticity of demand for necessities tends to be inelastic; the demand for **luxury goods** tends to be more elastic. Salt, bread, and milk are necessities for most people; and demand for these foods is very inelastic as compared with the demand for commodities such as steak, lobster, caviar, and crab meat, which most people view as luxury goods. If your income were to double or triple, you would probably not buy more beans, bread, or spaghetti but you might buy more steak, lobster, or other costly favorites. Also, many food-related services, such as those offered by fine restaurants, may be regarded as luxuries. The size of the restaurant industry is one index of affluence in a society.

Proportion of Income Spent on a Good

Goods responsible for a high percentage of consumers' budgets tend to be price elastic. In affluent countries meat is an important item in most consumers' food budgets, for example. This makes meat more price elastic. In comparison, bread requires only a small part of most consumers' food budget. A change in the price of bread will not have as much effect on the quantity demanded. The demand for salt tends to be highly inelastic for several reasons. A small amount of salt is necessary for good health, and adding some salt makes food more palatable; but even if this were not the case, salt is a negligible item in a food budget, and the price of salt has little effect on the quantity demanded.

Time

Generally, the longer the time for adjustment to a change in price, the greater is the price elasticity of demand, other things being equal. This is due partly to the fact that it takes time for habits to change or for people to change consumption patterns. Also, it may take some time for consumers to become aware of the available alternatives.

In a few cases, however, time has the opposite effect. For example, if there is a sharp increase in the price of meat, some people will make substitutions immediately, but over a longer time they may resume their former buying habits. Here, the short-run demand will be more elastic than the long-run demand. And people may react to a price decrease by building inventories of storable commodities, which makes their short-run demand more elastic. Hence, the consumer can gain from price fluctuations or price instabilities by taking advantage of periodically low food prices. Generally, however, you will find that the longer the time allowed, the greater will be the price elasticity of demand because it takes time to adjust to changes in prices.

Trade

Trade is discussed in detail in Chapter 8. However, trade does have an important effect on the price elasticity of demand. Goods traded internationally have a more elastic demand than if they were marketed only domestically, *ceteris paribus*. This is because there are many more consumers and there may be many more substitute commodities on the world market than in a domestic market.

INCOME ELASTICITY OF DEMAND

Income elasticity of demand is the percentage change in quantity of a good purchased resulting from a percentage change in income, *ceteris paribus*. The formula for income elasticity of demand (E_{dy}) is as follows:

$$E_{dy} = \frac{\text{percentage change in quantity demanded}}{\text{percentage change in income}} = \frac{Q_2 - Q_1}{Q_2 + Q_1} \div \frac{Y_2 - Y_1}{Y_2 + Y_1}$$

For most goods E_{dy} is positive. However, the positive income-elasticity coefficient varies widely among goods, especially among various food commodities. For **normal goods,** such as butter, milk, and bread, the income elasticities are low but positive, between zero and 1.0. For **superior goods,** for example, lamb chops, lobster, and fancy cheeses, income elasticity is very high, greater than 1.0. The income elasticity of demand is also high for the amenities associated with food—for example, fine wines, gourmet restaurants, and personal services. For these goods the demand increases by a larger percentage than the percentage increase of income.

There are still other goods, **inferior goods,** for which an increase in income results in a decrease in consumption. These goods have negative income elasticities of demand. Consumers tend to decrease their purchases of these goods when their income increases. For example, most poor families in the United States will reduce their purchases of potatoes or rice if their incomes increase. This means that for such people these are inferior goods. This is most commonly observed for less-preferred foods that the poor consume only because of their low incomes.

The income elasticity of demand for total food is different from the income elasticity of demand for each food commodity, or for the services and amenities associated with food. As per person income increases, less grain is consumed directly as human food but much more grain is consumed as livestock and poultry products. For example, recently, the annual direct consumption of grain for human food was about half as high in Canada and the United States as in India and Mexico. However, the total grain used per person as meat and other products was over four times as high in Canada and the United States. The percentage of total income spent on food generally declines as income increases resulting in an income elasticity of demand for the total quantity of food less than one, a relationship known as **Engel's Law.**

ENGEL'S LAW: INCOME ELASTICITY OF DEMAND FOR FOOD

Ernst Engel (1821–1896), a Prussian statistician, established the relationship between income and the quantity of food demanded (the percentage of total income spent on food generally declines as income increases), and demonstrated that a general rise in productivity results in a shift from agriculture to manufacturing. He also used his observations as an argument against the fears of overpopulation, which were popularized by Thomas Robert Malthus (1766–1834). Malthus argued that there is a tendency for the human race to reproduce until it reaches the level of bare subsistence. To overcome this, Malthus advocated sexual abstinence and late marriage. Engel popularized the view that such drastic measures were unnecessary because advances in productivity and trade could balance total food supply and demand without great price increases.

TABLE 2-5   CHARACTERISTICS OF PRICE ELASTICITY OF DEMAND
AND OF INCOME ELASTICITY

Price elasticity	Price and income elasticity	Income elasticity
	Elastic	
Goods with substitutes Food products purchased by the poor	Luxury goods Long time horizons for adjustment	All food in very poor countries
	Inelastic	
Goods that are complementary to other goods Food products purchased by the rich	Necessities Short time horizons for adjustment	All food in countries with adequate incomes for food purchases

The E_{dy} for the total quantity of food demanded is very low in developed countries, but it is generally much higher in poor countries. There are, of course, limits to the amount of food people choose to eat and how much of their income they spend on food.

Income elasticity coefficients help predict which industries, or sectors of an industry, will experience the greatest increase in demand when incomes increase. Industries that sell normal or superior goods can expect to realize increasing sales as incomes rise. Industries selling inferior goods must expect poorer sales performance.

SUMMARY

The primary goal of this chapter has been to illustrate the law of demand. We started by pointing out a basic characteristic of human beings, they have unlimited wants and desires. Goods are consumed to partially satisfy these wants. The term utility was used to describe this effect. As more of a good is consumed, the incremental contribution to utility of additional consumption declines. This is the law of diminishing marginal utility.

The next important concept is the budget constraint; it determines the affordable combinations of goods. We examined how changes in market prices and in the budget change the affordable combinations. Combining the concepts of utility and the budget constraint, we found the single, utility-maximizing combination of goods for a given budget. Utility is maximized when the costs of marginal utilities are equal for all goods. Next we found that utility maximization also occurs at the point where an indifference curve is tangent to the budget line.

We showed that when the price of a good increases and all else remains constant, the consumption of that good falls. This illustrates the law of demand. The information about the quantity demanded at various prices (a demand schedule) can be used to construct a demand curve.

Finally, we presented the concept of elasticity of demand and showed how elasticity is calculated. Table 2-5 summarizes some of the major determinants of price elasticity of demand and of income elasticity.

LOOKING AHEAD

In the next three chapters, we examine the behavior of producers of goods. The basic problem is to determine how a firm decides how much to produce. We show that producer behavior parallels consumer behavior. Just as utility maximization drives consumer behavior, profit maximization is the basic goal of a firm and drives its decisions. The firm operates with a production technology that determines how inputs are converted into goods. We will use information on the interaction of profit-maximizing behavior, production technology, and changing market prices to explain the law of supply, which is that the supply curve slopes upward.

IMPORTANT TERMS AND CONCEPTS

arc elasticity   40
budget   28
budget constraint   28
budget line   28
ceteris paribus   25
complements   43
consumer choice problem   31
consumer income   43
consumption   24
cross-price elasticity of demand   42
demand   24
demand curve   37
demand schedule   37
determinants of price elasticity of
  demand   43
elastic demand   40
elasticity   39
Engel's Law   45
income elasticity of demand   45
indifference curve   35
indifference map   35
inelastic demand   40

inferior goods   45
law of demand   25
law of diminishing marginal utility   26
luxuries   44
marginal rate of substitution   36
marginal utility   26
necessities   44
negative marginal utility   26
normal goods   45
own-price elasticity of demand   40
point elasticity   40
price elasticity formula   40
price elasticity of demand   40
substitutes   43
superior goods   45
tangency of the budget line and
  indifference curve   36
tastes and preferences   39
unitary elasticity of demand   40
utility   26
utils   26
wants   25

QUESTIONS AND EXERCISES

Name That Term

Read the following sentences carefully and fill in the missing term or terms.

1 The law of _____ says that, other things being equal, a higher price means less is purchased.
2 The Latin phrase for "all else equal" that clutters economics texts is _____.

3 The satisfaction consumers derive from goods is called _____.

4 A consumer's _____ is the amount of money the consumer has to spend during a certain period of time.

5 _____ is money set aside today for future consumption.

6 A consumer, when faced with a budget, allocates consumption choices in such a way as to _____ utility.

7 To maximize total utility within a budget constraint, consumers allocate expenditures so that the ratio of _____ to price is the same for all goods.

8 The line connecting all combinations of two goods that give equal utility is called an _____.

9 The line connecting all combinations of price and quantity consumed, other things being equal, is the _____ for a good.

10 Goods such as apples and oranges are called _____.

11 Consumers buy less of an _____ good at every price when their incomes increase.

12 The income elasticity of demand for _____ goods is positive.

13 The income elasticity of demand for _____ is positive but less than one.

14 The income elasticity of demand for _____ goods is greater than one.

15 A negative cross-price elasticity of demand between goods A and B indicates that goods A and B are _____.

True/False

Read the following sentences, then decide whether each statement is true (T) or false (F) and mark it accordingly.

T F 1 Consumers maximize their utility at combinations of goods where their indifference curves intersect.

T F 2 Consumers maximize their utility by finding the point on their budget line that enables them to reach the highest possible indifference curve.

T F 3 An increase of a consumer's budget, other things equal, means the consumer can reach a higher indifference curve.

T F 4 If the price of ice cream doubles while the price of chocolate chip cookies remains the same, then the budget line for ice cream and cookies shifts to the right in a parallel fashion.

T F 5 The budget line separates affordable combinations from unaffordable ones.

T F 6 If you have $100 to spend on pizza and soda, and the price of soda is $1.00 per cup and the price of pizza is $2.00 per slice, then you can purchase 30 slices of pizza and 50 cups of soda.

T F 7 If the price of good A is three times as much as the price of a similar good B, then a consumer maximizes utility subject to a budget constraint by consuming more units of B than A.

T F 8 A vertical demand curve indicates that, other things equal, a consumer will purchase the same quantity of a good whatever its price.

T F 9 If the price of good A goes up, other things equal, a consumer will demand fewer units of good B.

T F 10 Consumers maximize happiness by striving to assure that the marginal utility of each good they consume is equal.

T F 11 An increase in the ticket price for football games will increase revenues from ticket sales if the demand for football tickets is price inelastic.

T F 12 If the own-price elasticity of demand for widgets is 0.8, a decrease in the price of widgets will lead to a decrease in the amount consumers will spend on widgets.

T   F   13 Most goods are inferior.
T   F   14 The own-price elasticity of demand for goods, by the law of demand, is positive.
T   F   15 The own-price elasticity is constant along a linear demand curve.
T   F   16 Consumer goods with many good substitutes tend to have price elastic demand.

Multiple-Choice Questions

Choose the letter of the response that best answers the question or completes the statement.

1 If asked to choose between combination A or B on one of your indifference curves between milk and cookies, you would choose:
   a the combination with more milk.
   b the combination with the most cookies.
   c the combination with the largest number of total calories.
   d the combination with the smallest number of total calories.
   e I don't care; both would be equally good.
2 Suppose under your present consumption pattern, your ratios of marginal utilities to prices are equal for all goods, except widgets for which the ratio is higher. To maximize utility you should
   a purchase fewer widgets.
   b purchase the same number of widgets.
   c purchase more widgets.
   d refuse to purchase widgets.
   e ask for more information.
3 The law of demand says that
   a people want things they cannot have.
   b other things equal, consumers prefer expensive goods to cheap ones.
   c other things equal, a good's price and the quantity demanded are inversely related.
   d only a few consumers are lucky enough to satisfy all their wants.
   e other things equal, consumers with higher budgets purchase more units of a good.
4 *Ceteris paribus* is a Latin phrase for
   a other things equal.
   b let the buyer beware.
   c money can't buy love.
   d markets aren't always fair.
   e never put off until tomorrow what you can do today.
5 If a good is free (that is, if its price is zero), according to utility theory you would consume
   a every unit of the good you could get your hands on.
   b none of the good because you know, "there is no such thing as a free lunch" and are wary of handouts.
   c as many units of the good as you dare without seeming greedy to your friends and neighbors.
   d units of the good until the marginal utility of consuming it reached zero.
   e however many units everyone else you know thought was about right.
6 If the price of hamburgers increased while the price of milk remained the same and your budget remained constant, then your budget line for milk and hamburgers would
   a cross the hamburger axis at a higher point.
   b shift inward in a parallel fashion.

   c shift outward in a parallel fashion.

   d pivot about the milk intercept toward the hamburger axis.

   e pivot about the hamburger axis toward the milk axis.

7 If the price of pizza and the price of soda remained constant, but your budget increased, then your budget line would

   a remain unchanged.

   b shift out in a parallel fashion.

   c shift inward in a parallel fashion.

   d pivot about the soda intercept toward the pizza axis.

   e pivot about the pizza intercept toward the soda axis.

8 Demand curves

   a slope upward.

   b show how quantity purchased changes as its price changes, *ceteris paribus.*

   c show how quantity of a good changes as budgets increase, *ceteris paribus.*

   d show how quantity of a good changes as the price of other goods change, *ceteris paribus.*

   e show how quantity purchased is influenced by advertising and other factors that influence tastes and preferences, *ceteris paribus.*

9 Consumers maximize utility where

   a the cost of marginal utility is the same for all goods.

   b an indifference curve is tangent to a budget line.

   c no other affordable combinations of goods yield greater satisfaction.

   d all of the above.

   e none of the above.

10 To find the number of shirts purchased from an indifference map for shirts and shoes

   a find the points where the budget line meets the shoe axis.

   b find the points where the budget line meets the shirt axis.

   c find the points on the shirt axis associated with the points where the cost of marginal utility is the same for shoes and shirts.

   d find the points on the shirt axis associated with the points where each of the indifference curves cross the budget line.

   e find the points on the shirt axis associated with the points where each of the indifference curves is tangent to a budget line.

11 What happens to a consumer's spending on a normal good if income decreases, *ceteris paribus?*

   a increases

   b remains constant

   c decreases

   d uncertain

12 If the price of good X increases and good X and Y are complementary goods, what happens to the number of units of Y consumed?

   a increases

   b remains constant

   c decreases

   d uncertain

13 If the price elasticity of demand for widgets is 1.2 and presently 1000 widgets are consumed, how many widgets will be consumed if the price falls by 15%?

   a 50 fewer

   b 20 fewer

   c 120 more

d 150 more

e 180 more

14 If the income elasticity of demand for widgets is 1.2 and presently 1000 widgets are consumed, how many widgets will be consumed if income increases by 20%?

    a 240 fewer

    b 200 fewer

    c 120 more

    d 200 more

    e 240 more

15 If the cross-price elasticity of demand for widgets and gidgets is 0.6 and presently 1000 widgets are consumed, how many widgets will be consumed if the price of gidgets increases by 10%?

    a 600 fewer

    b 60 fewer

    c 60 more

    d 100 more

    e 600 more

Technical Training

The following questions ask you to solve consumption problems graphically. *Hint:* You will find it easiest to see your work if you use graph paper and create a graph that covers at least half a sheet.

1 Suppose your demand curve for basketball game tickets takes the form

$$P_g = 30 - 2G$$

where P_g = price per game

       G = games attended

    a Graph your demand curve.

    b How many games would you attend if tickets cost $12.00?

    c What is the most you would pay to attend 5 games?

    d What is the most you would pay to attend 13 games?

2 Consider the indifference curve presented in the following graph.

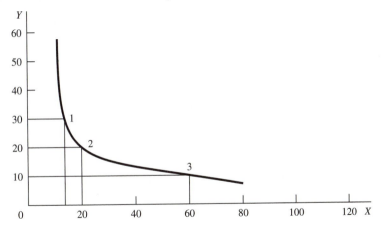

What combination of goods X and Y would the consumer choose in each of the following circumstances: 1, 2, or 3?

Case	Budget	Px	Py	Choice
a	120	1	6	_____
b	60	1	2	_____
c	90	1.5	3	_____
d	80	4	1	_____

3 Use the data provided to find the arc elasticity in the following questions. Of course, assume "other things equal."

a
Period	Units demanded	Income
1	100	$1200
2	110	$1600

What is the income elasticity of demand? Is the good a luxury, necessity, or inferior?

b
Period	Units demanded	Income
1	100	$1200
2	150	$1600

What is the income elasticity of demand? Is the good a luxury, necessity, or inferior?

c
Period	Units of X demanded	Price
1	200	$5.00
2	160	$7.00

What is the income elasticity of demand? Is the elasticity of demand elastic, unitary, or inelastic?

d
Period	Units of X demanded	Price of Y
1	60	$10.00
2	40	$12.00

What is the cross-price elasticity of demand of X with respect the price of Y? Are X and Y substitutes or complements?

Questions for Thought

The following questions are designed to help you apply the concepts you have learned to real-life consumer choices. Answering them will help prepare you for answering discussion questions based on the material in this chapter.

Use the following information to answer questions 1 through 6:

Lee Frosh is a recent high school graduate who is planning to attend the State University. His parents will provide him with $500 a month for his consumption needs. Since this is his only source of income, Lee will be cautious when deciding how to allocate the limited funds to his unlimited wants. Assume that Lee decides to allocate $100 of his monthly allowance between movies and bowling. Table 1 displays different quantities of each product that Lee could consume and the resulting utility.

1 Fill in the marginal utility columns in Table 1.

2 Assume that the price of a movie $P_m = 10 and the price of a game of bowling $P_b = 5. Plot Lee's budget constraint on the following graph.

TABLE 1 TOTAL AND MARGINAL UTILITY FROM BOWLING
AND MOVIES

Bowling		
Consumption/ month	Total utility	Marginal utility MU_b
1	50	———
2	80	———
3	104	———
4	125	———
5	143	———
6	155	———

Movies		
Consumption/ month	Total utility	Marginal utility MU_m
1	75	———
2	117	———
3	144	———
4	168	———
5	178	———
6	183	———

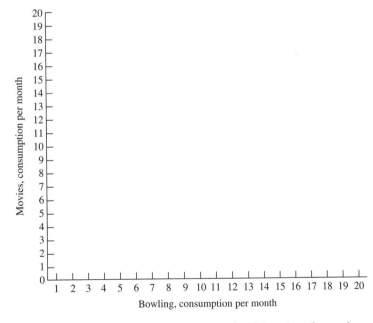

3 What would happen to Lee's budget constraint if the price of a movie were reduced
to $7? What would happen if his budget constraint were reduced to $50.00?

4 Would Lee's total utility increase or decrease from the price change in the previous question? Explain.

5 The following graphs give two possible indifference maps for the consumption of bowling and movies. Which map (Graph A or Graph B) indicates a stronger preference for the consumption of movies? Explain your answer. (*Hint:* Determine how many movies Lee would attend if the prices of movies and bowling were the same.)

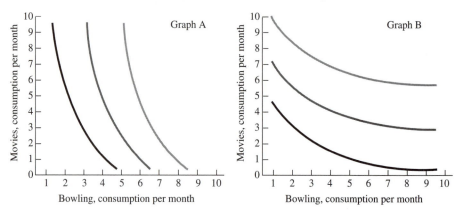

6 Use the following indifference map to derive the demand curve for bowling as the price of bowling P_b changes. Assume that the original budget $(B) = \$100$ and the price of movies $P_m = \$10$. Allow P_b to equal three different prices: $P_b = \$10$, $P_b = \$7$, $P_b = \$5$. Show the shifts in the budget constraint on Graph 1 and the derived demand curve on Graph 2.

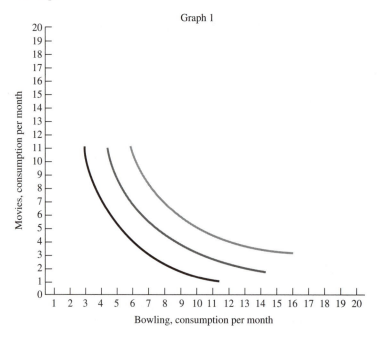

Graph 2

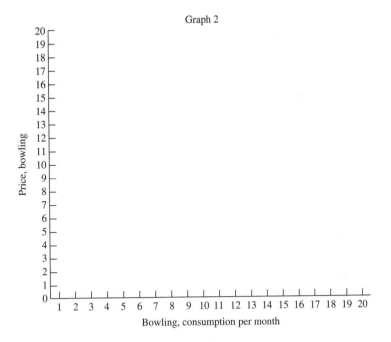

Bowling, consumption per month

7 Explain why marginal utility decreases as the consumption of a good is increased. Give an example from your own experience.

8 Sketch your indifference curve for a week's consumption of two kinds of sandwiches you enjoy. Now sketch your indifference curve for one of these kinds of sandwiches and your favorite beverage. Why are they different shapes?

9 How is the coefficient of price elasticity of demand computed? Is the elasticity the same at various points on a linear demand curve? Why or why not?

10 What are the determinants of the elasticity of demand?

11 How much do you guess you spend on table salt in a year? If the price doubled, would it change the amount you buy? How much do you spend on clothing in a year? If the clothing prices doubled, would it change the amount of clothing you would buy? Explain the difference in your demand for salt and clothing.

12 What is cross-price elasticity of demand? How is it computed?

13 If cross-price elasticity of demand is negative, are the two goods substitutes or complements?

14 Do you have everything you want? If not, list five goods you want but are unlikely to purchase in the next year. Why do these goods appear on this list? What would result in your deciding to purchase them this year?

15 Are your wants truly unlimited? Suppose you received a tax-free income of $10 million per year for the rest of your life. Would the consumption problem disappear for you?

16 Can we compare utility levels among consumers? Do you get more or less satisfaction, for example, from your consumption combination than people in other cultures who consume different things?

17 Cities may realize that monthly parking meter revenues increase after they increase the parking rates. What does this suggest about the price elasticity of demand for municipal

parking? Does it follow that the city can expect revenues to continue to increase indefinitely if they continue to raise the parking rates?

ANSWERS AND HINTS

Name That Term   **1.** demand; **2.** *ceteris paribus;* **3.** utility; **4.** budget; **5.** Savings; **6.** maximize; **7.** marginal utility; **8.** indifference curve; **9.** demand curve; **10.** substitutes; **11.** inferior; **12.** normal; **13.** necessities; **14.** luxury or superior; **15.** complementary

True/False   **1.** F; **2.** T; **3.** T; **4.** F; **5.** T; **6.** F; **7.** T; **8.** T; **9.** F; **10.** F; **11.** T; **12.** T; **13.** F; **14.** F; **15.** F; **16.** T

Multiple Choice   **1.** e; **2.** c; **3.** c; **4.** a; **5.** d; **6.** d; **7.** b; **8.** b; **9.** d; **10** e; **11.** c; **12.** c; **13.** e; **14.** e; **15.** c

Technical Training
1. a.

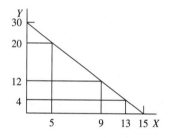

   b. 9 games; **c.** $20.00; **d.** $4.00
2. a. 3; **b.** 2; **c.** 2; **d.** 1
3. a. 0.33, necessity; **b.** 1.4, luxury; **c.** 0.67, inelastic; **d.** −0.2, substitutes

PRODUCTION FUNCTIONS AND PRODUCT CURVES

CHAPTER OUTLINE

OVERVIEW

The decision-making processes of consumers and producers are the fundamental building blocks of microeconomics. In Chapter 2, we discussed the basic economic theory of demand, and we showed how individuals make buying decisions to maximize their satisfaction or welfare. In this chapter and in the following two, we analyze production processes and show how producers make decisions to maximize profits. Using examples from agriculture, this chapter develops the basic physical relationships between the inputs used in production and the resulting output. Important measures of input and output are illustrated and discussed.

LEARNING OBJECTIVES

This chapter will help you learn:

- What a production function is
- How advances in technology affect the production function
- How average and marginal productivity are determined
- About the relationships among the several productivity measures

PRODUCTION FOR PROFIT

Production is the use of inputs to produce a good or service. Commodities such as wheat or cotton, goods such as automobiles or hamburgers, and services such as banking and waste disposal are produced when several resources, or factors of production, are combined in a production process. Even a plant growing in the wild requires sunshine, water, and chemicals from the soil. Agricultural crops generally require land, management, labor, seed, fertilizer, and machinery. Livestock require feed, water, shelter, and care. These are the inputs to the production process.

Farmers and other business operators manage production processes, which means making decisions about using resources to produce the products to be sold on the market. Before we analyze the production process, we stress that the analytical approach used in economics relies on a behavioral assumption about producers. The basic objective of the farmer, or of the operator of any business, is to maximize economic profit. While other objectives may come into play, involving such things as market share and social responsibility, the long-term "bottom line" is profit. One of the more important insights you will gain from this section of the book is that maximizing output does not maximize profits.

What is meant by economic profit? Economic profit is not the difference between the costs and returns or expenses and receipts. This difference is sometimes called *accounting profit,* because it is what an accountant would report. The problem is that some costs may be excluded. For example, several members of the family might work on a family farm without pay. Does this mean there is no value associated with their work? Of course not. The value of their work is the pay these individuals could receive in another position. This is their **opportunity cost**—the amount of money that they could have earned if they had taken the opportunity to work elsewhere. Or consider a farm operating on land purchased many years ago at a price well below the current land price. When assessing the costs of production the farmer must consider the current value of the land, perhaps best indicated by the amount of money that could be charged if the land were rented to another farmer. Similarly, the cost to you of a semester in college is more than the "out-of-pocket" costs of tuition and expenses. You must also consider the money you could earn if you were not in college, your opportunity cost of not working.

It is also worth noting that an individual may choose to work in a job or profession that pays less than some alternative. In this case, the opportunity cost associated with choosing a lower paying option may be explained by better working conditions, more security, or more satisfaction associated with the job performed.

Economic profit, then, refers to the difference between returns and the value of all the inputs used in producing the good or service sold. It is total returns minus the cash payments for purchased inputs and the opportunity costs of the firm-owned inputs. For example, a wheat farmer's economic profit is the value of the wheat sold minus all economic costs. Economic costs are the cost of all purchased inputs such as seed and fuel plus the opportunity costs. Opportunity costs include the family labor (the wages that could have been earned working off the farm) and land owned by the farmer (the rent that could have been received if the land were rented to another farmer).

We begin now to develop the economic model of the decision-making activities of a firm. We first consider the physical relationships; then we add costs in Chapter 4 and returns in Chapter 5, where we bring all of this together and identify the firm's profit-maximizing level of output.

THE PRODUCTION FUNCTION

The general purpose of studying the theory of production is to learn how to find the combination of *inputs* and *outputs* that will generate the most profit. Even for a relatively simple production process, it is difficult to determine the best combination if all the inputs and outputs are allowed to vary simultaneously. To simplify the problem, we start with the idea that the manager has a certain amount of land (or space), management, labor, and other inputs. Some of these inputs are **fixed,** and some are **variable.** Generally the distinction between fixed inputs and variable inputs is based on how long firms operate with a given amount of the input. A corn producer may vary the amount of seed, fertilizer, and labor each year (variable inputs) but will seldom buy or sell land or change machinery (fixed inputs). A restaurant will likely operate with the same building and equipment for several years (fixed inputs) but might adjust daily the amount of labor and the food purchased (variable inputs) to create the menus offered. The firm seeks to find the best combination of variable inputs to combine with the available fixed inputs.

There is a maximum amount of output that can be created with each combination of inputs or factors of production. This technical relationship between inputs to outputs is so important that economists have given it a name: the **production function.** *The production function is the technical relationship between inputs and outputs, indicating the maximum amount of output that can be produced with alternate amounts of variable inputs used in combination with one or more fixed inputs under a given state of technology.*

A production function is based on a given state of knowledge and technology. For example, a production function for rice gives many alternate combinations of inputs— land, seed, labor, fertilizer, water, and so on—that can be used to produce various amounts of rice using a given production technology. The production processes in every firm (including farm firms) can be represented by a specified, or functional, relationship between inputs and outputs, hence the term production function. Because of its importance to the decision processes of all firms, we discuss the production function in considerable detail.

> **VARIABLE AND FIXED INPUTS**
>
> While seed and fertilizer typically are considered variable inputs, once the seed has been planted and the fertilizer has been applied, they surely are fixed inputs, since changing the amount used is virtually impossible. Thus, after seed and fertilizer have been applied, they join land and the other inputs of planting the crop as fixed inputs. At that time, the variable inputs are those not yet applied, which might include additional fertilizer, pesticides, and the labor and fuel for harvest.

Fixed and Variable Factors of Production

Suppose you want to go into business. One of the first things you will have to do is buy or rent some land, a building, and some basic equipment. Even though you probably hope to expand—and you realize that you may lose it all—for the present you will operate with these inputs. These are your **fixed factors of production.** You will have to maintain them even if you don't produce anything, even if output is zero. The fixed factors of production remain in place, but the intensity of their use can change.

Once you have lined up the fixed factors of production, you must decide how much of the variable factors of production to use during your planning horizon. If you are farming, you will have to decide how much seed, fertilizer, fuel, and labor to use each season. If you are opening a restaurant, you will have to decide how much meat, vegetables, fruit, sauces, labor, utilities, and all other inputs to use each day or week. These are your **variable factors of production.** The amount of the variable inputs used changes as you change the amount of output. In much of the following, we will consider one of the inputs to be variable and all others to be fixed. Although this may seem too simple, it does reflect many typical decision processes. That is, the manager often frames a decision in terms of questions such as: Should I apply more fertilizer? Should I rent more land? Should I purchase a larger tractor? Of course, if the manager changes the amount of one of the inputs used, it is necessary to review the amount of other inputs used. But, by focusing on a single variable input and fixing the quantities of the other inputs, the physical and economic relationships can be presented more clearly.

Short-Run and Long-Run Time Horizons

Usually the distinction between fixed and variable inputs is based on the length of the planning horizon. In the short run, land, buildings and machinery usually are fixed inputs. Other inputs can be held fixed by the operator and the same concepts can be used to determine profitability. For example, given the available labor, machinery, and other inputs, land is a variable input in a decision about renting additional land. In the long run, all inputs are variable.

Production processes are classified according to the "length of run" or the length of time considered. The **short run** is so short that the business manager or owner does not change the use of the fixed factors of production. The **long run** allows sufficient time for the manager to change the use of all of the factors of production. In the long run, no inputs are fixed. There is no specific length of time that constitutes the short and long run. The short run would be at least one cropping season for a grain farmer since it is quite unlikely that the farmer would change the amount of land used during the crop-year. Similarly, a food processor would not vary the size of the plant during a processing season. Of course neither the farmer nor the food processor would likely change the size of the operation often. For a citrus producer, the short run is at least four years—the time it takes for citrus trees to start producing.

The vast majority of farms and agribusinesses produce more than one product. A farmer may produce several grain crops and cattle or hogs. A food processor may produce dozens of different food products. A restaurant will serve numerous items to breakfast, lunch, and dinner clientele. Each of these activities can be considered separate enterprises. We simplify the presentation by focusing on how much the production of one good or service produced changes as the use of one variable input is changed while holding all other inputs fixed. In Chapter 6 we discuss changing more than one input and output.

QUANTITY AND YIELD

We sometimes use the term *yield* or output per unit of land instead of referring to the quantity of output produced. Grain yields are generally reported in terms of bushels per acre or metric tons per hectare, milk production is reported in hundreds of pounds or kilograms of milk produced per cow, forage crops are reported in tons per acre, and cotton in bales per acre or hectare. In each case the quantity of output is reported in terms of the amount produced per unit of land or other fixed inputs, when in fact there are several fixed inputs.

Total Physical Product

A total physical product (TPP) curve gives the relationship between a single variable input and output. It is a portion of the production function, and it indicates how much of a good or service can be produced with increasing amounts of one variable input while holding constant the amount of all the other inputs.

A graph of a total physical product curve is given in Figure 3-1. As the use of the variable input increases, the quantity produced increases, as shown by the increase in TPP. At low levels of input, TPP increases at an increasing rate. With more of the input, the production function does not go up as fast, it increases at a decreasing rate. The

FIGURE 3-1   The total physical product curve.

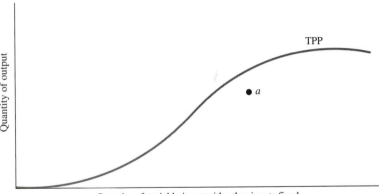

Quantity of variable input with other inputs fixed

quantity produced reaches a maximum at some point. Beyond this maximum point, further increases of variable input used result in a decrease in output. The manager has the option to operate anywhere on, or below, the TPP curve.

WHY THE TPP CURVE IS CURVED

Think of how a plant, your favorite houseplant or an agricultural crop, responds to water. Without water the plant cannot live. After you reach a level that is adequate to support plant growth, as small amounts of water are added you see dramatic responses in the plant. In the case of a food-grain crop, the plant may grow at low levels of water input, but the amount of grain produced would be quite low. As additional water is added and other requirements are met, the amount of grain harvested increases dramatically. With further additions of water, there is some increase in harvest, up to the maximum possible yield level. Beyond this point additional water reduces the amount of harvest. At some point additional water kills the plant, and there is nothing to harvest.

If the TPP function were not curved, it would imply that you could feed the world from a flower box—"the flower box phenomenon." If the amount of output from a fixed amount of land did not decrease at some point you could simply add more and more seed, fertilizer, water, and so on, to your flower box and produce more and more output. Of course this will not happen, even if you have a very, very green thumb! The amount of food you can produce in a flower box is limited by the amount of soil (land) in the box.

A farmer with a certain amount of land and equipment decides how much of each of the variable inputs—such as seed, plant nutrients, and labor—to use in producing a crop. This is analogous to the owner of a tomato processing plant with several production lines deciding how much labor and other variable inputs to use per processing line in processing a given quantity of tomatoes.

EFFICIENCY

The TPP curve gives the maximum quantity of output that can be produced with different amounts of the variable input. It is possible, but seldom desirable, to produce below the production function. *A firm operating on the TPP curve is efficient; it is producing as much output as possible given the quantity of inputs used.* A firm is inefficient if it produces less than the maximum and operates below the curve. For example, if the firm shown in Figure 3-1 operated at point *a*, it would not be efficient.

Effects of Technological Change

Next we discuss the effects of **technological change** on the production function. The following discussion indicates the effects of technological change and provides a better understanding of a production function and its importance to the firm.

A production function gives the maximum amount of output that can be produced by a firm at any given time using a given technology. The production function can shift over time as a result of research and development. As indicated in Figure 3-2, some changes in technology shift the production function so that more can be produced from the same quantity of inputs at any point on the function. That is, more of a good or service can be

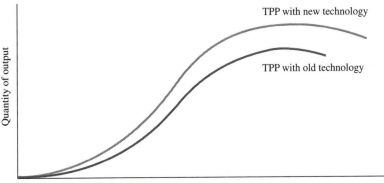

FIGURE 3-2   Effects of technology improvements on the production function.

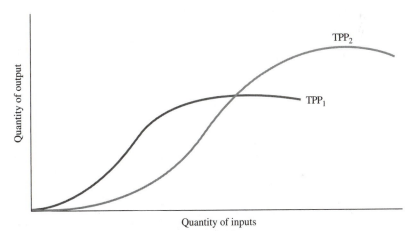

FIGURE 3-3   TPP curves representing alternative production technologies.

produced from a given level of input use, or the same amount can be produced using less of the inputs. For example, the corn production function shifts upward if an improved corn variety increases the response of the crop to fertilizer. Or, improved gas mileage implies an increase in automotive technology.

Figure 3-3 illustrates another effect of technology on a production function. The technology represented in curve TPP_2 is superior at high levels of inputs, and the technology represented in TPP_1 is superior at low levels of inputs. These curves might represent the production of a crop with varying quantities of fertilizer using different varieties and production techniques.

Generally speaking, technological advances tend to "come together" as "packages" of new technology. When they do, the inputs are complementary in production. The increased productivity of one depends upon using more of the other. The Green Revolution in developing countries, for instance, was based on a package of new technologies including new crop varieties that could use fertilizer more efficiently.

THE GREEN REVOLUTION

The **Green Revolution,** which began in the 1960s, was at least in part the result of an international effort to improve the productivity of agriculture in the developing nations. New wheat and rice varieties, new means of controlling plant diseases, and other changes necessary to improve crop yields were developed and made available to farmers. Major increases in production of wheat and rice occurred in many areas of the world. There is considerable debate about whether we can continue to increase yields to keep up with growing world demand.

Finally, it is important to note that some changes in production technology have adverse as well as beneficial effects—damaging the environment, for example. These adverse effects reduce the social value of the technology. If they are significant, the desirability of the technology is likely to be questioned. If there are fears of unknown adverse consequences, there may be objections to making the technology available to producers. Society has, on occasion, feared technologies that proved benign and at other times embraced technologies that proved damaging.

We turn now to the development of the set of product curves derived from the firm's TPP curve.

PRODUCT CURVES

We begin the discussion using one of the simplest production functions possible. One product, corn (maize), is produced using a single variable input, fertilizer. To simplify the analysis, other inputs are treated as fixed. Thus, the question is, How many units of the variable input, fertilizer, are used with the fixed inputs to produce a given amount of corn?

Total Physical Product Curve

The TPP curve shows how much of a product is produced with varying amounts of the variable input *given the quantity of fixed inputs used in the production process*. For example, farmer Jones visualizes the crop response to fertilizer based on knowledge of the production function, as shown in Table 3-1 and in Figure 3-4. The total physical

TABLE 3-1 A HYPOTHETICAL PRODUCTION FUNCTION SCHEDULE ILLUSTRATING TOTAL PHYSICAL PRODUCT

Input	TPP
0	0.00
1	9.00
2	27.00
3	50.00
4	70.00
5	85.00
6	96.00
7	101.00
8	101.00
9	95.00
10	85.00

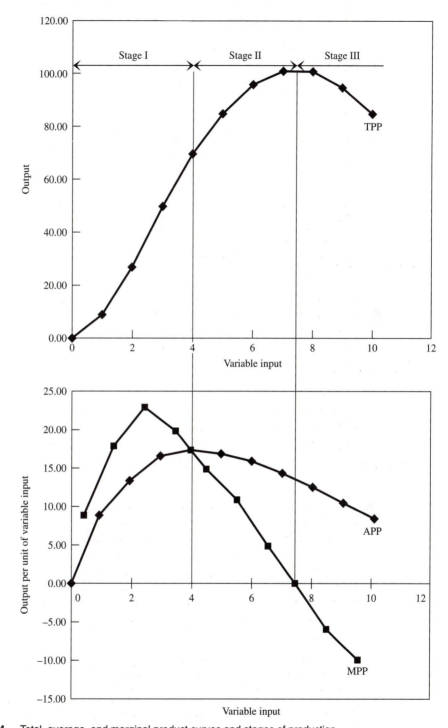

FIGURE 3-4   Total, average, and marginal product curves and stages of production.

Note: The marginal data are plotted at the midpoints of the two input values. This is because the MPP values measure the change in output associated with these two input levels. Also, note that the units on the two vertical axes are different.

product (TPP) curve describes the total amounts of a product (corn) produced from different amounts of the variable input (fertilizer) given the fixed inputs. It is a physical relationship because it describes the relationship between quantity of inputs and the quantity of outputs. (Cost information is added in the next chapter.) Farmer Jones's task is to select the profit-maximizing amount of input and associated amount of output.

The TPP curve in Figure 3-4 rises very little with the addition of the first few units of variable input. It rises more with the addition of the second unit, and still more as the third unit is added. Then it *rises less and less* with each successive unit added until it reaches its maximum. The maximum will occur somewhere between 7 and 8 units of input. If too much fertilizer is added, the corn is damaged ("burned") by excessive nutrients and the TPP declines. (If you do not believe that too much fertilizer will damage a plant, add about an inch of fertilizer to a plant you are not too fond of!)

Average Physical Product

The **average physical product (APP)** is the total physical product produced per unit of variable unit used (the total physical product divided by the number of units of the variable input). The APP for Jones's farm is shown in Table 3-2 and is graphed in Figure 3-4 (MPP is also shown, it is discussed in the next section). The APP curve shows all possible APPs. Jones will choose one level of input and that will determine the APP (and TPP).

Table 3-2 and Figure 3-4 give the input–output combinations from which Jones may choose. Notice that at low levels of fertilization, as the amount of fertilizer increases, APP increases. It reaches a maximum of 17.5 units of corn per unit of fertilizer when 4 units of fertilizer are used with the fixed inputs. At higher fertilizer levels APP decreases as more fertilizer is applied (TPP continues to increase). The APP curve shows how much production is possible *per unit of the variable input* at varying levels of the variable input, with a fixed amount of other inputs. Remember, Jones (and we) eventually will choose just one of the possible input/output combinations.

TABLE 3-2   A HYPOTHETICAL SCHEDULE
ILLUSTRATING AVERAGE PHYSICAL
PRODUCT PER UNIT OF FERTILIZER

Input	TPP	APP
0	0.00	0.00
1	9.00	9.00
2	27.00	13.50
3	50.00	16.67
4	70.00	17.50
5	85.00	17.00
6	96.00	16.00
7	101.00	14.43
8	101.00	12.63
9	95.00	10.56
10	85.00	8.50

Stages of the Production Function

With only the TPP and APP information, it is possible to begin to answer the question of how much fertilizer Jones should apply. Jones could decide to maximize yield per acre. This would mean applying 8 units of fertilizer to the fixed inputs to maximize output, TPP. Jones would certainly not apply more than 8 units of fertilizer because output per unit of fertilizer and land would both decline. Or, Jones might consider trying to maximize the amount of corn produced per unit of fertilizer. Applying 4 units of fertilizer per acre would maximize the APP, or maximize output per unit of variable input. Jones would not use less than 4 units of fertilizer. Comparing the results using 3 and 4 units of fertilizer, Jones would find that with 4 units of fertilizer the amount of corn produced per acre and per unit of fertilizer are both higher. If production is worthwhile with 3 units of fertilizer, it is better to use 4.

There are three **stages of the production function.** Stage I ends and Stage II begins where APP peaks. Stage II ends and Stage III begins where TPP peaks. In Stage I, as more fertilizer is applied, the output per unit of both the variable and fixed inputs increases. Both the TPP and the APP are increasing. If it pays to use any fertilizer, it will pay to increase fertilizer use until the end of Stage I is reached. Therefore, operating in Stage I of the production function is not rational because the amount of output per unit of both the fixed and variable input is increasing.

EFFECT OF LIMITED RESOURCES

It is possible that a manager might not be able to reach the end of Stage I because of limited resources, including borrowing funds to purchase variable inputs. What then? With a limited amount of the variable factor, more output can be obtained by not using some of the fixed factor, as opposed to spreading the variable factor too thinly. If, for example, a farmer cannot find any additional help and must operate a farm alone, it is better to leave part of the farm idle rather than produce on all available acres. Similarly, if the farmer cannot buy all the fertilizer needed to reach yields at least as high as the maximum APP, the total product will be higher by producing on only part of the farm. The alternative of spreading the fertilizer more thinly over the entire area will not produce as much corn.

Stage III begins at the peak of the TPP curve. Using more fertilizer than necessary to maximize TPP will result in a lowering of the TPP and the APP. In Stage III of the production function the manager can produce more corn by applying less fertilizer. Thus, it is clear that Jones will not operate in Stage III. Operating in either Stage I or III is irrational and we will find that profits cannot be maximized in either of these stages.

Within Stage II, if more units of the variable input are used, TPP is higher but APP is lower. The average output from the fixed inputs used increases, but the average output per unit of the variable input falls. Thus in Stage II there is a tradeoff. The resolution of this tradeoff and the profit-maximizing rate of use of the variable input depend on the prices of the variable input and output. The optimum will occur somewhere within Stage II. This is where Jones will find a profit-maximizing level of production. It is the only rational stage of production.

Marginal Physical Product

To determine the profit-maximizing level of production, we use one more physical relationship. The **marginal physical product (MPP)** is the addition to total physical product from an additional unit of variable input. In the example in Table 3-2, the variable input used is changed in one-unit increments. The MPP is the addition to TPP resulting from one more unit of fertilizer. In mathematical terms:

$$\text{TPP}_x - \text{TPP}_{x-1} = \text{MPP}_x$$

More generally, it is:

$$\text{MPP}_x = \Delta\text{TPP} \div \Delta\text{X}$$

EXAMPLES OF PRODUCTION

Have you ever waited in a long line at a fast-food restaurant? If so, you almost certainly have witnessed a firm operating at a non-optimal point on its TPP. The restaurant is clearly shorthanded—the small number of workers are scurrying from task to task and are not able to serve many customers quickly. With more workers, a fast-food restaurant can be just that! The workers serve numerous customers quickly, TPP is higher than if the restaurant is shorthanded. You also may have seen a restaurant that prepared for a rush period with such a large number of workers that there was mass confusion. Workers get in each other's way—with lots of bumping, spilling, and "pardon-me's." The restaurant has reached the point where it has more workers than can be used efficiently in an establishment of its size. At some point, adding still more workers would reduce the number of customers served.

Here is another example. Think about how your grades are affected by the amount of time you devote to your classes. If you don't spend any time on the classes, you will very likely receive failing grades. Spending small amounts of time greatly improves your understanding of the material and your grades. Additional time spent improves your grades, but the improvement is not as dramatic. At the extreme, additional time will reduce the amount of sleep you get and your grades will fall.

Table 3-3 repeats the TPP data from Table 3-1 and adds the MPP data. For example, the MPP of adding the sixth unit of input equals the TPP produced with 6 units of variable input minus the TPP produced with 5 units of input ($96 - 85 = 11$). (If the amount of the variable input is changed by more than one unit, say by 10 pounds or 10 kilograms, MPP is calculated by dividing the change in TPP by the change in quantity of the variable input, or 10.)

Figure 3-4 shows the general shape of the MPP curve and its relationship to the TPP and APP curves. There are several important relationships among the three functions. In each case below, think of the changes as occurring as more of an input is used to produce a higher level of output.

• Within Stage I, the TPP curve increases first at an increasing rate and then at a decreasing rate. The changeover point is called the *point of inflection.*

• The MPP curve is above the APP curve until the APP curve reaches its peak. At the peak of APP, MPP equals APP because the change in TPP is equal to the APP. Beyond that point, at higher levels of input use, the MPP curve is below the APP curve and

TABLE 3-3 A HYPOTHETICAL PRODUCTION
SCHEDULE ILLUSTRATING
MARGINAL PHYSICAL PRODUCT PER
UNIT OF FERTILIZER

Input	TPP	MPP
0	0.00	
		9.00
1	9.00	
		18.00
2	27.00	
		23.00
3	50.00	
		20.00
4	70.00	
		15.00
5	85.00	
		11.00
6	96.00	
		5.00
7	101.00	
		0.00
8	101.00	
		−6.00
9	95.00	
		−10.00
10	85.00	

Stage II begins. You can think of this relationship between the average and marginal curves in mathematical terms. If the last unit added is larger than the average, the average must increase. If the added unit is less than the average, the average must fall. For example, if you receive a grade higher than your GPA (grade point average) your GPA increases. A grade lower than your GPA will reduce it.

• The MPP equals zero (crosses the horizontal axis) at the point where the TPP curve is at its maximum. Adding zero does not change the total!

• If MPP is positive, the TPP is increasing; if MPP is negative, the TPP is decreasing. And at just the point where the marginal physical product is equal to zero, the TPP must be at its maximum.

• TPP and APP are never negative; MPP becomes negative where the TPP curve reaches its maximum.

• If the APP curve is increasing, MPP is greater than APP; if the APP curve is decreasing, MPP is less than APP. Or, if the average is increasing, an amount larger than the average is added by each unit of input. If the average is falling, an amount smaller than the average is added by each unit of input.

Diminishing Marginal Physical Product

The shape of the marginal physical product curve demonstrates another important relationship in economics, the law of **diminishing marginal physical product.** This law states that, *ceteris paribus,* the contribution to output of each added unit of the variable input will eventually be less than the contribution of the previous unit. As more of a variable input is used, eventually the rate of increase in total output will decrease.

In Table 3-3, the MPP of the first unit of variable input is 9 units of corn. The second unit adds 18 units; the third, 23 units of corn. Thus, there is increasing marginal productivity up to the third unit of variable input. The fourth unit of variable input adds only 20 units, and the fifth unit adds 15 units of output. The marginal product is diminishing. Beyond the eighth unit of input the MPP becomes negative and the firm returns will fall as costs continue to rise.

How typical is this illustration? Think about the number of sales associates in a clothing store. As the number of clerks increases from one to two, and then to three, there is a rapid increase in the number of customers who can be served. The MPP is increasing, and it is above the APP. When the number of associates reaches an efficient crew size, further increases in workers will increase the number of customers served but at a much smaller rate. Diminishing productivity occurs. MPP falls, APP will fall as well, but TPP continues to increase. If there are so many workers that they get in each other's way, the number of customers served will fall. Productivity continues to fall until MPP is negative and then TPP decreases.

SUMMARY

The production function is the technical relationship between the inputs and outputs of the production process. Research and development activities lead to technological change, which allows the production of more output from a given quantity of inputs. Some inputs are used in quantities that vary in the short run, while the quantity of other inputs used is fixed in the short run. In the long run all inputs are variable.

Production functions cover three stages of activity. In Stage I, both the TPP and APP curves are increasing, and the MPP reaches its maximum at the point of inflection of the TPP curve. Since the MPP curve is above the APP curve until the end of Stage I, if it pays to produce it will pay to expand production at least to the end of Stage I. In Stage II, the MPP curve falls below the APP curve, but the TPP continues to increase until the MPP falls to zero and the TPP is at its maximum, which marks the end of Stage II. Profits are maximized at some (thus far undetermined) point within Stage II. The location of the point of maximum profits depends on the relationships between the product curves and the prices of the inputs and output. Since all the product curves are falling in Stage III, production in Stage III is irrational, as is production in Stage I.

IMPORTANT TERMS AND CONCEPTS

accounting profit   58
average physical product (APP)   66
diminishing marginal physical
  product   69
economic profit   59
efficient   62
fixed factors of production   60
fixed inputs   59

Green Revolution   64
inputs   59
long run   60
marginal physical product (MPP)   68
opportunity cost   58
outputs   59
production   58
production function   59

short run 60

stages of the production function 67

technological change 62

total physical product (TPP) 61

variable factors of production 60

variable inputs 59

QUESTIONS AND EXERCISES

Name That Term

Read the following sentences carefully and fill in the missing term or terms.

1 _____ is the use of inputs to produce a good or service.

2 Receipts minus expenses equal _____ profit.

3 _____ profit is the total returns minus the cash payments for purchased inputs and the opportunity costs of the firm-owned inputs.

4 The _____ is the technical relationship between inputs and outputs, indicating the maximum amount of output that can be produced with alternative amounts of variable inputs used in combination with one or more fixed inputs under a given state of technology.

5 The amount of _____ used changes as a producer changes the amount of output in the short run.

6 The _____ physical product curve gives the maximum quantity of output that can be produced with different amounts of the variable input.

7 TPP divided by the number of units of variable inputs used in production is the _____ product.

8 The law of _____ says that, all else constant, the contribution to output of the variable input will eventually be less than the contribution of the previous unit.

9 _____ product is the change in output resulting from one additional unit of a variable input.

10 The _____ is a period of time long enough for producers to change the quantities of all resources used in production.

True/False

Read the following sentences, then decide whether each statement is true (T) or false (F) and mark it accordingly.

T F **1** Marginal physical product of an input is positive but declining in Stage II of the classical production function.

T F **2** A technological improvement causes the TPP curve to shift up.

T F **3** Stage III of a production function featuring one variable input begins where MPP of the variable input is zero.

T F **4** If the MPP of a variable input is decreasing then the APP of that input also must be decreasing.

T F **5** Stage II of a classical production function for a single variable input begins where the marginal physical product of an input reaches a maximum and begins to decline.

T F **6** A profit-maximizing corn farmer should use the level of fertilizer that maximizes total output of corn per acre.

T F **7** A firm can recoup part of its fixed costs by ceasing production.

T F **8** An 800-acre farm that yields 120 bushels of corn per acre generates 96,000 bushels of corn.

T   F   **9** An efficient firm operates below its TPP curve.
T   F   **10** The Green Revolution involved developing and making available to farmers new wheat and rice varieties, new means of controlling plant diseases, and other changes necessary to improve crop yields.

Multiple-Choice Questions

Circle the letter of the response that best answers the question or completes the statement. (Use the following graph to answer questions 1–7.)

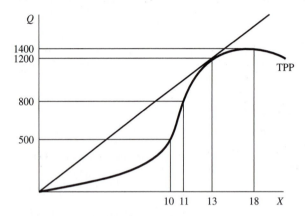

1 What is the TPP at X = 11?
   a 0
   b 500
   c 800
   d 1200
   e 1400
2 What is the APP at X = 10?
   a 80
   b 50
   c 20
   d 10
   e not defined
3 What is the MPP at X = 18?
   a 0
   b 10
   c 20
   d 30
   e 40
4 At X = 13, MPP is _____ APP.
   a above
   b same as
   c below
5 MPP is at its maximum at _____?
   a 10
   b 11

    c 13
    d 18
    e not defined
6 Stage II begins at X = _____?
    a 0
    b 10
    c 11
    d 13
    e 18
7 Stage III begins at X = _____?
    a 0
    b 10
    c 11
    d 13
    e 18

Technical Training

1 Professor William Saupe of the Agricultural Economics Department of the University of Wisconsin has a trophy in his office certifying that, in an earlier life, he won the Northwest Iowa Corn Yield Contest. When Professor Saupe brings his trophy to farm management courses, holds it up to the class, and says, "Look what I won for driving the marginal physical product of nitrogen to zero!", what does he mean?

2 Suppose scores on an exam (E) were related to hours studied (H) according to the following production function: E = H. Graph the TPP, MPP, and APP of hours studied. Do you think that this is an accurate relationship between hours studied and grades? Why or why not.

3 Find the missing values in the following table. Let X = number of variable inputs, TPP = total physical product of output, MPP = change in output/change in input, and APP = output ÷ input.

X	TPP	MPP	APP
10	500		_____
		_____	
20	1200		_____
		50	
30	_____		_____
		_____	
40	2000		_____
		_____	
50	_____		42
		−10	
60	_____		_____
		_____	
70	1700		_____

Use the above table to answer questions 4–7.

4 At what level of X does TPP reach the maximum?

5 At what level of X does Stage III of the production function begin? Remember Stage III begins where MPP first equals zero.

6 At what level of X does Stage II of the production function begin? Remember Stage II begins where MPP = APP.

7 Would it ever make sense to purchase and employ 60 units of input?

Use the production function $Q = -400 + 100X - X^2$ to answer questions 8–9.

8 Graph TPP (or Q) for input (X) from 10 to 70.

9 Graph APP and MPP for inputs from 10 to 70. Remember APP $= Q \div I = -400 \div I + 100 - I$ and, in this case, MPP $= 100 - 2X$. Use this graph and the graph from question 8 to answer questions 9–18.

10 At what level of X does APP reach a maximum?

11 What is the maximum level of APP?

12 At what level of X does MPP = APP?

13 Where does Stage II of the production function begin?

14 What is the maximum level of Q?

15 At what level of X does MPP = 0? What is the Q associated with that level of X?

16 At what level of input does Stage III of the production function begin?

17 Are the graphs you drew based on the production function consistent with the data presented in the table?

18 What are the values of TPP, APP, and MPP at X = 32?

19 Could you have used the table to answer question 18? Why or why not?

Questions for Thought

Use the following information to answer questions 1–8.

You and several of your closest friends have decided to go into the restaurant business. Upon evaluating the customer base you conclude that the public's demand for Hungarian food is not being met. Hence, you and your new business partners rent a building on Main Street and begin selling Hungarian chicken wings. Table 1 displays different quantities of variable input (hours of labor) you could employ and the resulting weekly production. To make it simpler, all other inputs are constant.

1 Complete Table 1 by entering the missing values.

TABLE 1

Input (hours)	TPP (chicken wings)	MPP	APP
5	100	———	———
10	300	———	———
15	———	50	———
20	750	———	———
25	———	———	34
30	———	10	———
35	800	———	———
40	650	———	———

Note: APP = average physical product and output/input; MPP = marginal physical product and change in output/change in input; TPP = total physical product (chicken wings).

2 Plot the TPP on Graph 1 and the MPP and APP on Graph 2. Show the three stages of production on the graphs. Remember to label your graphs.

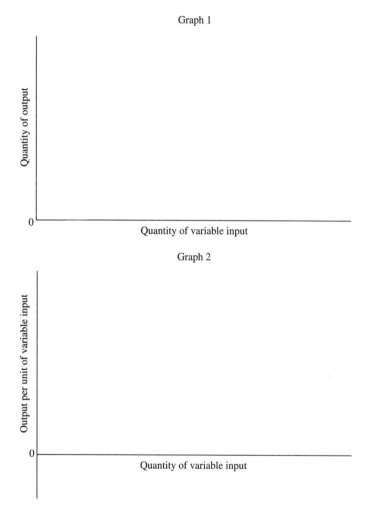

Graph 1

Quantity of output

0

Quantity of variable input

Graph 2

Output per unit of variable input

0

Quantity of variable input

3 What quantities of labor, in hours, would be considered rational to employ?

4 Would you ever want to employ 10 hours of labor in this example? Explain why or why not.

5 The City Inspection Agency discovers that your frying oil emits a gas that is killing the local squirrel population. You must use a different type of cooking oil. To your dismay, cooking with the new oil is not as efficient because the chicken requires a longer cooking time. Discuss the likely effects of the new oil on the production function, and show the shift in TPP on Graph 1.

6 Give an example of fixed and variable factors of production other than those mentioned in the chapter.

7 Examine Graph 1 and estimate the quantity of input at which the law of diminishing marginal returns begins to take effect.

8 Under what circumstances would your Hungarian chicken wing restaurant make a decision that would be classified as a long-run decision? Give an example of such a decision, and explain how it differs from short-run decisions.

9 Is maximizing profits the only objective of the operator of a business in the short run? In the long run? What other objectives are possible? What effect does pursuing these other objectives have on profit in the short run? In the long run?

10 Consider the following production function: $G = f(H)$, where G is the number of points scored on the most recent exam you took and H is the number of hours you spent studying for that exam. Sketch the TPP, APP, and MPP of H, using numbers appropriate to your own case.

11 Suppose you rent an apartment for $600 per month. Would you ever choose to spend a day in your apartment just because, "this place cost me about $20 per day?" Why or why not?

12 In Chapter 1, you learned how the declining number of farmers has adversely affected many rural communities. One argument used to justify spending public money to maintain these communities is that it doesn't make sense to abandon perfectly good infrastructure. Analyze this argument using fixed, variable, and marginal cost concepts.

13 What are the fixed and variable costs associated with your taking this class?

ANSWERS AND HINTS

Name That Term   1. Production; **2.** accounting; **3.** Economic; **4.** production function; **5.** variable inputs; **6.** total; **7.** average physical; **8.** diminishing marginal productivity; **9.** Marginal physical; **10.** long run

True/False   1. T; **2.** T; **3.** T; **4.** F; **5.** F; **6.** F; **7.** F; **8.** T; **9.** F; **10.** T

Multiple Choice   1. c; **2.** b; **3.** a; **4.** b; **5.** b; **6.** d; **7.** e

Technical Training   1. He means that nitrogen (a fertilizer) is a variable input in corn production and that, in order to win a trophy for maximum corn yield, he applied nitrogen (and all of the other variable inputs) to the point where the MPP was close to zero.

2.

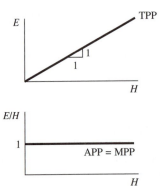

3.

X	TPP	MPP	APP
10	500		50
		70	
20	1200		60
		50	
30	1700		57
		30	
40	2000		50
		10	
50	2100		42
		−10	
60	2000		33
		−30	
70	1700		24

4. 50; **5.** somewhere between 40 and 60; **6.** somewhere between 10 and 30; **7.** no, you could get more output (and revenue) for fewer inputs (and costs) by cutting back to 50 units of input;

8.

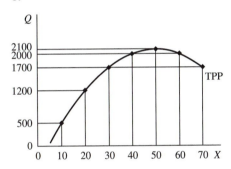

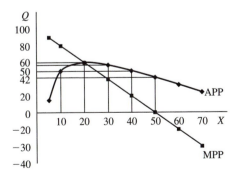

9. 20; **10.** 60; **11.** 20; **12.** 20; **13.** 2100; **14.** 50; **15.** X = 50, Q = 2100; **16.** 50; **17.** yes; **18.** TPP = 1776, APP = 55.5, and MPP = 36; **19.** Not really. You could obtain approximate answers (good guesses) from the table but you need the function to get precise answers.

4

COSTS, RETURNS, AND PROFIT MAXIMIZATION

CHAPTER OUTLINE

OVERVIEW

This chapter combines production function and price of input information to show the relationship between the firm's costs of production and level of output. It also shows how cost information and output prices combine to determine the profit-maximizing level of production. We then examine how the quantity of output produced changes with the output price to determine the supply curve. Finally, we present information on the elasticity of supply.

LEARNING OBJECTIVES

This chapter will help you learn:

- What influences the firm's choice of combinations of inputs and outputs
- What determines the cost curves of the firm
- The profit-maximizing level of output
- The causes of economies and diseconomies of scale
- How to find the elasticity of supply and its determinants

COST AND REVENUE CONCEPTS

In this chapter, we build on the physical relationships between input and output developed in Chapter 3. We examine the costs associated with various production levels. A firm manager uses cost data to determine whether the firm should operate and to determine the profit-maximizing (or loss-minimizing) combination of fixed and variable inputs. We begin by explaining the difference between explicit and implicit costs. Next we consider the factors determining the optimum combination of the fixed and variable inputs. We add revenue information and combine it with the cost data to find the profit-maximizing output. Next, we analyze how costs change as the level of output is changed, both in the short run and the long run. We conclude the chapter with an examination of the elasticity of supply.

COSTS OF THE FIRM

The reason we, and owners of firms, are concerned with costs is that they must be paid out of the revenues received as the firm tries to earn a profit. But simply paying the out-of-pocket costs is not an accurate way of determining the actual costs incurred by the firm. Why? Because this may miss the costs of some of the resources. For example, members of the family might work without pay in a family business or farm. Does the fact they work without pay mean their work has no value? Of course not, but how much is their work worth? The value of their work is the pay they could receive working elsewhere. This is called their opportunity cost. It is the pay that they could have earned if they had taken the "opportunity" to work elsewhere. (Individuals may accept the opportunity cost of a lower-paying option if the working conditions are better or there is more job security.)

*In general terms, the **opportunity cost** of using a resource (input) is the value of the contribution that resource could generate in the highest priced alternative use.* Any resource that could be used in some other way has an opportunity cost, whether or not a specific payment is made. An opportunity cost applies to leisure, recreation, and even sleep, as well as to the production of goods and services. It is a universal phenomenon. For example, consider a business that uses land and buildings purchased many years ago at a price well below the current value. To correctly assess the costs of production the business operator must consider the current value of the contributions of the land and buildings, perhaps best indicated by the amount of money that could be charged if the land and buildings were leased to another business. As we noted in Chapter 1, the cost to you of college is more than the "out-of-pocket" costs of tuition, books, and other expenses. You must also consider the money you could earn if you were not in college, your opportunity cost of not working (but not the cost of housing and food because you need those regardless of whether you are in college).

The firm's economic costs are the payments that it must make to attract resources and keep them from being used to produce other products. These payments can be explicit or implicit. **Explicit costs** are the payments made for such inputs as hired labor, rented land, fertilizer, pesticides, and fuel. These are the normal "out-of-pocket" costs of the inputs purchased for use in producing a product.

An **implicit cost** is the opportunity cost of an input that the firm or a family business owns or controls. Since the firm owns or controls these resources, it need not make a cash payment for their use, but such resources are not free.

DEFINITION OF GRAPH AXES

In graphs of production costs economists graph output on the horizontal axis, and costs and returns on the vertical axis. This is different from the productivity graphs in which output is on the vertical axis. It is important that you are aware of this difference—it is a major source of confusion and can cost many points on tests!

Next we consider the three categories of costs: fixed, variable, and marginal. In the discussion of costs we continue the Jones farm corn production example we used in the previous chapter. In Table 4-1 we give the total fixed and total variable costs for this

TABLE 4-1   HYPOTHETICAL TOTAL COSTS OF CORN PRODUCTION

INPUT (units)	TPP (bu)	TFC ($)	TVC ($)	TC ($)
0	0	80	0	80
1	9	80	25	105
2	27	80	50	130
3	50	80	75	155
4	70	80	100	180
5	85	80	125	205
6	96	80	150	230
7	101	80	175	255
8	101	80	200	280
9	95	80	225	305
10	85	80	250	330

Note: Hypothetical data.

corn producer. The TPP curve in Chapter 3 gave the maximum amounts of product that could be produced with differing amounts of the variable input. The total costs are the minimum cost of the differing amounts of the variable input used in producing various levels of output. For example, the minimum cost of all inputs needed to produce 85 bushels of corn is $205. Of this cost, $125 is for five units of the variable inputs and $80 is to pay the fixed costs. We begin our review of these costs by considering in more detail total fixed costs.

Total Fixed Costs

Total fixed costs (TFC) are costs that do not change as output changes. These costs include both explicit and implicit costs of inputs that are fixed. For example, they include the explicit costs of property taxes, interest on farm mortgages, insurance, and contract payments for labor. They also include implicit charges, such as the potential income from employment opportunities forgone, and the rental rate for owned assets. Notice that the firm pays fixed costs even if it doesn't produce any corn. In this example, as shown in Table 4-1, total fixed costs are a constant $80.00. Figure 4-1 shows total fixed costs as a horizontal line. Regardless of how much Jones decides to produce, fixed costs will be $80.

Total Variable Costs

Total variable costs (TVC) shown in Table 4-1 are the costs of inputs that change as the level of output is changed. If no variable inputs are used, no output is produced and the total variable cost is zero. To produce a product, the firm must use variable inputs and

FIGURE 4-1   Total fixed costs.

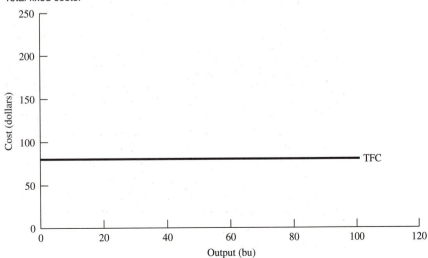

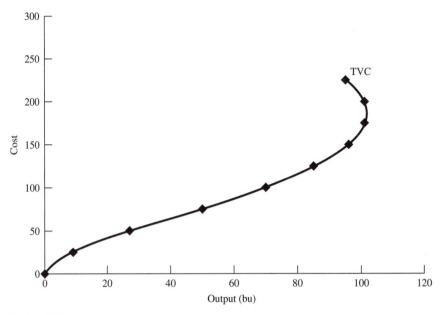

FIGURE 4-2   Total variable costs.

incur variable costs. Variable costs include explicit costs of inputs such as fertilizers, chemicals for control of crop pests and disease, cultivating, harvesting, drying, and fuel. They also include implicit costs such as the opportunity cost of family members who work for the firm but do not receive pay. In a restaurant, typical examples of variable costs include labor, food ingredients, and utility payments for cooking and cleaning.

Figure 4-2 shows Jones's total variable cost (TVC) curve. Variable costs are zero if no output is produced. To produce higher levels of output Jones must use more variable inputs and the total cost of these variable inputs will, of course, increase. In this case each unit of input costs Jones $25.00. The reason the curve turns back toward the vertical axis is that at high levels of the variable input use, less is produced (you may want to look back to the discussion of Stage III of the production function in Chapter 3).

Total cost is the sum of total fixed and total variable costs at each level of output.

$$TC = TVC + TFC$$

Figure 4-3 displays total fixed, total variable, and total costs. Graphically, **total costs (TC)** are the vertical sum of the fixed and variable cost curves. At higher levels of output, total variable costs (TVC) are higher, but by definition total fixed costs (TFC) remain the same. All the changes in total costs are explained by the changes in variable costs. Note that the shape of the variable cost curve and the total cost curve is exactly the same; the vertical distance between the variable cost curve and the total cost curve is fixed costs.

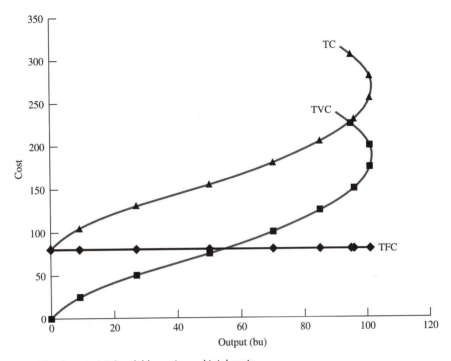

FIGURE 4-3   Total fixed costs, total variable costs, and total costs.

CAPACITY

Not all texts show the "backward bending" portion of the cost curves. What the backward bending portion suggests is that the firm is operating in Stage III of the production function. Adding more of the variable input increases the cost but reduces the amount of output produced. Since a rational producer would never go beyond the point where the curves bend backward, this is the firm's capacity—given its fixed resources.

Average Fixed Costs, Average Variable Costs, and Average Total Costs

Table 4-2 shows the average costs per bushel for the three average cost curves: **average fixed cost, average variable cost,** and **average total cost.** In each case the average cost is equal to the total cost divided by output.

Average fixed cost is equal to total fixed cost divided by output.

$$\text{AFC} = \text{TFC} \div \text{TPP}$$

Since the total fixed costs do not change as total output increases, average fixed costs (AFC) decline as output (TPP) increases. Fixed costs are sometimes called overhead costs. Farm managers and other business managers sometimes refer to declining AFC when output increases as "spreading the overhead." For example, most businesses own or have a long-term lease for a production facility (land, buildings, and machinery). As output increases by using more of the capacity of the facility, perhaps by operating with two and then three shifts of workers, the fixed cost of plant and equipment is "spread" over the larger total output produced by more workers.

Average variable cost is equal to total variable cost divided by output.

$$AVC = TVC \div TPP$$

Table 4-2 gives the average costs calculated from the total cost data in Table 4-1. Both average variable costs (AVC) and average total costs (ATC) decline initially, reach a minimum, and then increase, generating U-shaped curves as indicated in Figure 4-4. For many production processes the AVC and ATC curves are rather flat over a wide range of outputs, as in our example. Because of initially increasing returns, as discussed in Chapter 3, it takes fewer and fewer of the variable inputs to produce each additional

TABLE 4-2   HYPOTHETICAL AVERAGE COSTS OF CORN PRODUCTION

TPP (bu)	AFC ($)	AVC ($)	ATC ($)
0	Undefined	Undefined	Undefined
9	8.89	2.78	11.67
27	2.96	1.85	4.81
50	1.60	1.50	3.10
70	1.14	1.43	2.57
85	0.94	1.47	2.41
96	0.83	1.56	2.40
101	0.79	1.73	2.52
101	0.79	1.98	2.77
95	0.84	2.37	3.21

Note: Hypothetical data.

FIGURE 4-4   Average fixed, variable, and total costs of production.

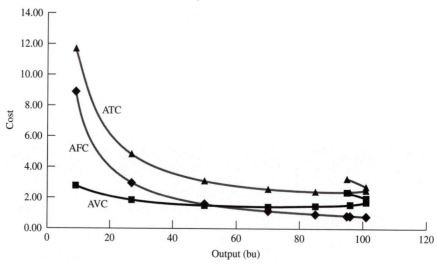

bushel of corn. The AVC declines to $1.43 per bushel. Between 70 and 100 bushels, the AVC tends to level out. Beyond 100 bushels, AVC rises rapidly as diminishing returns necessitate the use of more and more of the variable inputs to produce each additional bushel of corn. In Stage III, when the MPP has become negative, AVC curls back as additional variable inputs cause production to fall. For example, it is possible to produce 95 bushels of corn at an average variable cost of either $1.56 or $2.37 per bushel. This is not a hard choice for Jones!

Average total cost is equal to the sum of average fixed plus average variable cost.

$$ATC = AFC + AVC$$

It is also equal to the total cost divided by the total output.

$$ATC = TC \div TPP$$

Graphically, ATC is found by adding the AFC and AVC curves vertically as in Figure 4-4. The vertical distance between the ATC and AVC curves is equal to AFC for all levels of output. At higher levels of output, ATC is closer to AVC because AFC declines at high levels of output. The average total cost, or average cost, is the cost per unit of output and is often termed the "unit cost of production" by business operators.

Marginal Cost

The final cost concept, marginal cost (MC), is shown in Table 4-3. Understanding marginal cost is crucial to the operator of a firm. Jones will use marginal cost data to answer

TABLE 4-3   HYPOTHETICAL COSTS OF CORN PRODUCTION

TPP corn (bu)	TVC ($)	TC ($)	AVC (TVC/TPP) ($)	MC Δcost/ΔTPP ($)
0	0	80	Undefined	
				2.78
9	25	105	2.78	
				1.39
27	50	130	1.85	
				1.09
50	75	155	1.50	
				1.25
70	100	180	1.43	
				1.67
85	125	205	1.47	
				2.27
96	150	230	1.56	
				5.00
101	175	255	1.73	
				Undefined
101	200	280	1.98	
				Undefined
95	225	305	2.37	

the important question, how much more must I spend to produce additional output? (Looking ahead, to make the profit-maximizing decision Jones will also need revenue information.)

Marginal cost is the increase in cost associated with a one-unit increase in total output. *It can be computed by dividing the change in either the total costs or the total variable costs by the change in the quantity produced (TPP).*

$$MC = \Delta TC \div \Delta TPP$$

or,

$$MC = \Delta TVC \div \Delta TPP$$

The reason MC can be calculated from either of these cost measures is that each reflects the entire change of total costs. The total fixed costs do not change as output changes.

In our example and in most practical applications, marginal cost is calculated on the basis of changes in output larger than a single unit. In the example in Table 4-3, increasing production from 50 to 70 units of corn requires spending $25.00 more on variable inputs. Dividing $25.00 by 20 equals $1.25, the (representative) marginal cost of increasing production from 50 to 70 bushels. (It is representative since the value is in mid-range of the MC values that would result if the MC of each incremental unit of output could be calculated.) If Jones considers output levels of 95 and 100, the additional variable inputs required to increase production by the last 5 units will increase variable and total costs by $25.00. The (representative) marginal cost is $25 ÷ 5, or $5 per unit of corn. Note that since fixed costs do not change, they do not affect the marginal cost.

In Figure 4-5 the marginal cost and the average cost curves from Table 4-3 are graphed. The shape of the marginal cost curve is a direct result of increasing and then

FIGURE 4-5   Average fixed, average variable, average total, and marginal costs of production.

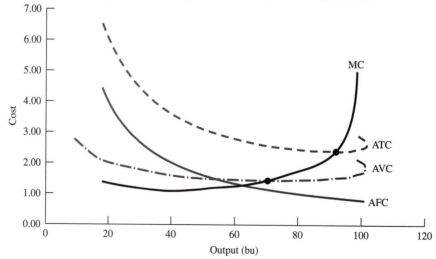

diminishing physical returns of the variable input. This can be seen by comparing the marginal cost (MC) curve in Figure 4-5 with the marginal physical product (MPP) curve for the variable input in Figure 3-4. (To convince yourself of this, return to Figure 3-4, turn the page, hold the page to the light so you can see through it, and turn it upside-down and you will see curves that look like the MC and AVC curves.) At low output levels, MPP is increasing because of increasing returns, resulting in falling marginal costs. The output level that results in the maximum MPP in Figure 3-4, also results in the minimum MC. At higher levels of output, diminishing returns occur, the marginal physical product is lower, and the marginal cost is higher. At high levels of output where the marginal product approaches zero, the marginal cost shoots upward. If Jones attempts to push output to a very high level, the marginal cost of the last few units of output is very high. This is because the last units of input result in very small increases, and even decreases, in output. Thus, a rising marginal product (increasing returns) generates a decreasing marginal cost. A falling marginal product (diminishing returns) produces an increasing marginal cost.

Relation of MC to AVC and ATC

A number of important relationships among these cost curves can be observed in Figure 4-5. The marginal cost curve cuts the average variable cost and average total cost curves at the minimum point on each of these average cost curves. This is a mathematical relationship, there are no exceptions. As long as MC is below AVC and ATC, both of the average cost curves must be falling because less is being added to the total cost for each successive unit of output than the average of all the previous units. Beyond the point where the MC curve cuts the AVC and ATC curves, each of the latter curves must rise because the MC of each added unit of output is larger than the average of the previous units. (If you have a "B" average and get an "C" your grade point average will go down, if you get an "A" it will go up.)

Remember that a firm would not operate in Stage I because, with each additional unit of input used, output per unit of both the variable and the fixed input increases. That one would not operate in Stage I is also apparent from the AVC curve. If it is profitable to produce at all, it will be profitable to produce at least up to the point where AVC is at its minimum, where Stage I of the production function ends. For example, if producing 27 bushels at a cost of $4.81 each is good for Jones, producing 50 bushels at $3.10 each is even better.

DETERMINING HOW MUCH TO PRODUCE

Thus far we have focused on the physical and cost aspects of production. We turn now to an analysis of a producer's revenue. Producers use the revenue information in combination with production-cost information to determine the optimum quantity to produce. The firm's revenue is determined by the price received for the product and the quantity of the product sold. Jones and other farm operators are **price takers.** That is, Jones cannot control the market price received for the output. Jones develops production plans based on the current or expected price. For this reason, the analysis in this chapter deals with a firm that is a price taker. Some firms produce a unique or differentiated product or operate in markets that are not purely competitive. Such firms can, to some degree, set their output prices. We consider this situation in Chapters 5 and 7.

Determining How Much to Produce to Maximize Profit

We now turn to the important issue of profit maximization. Here we are referring to economic profit, the difference between returns and the value of all the inputs used in producing the good or service sold. Economic profit is total returns minus the explicit cash payments for purchased inputs plus the implicit opportunity costs of the firm-owned inputs. For example, the economic profit of a wheat farmer is the value of the wheat sold minus the cost of all the purchased inputs and the opportunity costs of farm owned resources. Purchased inputs include things such as seed and fuel. Opportunity costs include family labor (the wages that could have been earned working off the farm) and land owned by the farmer (the rent that could have been received if the land were rented to another farmer).

We present two methods of finding the profit-maximizing output level. The first compares total revenues and total costs. The second compares marginal revenue and marginal cost. Both of these approaches are accurate means of determining the profit-maximizing output. However, the marginal revenue—marginal cost approach is preferable, because the focus is on the changes in costs and returns. Therefore, we present the total approach briefly, but concentrate on the marginal approach.

Total Revenue—Total Cost Approach

How does a manager determine the level of output to produce to maximize profits? Since **profit** is the difference between total revenue and total cost at each level of output, we must have revenue information as well as the cost information considered here. When the firm is a price taker, it can sell any amount of output at the market price. *Total revenue is the product of the quantity sold* (Q) *times the price received* (P).

$$TR = Q \times P$$

Total revenue is often called total sales, or the gross income of the business.

Profit (π) *(loss) is the difference between total revenue and total cost at a given level of output.* (The Greek letter π is used to represent profit.)

$$\pi = TR - TC$$

The first two columns of Table 4-4 reproduce the TPP data from Chapter 3 and the TC data from Table 4-3. The third column is the total revenue assuming a corn price of $3 per bushel. The fourth column is the profit or loss for each level of production. From Table 4-4 you can see that with a corn price of $3 per bushel, Jones maximizes profits by producing 95 bushels per acre, or slightly higher. Also note that at 95 bushels per acre Jones's profit is $10 higher than it would be at the maximum output of 101.

The relationships among total revenue, total cost, and profit are graphed in Figure 4-6. For a firm operating in a competitive market in which it can sell as much output as it

TABLE 4-4   HYPOTHETICAL COSTS AND RETURNS OF
CORN PRODUCTION

TPP (bu)	TC ($)	TR ($)	Profit ($)
0	80	0	−80
9	105	27	−78
27	130	81	−49
50	155	150	−5
70	180	210	30
85	205	255	50
96	230	288	58
101	255	303	48
101	280	303	23
95	305	285	−20

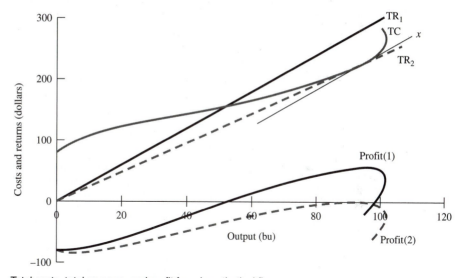

FIGURE 4-6   Total costs, total revenue, and profit for a hypothetical firm.

can produce at a given market price, the total revenue curve is a straight line from the origin because the firm receives the same price for all units of output (the firm is a price taker). Profit is the difference between total revenue and total costs.

The optimum level of production is easy to determine from the profit curve. Look for the point where the profit curve is at the maximum. With some effort, the optimum level of production can be determined from the total revenue and cost curves. Profit is the vertical distance between the total revenue, TR_1, and total cost curves, TC. If Jones considers a very low level of output, TR_1 is less than TC and Jones loses money. At higher output levels, where TR_1 is higher than TC, Jones earns a profit. We focus on that area.

Think of Jones as deciding whether to increase production by one unit. Jones will choose the higher level of production only if the addition to revenue is greater than the

addition to costs. If the revenue from the additional unit is greater than the cost of producing it, the firm will produce at the higher level. If the revenue from the last unit is less
than the cost to produce it, the firm will produce at the lower level. Since this firm is a
price taker, Jones sells each unit at the same price and the price is the added revenue from
selling an additional unit. Therefore, if the cost of producing the additional unit is less
than the price at which it can be sold, Jones will go for the higher level of production.

The added cost of the last unit determines the slope, at that point, of the total cost
curve. The slope of the total revenue curve is determined by the price of the product.
Since the output price is the same, the total revenue curve is a straight line. Thus, the optimum level of production must occur at the point where the slope of the TR curve
equals the slope of the TC curve. This occurs where a line parallel to the TR curve, such
as line x in Figure 4-6, is tangent to the TC curve. This is the point of maximum profit,
break even, or minimum loss. So, this is the best Jones can do in the short run, with the
firm's fixed resources and the current prices. In Figure 4-6, the maximum profit occurs
at output Q_1 when the price of output results in TR_1. Increasing production beyond the
point where the slopes of the TR and TC are equal reduces the profits because costs go
up more than revenues with each additional unit produced.

A change in the price of the product changes the slope of the total revenue line. In
Figure 4-6, the TR_2 curve results from a lower price than TR_1. If, as in the case of TR_2,
the price is at a level that results in the TR curve being tangent to the TC curve, the firm
will produce Q_2 and will break even. Jones is not making profits or realizing losses.

WHAT DO WE MEAN BY PROFIT?

Economists refer to normal, abnormal, windfall, and excess profits. What is the difference? As
indicated above, profit is the difference between total costs and total returns. The firm earns
normal profits when all costs, including the opportunity costs of family management, family
labor, and invested capital that could earn a higher return in an alternative use, are just covered by the revenue received. (If an operator does not include the costs of family labor and
management inputs, profits are overstated.)

Abnormal profits or excess profits occur if higher rates of profit are realized. These are
sometimes called windfall profits if a firm realizes higher than normal profits because of a for-
tunate set of circumstances; that is, "being in the right place at the right time."

Marginal-Revenue—Marginal-Cost Approach

An alternative approach to determine how much to produce is to compare the amounts
that each *additional* unit of output adds to the total revenue and to the total cost, that is,
by comparing MR and MC. First, we define two additional concepts, **average revenue
(AR)** and **marginal revenue (MR).**

Average revenue is total revenue divided by the quantity sold (AR = TR ÷ Q). *Marginal revenue is the addition to total revenue from producing one more unit of output*
(MR = ΔTR ÷ ΔQ). Marginal cost, defined earlier, equals ΔTC ÷ ΔQ.

If the firm sells all of its output at the same price, marginal revenue, price, and average revenue are the same (MR = AR = P). If all units of output sell at the same price,

then clearly this is the average price (AR). Similarly, if all units sell at the same price, the revenue received from selling one more unit (MR) will be the price.

We can find the point of maximum profit or minimum loss by comparing MR and MC at various levels of production. *If MR is greater than MC; profit will be increasing with production.* Why? For each unit where MR > MC the amount added to total revenue is more than the amount added to total cost. Each successive unit adds to profits, or subtracts from losses. Similarly, *if marginal cost exceeds marginal revenue, increasing the amount produced will reduce profit.* Each unit adds less to total revenue than it adds to total cost. Thus, the firm's profit is maximized where MR = MC.

Managers base production decisions on marginal revenue and marginal cost because they need to consider only the change in revenue and cost from each added unit of output. The manager does not have to reappraise the entire production plan. For example, Jones begins a growing season with a given amount of land, a set of equipment, a certain amount of family labor, and other fixed factors of production. Jones considers the profit potential from various possible yield levels. Higher yields mean higher revenues as well as higher costs because of higher fertilizer requirements, higher seeding rates, perhaps higher levels of irrigation, and more spending to control weeds and insects. Also, the total costs of harvesting and handling the increased output will be higher. Thus, Jones's decision about how much to produce depends on whether the additional revenue associated with higher yields is greater or less than the additional costs. If the marginal revenue from the last bushel produced is $2.10 (MR = $2.10), and the marginal cost (MC) is $2.00 (MC = $2.00), then producing the last unit will increase profits by 10 cents. It will pay Jones to increase output as long as the increase in revenue is more than the increase in total cost. However, if producing the last bushel adds $2.20 to the cost, Jones would reduce profit by producing it.

Other firms use this approach of equating marginal revenue and marginal cost as well. For example, when deciding whether to expand output by hiring an additional employee (perhaps you!)—along with associated expenses such as a desk, a computer, and training programs—the manager of an agribusiness firm determines whether the additional income produced by that one more employee is more or less than that employee's salary and benefits plus all the associated costs.

The intersection of the MR and MC curves is a unique point. It is the basis for the general MR = MC rule: *A firm maximizes profits or minimizes losses by producing at the level of output where marginal revenue equals marginal cost.*

Determining Profit-Maximizing Output

We use the MR = MC rule to determine *profit-maximizing output.* Continuing with the same example Table 4-5 gives the MC and MR information. We assume that Jones is operating in the short run. There are three related questions: (1) Shall I produce? (2) If so, how much? (3) What is the maximum profit, or the minimum loss? The answer to the first question (whether to produce) is *not* determined by whether there will be a profit or a loss. Regardless of whether Jones produces, the fixed costs—such as insurance, taxes, and the opportunity cost of the land—must be paid. Although a firm may incur losses, it should continue to produce if it loses less money than it would lose by closing down.

TABLE 4-5 HYPOTHETICAL TOTAL, AVERAGE, AND MARGINAL COSTS AND RETURNS OF CORN PRODUCTION

TPP (bu)	AVC ($)	ATC ($)	TC ($)	TR ($)	MC ($)	MR ($)
0	0	0	80	0		
					2.78	3.00
9	2.78	11.66	105	27		
					1.39	3.00
27	1.85	4.81	130	81		
					1.09	3.00
50	1.50	3.10	155	150		
					1.25	3.00
70	1.43	2.57	180	210		
					1.67	3.00
85	1.47	2.41	205	255		
					2.27	3.00
96	1.56	2.39	230	288		
					5.00	3.00
101	1.73	2.52	255	303		
					Undefined	3.00
101	1.98	2.77	280	303		
					Undefined	3.00
95	2.37	3.21	305	285		

The decision rule in the short run is; *produce if the firm will make a profit or lose less than its fixed costs.* This means that the firm must at least cover its variable costs. If the firm doesn't cover all its variable costs, producing nothing is the best short-run strategy. Table 4-5 shows that Jones will operate since average revenue of $3.00 is greater than the minimum average variable cost of $1.43.

The answer to the second question (how much to produce) is answered with the following decision rule: *In the short run, produce the output level where MR = MC in order to maximize profits or minimize losses.* Suppose Jones considers whether to produce 70 or 85 bushels. The average net additional revenue for each bushel is $3.00. The additional cost is $1.67 per bushel. Therefore, Jones will produce 85 bushels. The same analysis of whether to produce 85 or 96 bushels will lead Jones to produce 96 bushels, at which point the marginal cost rises to $2.27 per bushel. It pays Jones to increase production slightly beyond 96 bushels because MR > MC at 96 bushels. It does not pay Jones to increase production to 101 bushels per acre because the MC of the last bushel is $5.00, which is more than the marginal revenue of $3.00. The most profitable level of output, or the short-run equilibrium level of output, is about midway between 96 and 101 bushels per acre.

To confirm the validity of the MR = MC rule, suppose as in Table 4-5 we are operating the farm at 96 bushels per acre. We can change the revenue and cost relationship by either expanding or reducing the output level. If we reduce the level of output to 85 bushels per acre, profit from each acre will fall by $5 because the cost of producing 11 fewer bushels is $25 less ($230 − 205), but the return falls $33 ($288 − 255). If we increase the level of output to 101 bushels per acre, profit falls by $10 because the

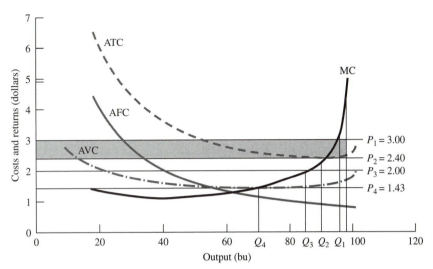

FIGURE 4-7   Utilizing average and marginal costs to determine profit-maximizing output at various prices.

revenue increase is only $15 ($303 − 288), and costs increase by $25 ($255 − 230). However, since MR > MC at 96 bushels per acre, there is an output level between 96 and 101 bushels per acre where MR = MC at $3 per bushel. Profits would be maximized at that point.

Now we can address the question of the amount of profit or loss realized. We use the cost and revenue curves in Figure 4-7 to determine graphically the quantity of output produced at different output price levels. This figure shows the costs of producing various possible levels of output. The challenge Jones faces is to choose the profit-maximizing level of production. This level will depend on the price of the product being produced, in this case corn, and the firm's costs. With an output price of $3.00, the profit-maximizing firm produces 98 bu because MR = MC = P at 98 bu. The vertical line from 98, labeled Q_1, on the output axis gives the values for ATC, AVC, and AFC. Thus, on the vertical line projected above Q_1, the difference between per unit revenues (AR) and per unit costs (ATC) is the average profit per unit of the 98 units of output. The shaded rectangle is the profit per unit times the quantity, the number of units produced, which is the total profit associated with this level of production. Remember that the firm's costs include all fixed and variable costs of production (including a return to management and to investment equal to returns in alternative economic activities).

At prices even higher than $3.00, the firm would operate where MR = MC = P and would produce slightly more than 98 units of output. Also the profit per unit would be higher since the distance between the MC and ATC increases as output increases. At prices below $3.00, less output is produced and less profit would be realized.

Break-Even or Equilibrium Price   If price falls to P_2, about $2.40, the firm is operating at the **break-even point.** At the break-even point the firm's average revenue (P = AR) is equal to its average total cost (ATC) at the minimum point on the ATC curve, and it is not earning profits. This is because, on average at Q_2, each unit costs just as much to produce as the sale price. Looking ahead, we will find that $2.40 is a long-run equilibrium price because there is no incentive for this firm to change its level of production, or for firms to enter or exit the industry.

Loss Minimization   How much will the farm operator produce if the price falls even lower? At any price below $2.40 there are losses, but the firm may continue to produce in the short run. If the firm ceases production, all fixed costs must still be paid. For example, at a price Q_3 of $2.00 the choice is between producing nothing and paying the $80.00 fixed costs, or producing 85 units and paying the $124.95 ($1.47 × 85) of variable costs and $45.05 ($2.00 × 85 − $124.95) of the fixed costs. So the firm's loss is $34.95 and not the full $80.00 of fixed costs that must be paid if nothing is produced. Note that losses are minimized by producing where MR = MC. Given a choice between losing all of fixed costs or a lesser amount, the choice is clear. If an operator cannot make a profit, **loss minimization** is the next best choice.

While loss minimization is the clear best choice, it is a problem. The losses are real money and must be paid out of savings, by borrowing or some other source. If the funds are borrowed, the lender will go only so far before pulling the plug and demanding payment. If output prices do not increase, costs must be cut to return to profitability. If this is not possible, the firm will close and exit the industry because it will not incur losses in the long run.

Closedown Case   In this case, it pays to shut down in the short run even though the firm retains and pays for the fixed factors of production. (In the extreme case, if there is no market for the product, P = MR = 0, the firm will certainly not produce.) Producing in this situation would require incurring additional costs—the variable costs—which would also be lost. The firm loses less if it closes down and pays the fixed costs. If the price is $1.43, the firm is indifferent as to whether to operate. If it shuts down, the fixed costs must be paid. If the firm operates, it produces 70 units, which occurs at the minimum point on the AVC curve. It is just able to cover its variable costs. It still loses the $80.00 of fixed costs. This is because MR = MC at an output level at which AR = AVC.

In Figure 4-7, at any price (AR) less than $1.43 the firm shuts down. This is because the firm's AVC is greater than its AR. Its returns per unit are less than its variable costs per unit and it loses more than the fixed cost it would lose if it shut down. The firm operates in the short run at any price above the minimum point on the AVC curve. At prices between the minimum point on the AVC and the minimum on the ATC curve the firm will incur losses. At prices above the minimum of the ATC curve it earns profits.

Figure 4.8 is constructed to show more clearly the relationship among the cost and revenue curves. At each price (P_1, P_2, P_3, and P_4) the firm operates at the quantity where MC = MR, where MR = price. At P_1 the firm produces quantity Q_1 and earns profits equal to the shaded area. At P_2 the firm produces at Q_2 and breaks even. At P_3, output is Q_3, and the firm realizes a loss, but operates in the short run. The firm will not continue to operate in the long run with losses as indicated by the cross hatched area. The firm will close. At any price below P_4 the firm will not operate, even in the short run because its revenue will be less than its variable costs.

Implications   A farm, or any firm, will earn profits only if the price of the product produced is above the lowest point of the ATC curve. However, even if the price in a given

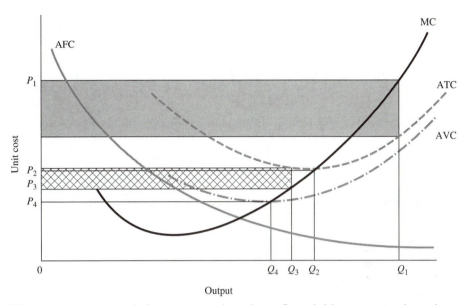

FIGURE 4-8   Utilizing average and marginal cost curves to determine profit-maximizing output at various prices.

year goes down to a level that results in short-run losses, production should continue at the quantity where MR = MC as long as the output price is higher than the minimum AVC. The firm may draw down reserves (savings), assuming there is reason to hope for better prices. Of course if the losses are projected to continue, eventually the firm must close.

In production agriculture the proportion of costs that is fixed varies among types of operations. As a result, they have different shutdown prices. For example, a grain producer has a high proportion of fixed costs in the form of land and machinery. Grain prices can be quite low and the grain producer will continue to operate because the firm can cover variable costs and some portion of fixed costs. Many cattle ranchers have a large number of acres and breeding herds that have large fixed costs. Thus they continue to operate in the face of low prices. Likewise, most agribusinesses have high fixed costs in plant, equipment, and personnel, and thus will continue to produce in the short run while incurring losses.

In contrast, livestock feeders who buy and feed cattle, hogs, or poultry usually operate with a small amount of land and relatively inexpensive buildings. Therefore they have relatively low fixed costs. The two major costs for a livestock feeder are the animals purchased to be fed and the feed that they eat. A relatively small drop in price generally will result in the livestock feeder shutting down in the short run because AR is less than AVC at the optimum level of production.

Marginal Cost and the Supply Curve

We can use the cost and return information to state a generalization concerning the purely competitive firm. The portion of the firm's marginal cost curve above the average variable cost curve is the firm's **short-run supply curve.** Supply is a schedule

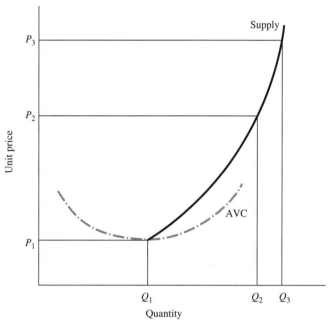

FIGURE 4-9   Supply curve for a firm.

showing the amount of output a firm produces at different prices during a specified period of time. Note in Figure 4-7 that, at a price equal to or above the AVC, the firm determines how much to produce by selecting the output level where the MR curve crosses the MC curve. This is true whether the firm is maximizing profits or minimizing losses. If the output price is below the minimum point of AVC ($1.43 in this case), the firm generally ceases production and may exit the industry. At prices above this level, production levels are higher. The MC curve can be derived from either the TC or the ATC curve, which are a function of the production function and the input prices. So the supply curve is determined by the combination of the technical relationships that determine the production function and the prices of the inputs and outputs of the firm.

Figure 4-9 shows the general shape of the supply curve for a firm. At a price below P_1, production is zero. As the price moves above P_1, the quantity supplied increases. The positive relationship between increasing price and quantity supplied is called the **law of supply.**

LONG-RUN PRODUCTION COSTS

As shown in Chapter 3, firms can make adjustments in the long run that they cannot make in the short run. Firms can alter the capacity of a plant, and they can buy and sell resources such as land and other capital inputs. They can merge and grow, cut back, close down, or even go out of business. The number of firms in the industry can increase or decrease as firms enter or leave. We show how firms make these long-run adjustments and indicate the long-run equilibrium position of firms.

The **long run** is a time period long enough to allow farmers to quit farming, or to change from one enterprise to another in whatever way might be profitable. It also permits others to enter farming. In the agribusiness context, it includes decisions to add production capacity, add or drop product lines, merge or divest, or expand into foreign markets.

A Firm's Long-Run Unit Costs

We now consider long-run unit costs. In the long run, firms can change the size of plants, the mix of inputs, and the combinations of products to achieve the highest possible profits. Farmers can purchase or sell land or other fixed assets. Since any resource adjustments can be made, all costs are variable; there are no fixed costs. Further, since there are no fixed resources or fixed costs, the law of diminishing returns does not apply. The individual firm's long-run average cost curve is U-shaped, not because of increasing and diminishing returns but rather because of the way costs change when the size of the firm changes. This is illustrated in Figure 4-10.

The level of a firm's short-run cost curve depends on the size of the firm. As shown in Figure 4-10, as the firm increases in size, the ATC and MC curves both shift downward as the firm grows to the size shown by ATC₃, the lowest-cost firm size. If the firm expands further, costs increase. Consider the example of a movie theater where output is the number of moviegoers. A theater with one screen will incur high costs per moviegoer because the employees are not productive while the film is showing. With several

FIGURE 4-10   Long-run average costs for a hypothetical firm.

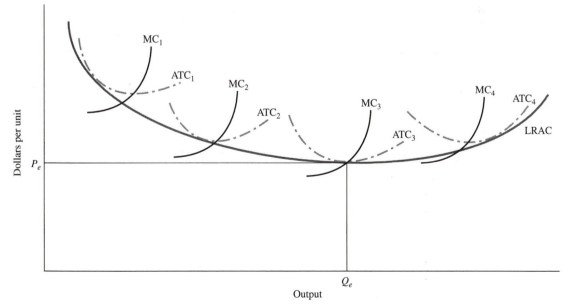

more screens, the management can schedule the showings so that the employees are productive at nearly all times as they sell tickets and concessions, collect tickets, clean theaters, and so on. At some point, adding more screens does not reduce costs by allowing more efficient scheduling. Further, with a very large number of screens, crowd control may become a problem, requiring additional staff and rising costs.

The long-run average cost (LRAC) curve envelops the several SRAC curves. The LRAC envelopes the SRAC cost curves of operating at various firm sizes. In Figure 4-10, a firm operating where the output price P_e equals the minimum of both the short-run average total cost (ATC_3) and the long-run average cost is in both short-run and long-run equilibrium. The firm has no incentive to change the amount produced by adjusting the quantity of fixed or variable inputs. Also, if the industry consisted entirely of firms of the size depicted by ATC_3, there would be no incentive for firms to enter or leave. All firms would be earning a rate of return on their investment equal to that earned in other industries.

If the price is P_e, then any firms operating at size one, two, and four will lose money in the short run because their ATCs are above P, which equals MR. These firms will operate in the short run if they are covering their AVCs. However, in the long run, they will either adjust the size of their firm so that their ATC and MC curves are like those of firm three or they will go out of business.

Industry Long-Run Equilibrium

To discuss conditions for a long-run equilibrium, we take an industry, rather than a firm, perspective. To simplify the problem, we assume that all firms in the industry have adjusted to a point such as Q_e in Figure 4-10 and have identical cost curves. This cost structure is "representative" or "typical" of all firms and can be used to demonstrate what happens in an industry. Second, changes in production levels are made by the entry and exodus of firms, rather than by increasing and decreasing the size of existing firms. This permits us to ignore the short-run adjustment of the firm and concentrate on what happens in the long run. Third, neither firm entry nor exit affects input prices.

The **long-run equilibrium** in a purely competitive industry occurs when the product price is equal to the minimum average total cost of all firms. This occurs at the long-run equilibrium point in Figure 4-10. What applies to the representative firm applies to all firms; that is (1) all firms seek profits and avoid losses, and (2) all firms can enter or leave the industry. If the price of the product exceeds the minimum point of the average total cost, the resulting excess profits attract new firms into the industry. Entry continues until the increase in the quantity supplied drives the price of the product back down to equal the average total cost. Simply stated, if firms in an industry are making higher than normal profits, others will decide to enter. Conversely, if MR = MC at a point below the minimum of the average total cost curve, the resulting losses will cause some firms to leave the industry. This reduces the total quantity supplied and drives the product price up to P_e.

This adjustment process implies that, with fixed output prices, in the long run the average cost of production will equal the selling price. This means among other things, as will be discussed much later, that the long-term maintenance of farm price supports, or target prices, will result in farmer's average cost of production equaling the support price.

Economies and Diseconomies of Scale

Thus far, this chapter has dealt with the costs of and returns to production and the interaction with market prices. The size of the firm was mentioned in the discussion, but the factors determining the size were not explained. We start by clarifying the distinction between increasing the output produced by a firm of a given size (e.g., a farm of a given acreage) and increasing the size of the firm (increasing the acreage of the farm). The term **economies of scale** refers to changes in the cost of production associated with a change in the amount of the fixed factors of production.

The two primary factors that determine optimum firm size are technical and pecuniary. The technical relationships among inputs and outputs determine the shape of the firm's production function. These relationships cause the **long-run average cost curve** to decrease and then increase as the size of the plant increases, as shown in Figure 4-10.

The optimum size varies widely among industries and farming systems. Ranches are commonly thousands of acres in size, while some fruit farms may be only a few acres. However, some fruit farms are quite large in terms of dollar value of sales, and some ranches with large acreages realize rather small total revenues. A large farm may be very small when compared with an average-sized agribusiness. Thus, comparisons of firm size should be made relative to other firms in the same industry.

Technical Advantages and Disadvantages of Large Scale   The *technical advantages of large scale* come from savings in inputs such as management, labor, fuel, and machinery. There are labor-related economies of scale associated with operating machinery of larger size. (One person may operate either a large or a small tractor, for example.) The marginal cost of labor for machinery operation decreases as the size of the farm increases because the larger farm allows more efficient use of large tractors and other large machines. For example, generally it does not take twice as much labor to operate a 1400-acre corn and soybean farm as it does to operate a 700-acre farm. Another technical factor leading to economies of scale is the ability of management to specialize in performing some tasks, for example, purchasing, producing, marketing, and financing.

The relationship between the size of the operation and the nature of ownership also varies widely. Generally speaking, in the United States, a family farm (a farm on which the family provides the management and most of the labor) can achieve most of the technical efficiencies such as from using large machinery. Thus, over a wide range of crop and livestock enterprises, the family farm may be as technically efficient as a larger farm employing many labor and management employees. There are of course many exceptions. For example, in fruit and vegetable growing, producers employ many seasonal workers. The cost advantages of large-scale, automated enterprises are evident in cattle feeding, in hog production, and in broiler and egg production, especially where large-scale feed-handling equipment is technically efficient. Operations of this type are operated as family farms, but many are operated under contract with large-scale corporate feed or packing firms.

However, at some size, the operation encounters technical inefficiencies. For example, if farm machinery becomes too large, it is difficult to move to the field or to operate. At this point the per-unit cost of operation increases because of **diseconomies of scale.** There are technical disadvantages to large scale which arise out of management's inability to coordinate more resources within a single unit. This is one of the reasons why large corporations often operate with several types of production operations. Problems associated with organizing production and arranging for financial resources increase as farms and agribusinesses become larger. Also, some machines are less efficient beyond a certain size. The multiplication of such problems tends to offset other technical advantages so that beyond a certain size the main advantages of large scale are usually pecuniary rather than technological.

Pecuniary Advantages and Disadvantages of Large Scale   The **pecuniary economies** refer to the prices paid and received. Pecuniary advantages of large scale, often more so than technical advantages, are the reason for the scale advantages of very large farms and agribusiness firms. These advantages take several forms, such as bulk and quantity discounts on purchased inputs, leverage to negotiate lower rates on borrowed capital, lower prices on equipment purchases, and the ability to use bulk transportation facilities. They "spread their overhead" costs over the larger number of units of output produced by a larger firm. Also, some large firms can sell at premium prices. Because of these pecuniary advantages some farms and agribusinesses grow to very large sizes, well beyond the size at which technical economies of scale are exhausted. Thus some very large firms, including very large family-operated businesses, may not be more technically efficient than the typical-sized family farm. But they may be more profitable because of pecuniary advantages. On a farm as well as in agribusinesses, the minimum average cost of production occurs when a manager combines the technical and pecuniary factors into the optimum arrangement.

PRESSURE FOR OPTIMAL SIZE

As noted above, in some industries firms of various sizes can operate at approximately the same cost per unit of output. In these industries, various sized firms can remain in the industry. However, in many industries scale economies are significant. In these industries there is great economic pressure on the firms to become more efficient by adjusting to the optimal size or to exit the industry.

PRICE ELASTICITY OF SUPPLY

We turn now to elasticity of supply. This concept is as important for understanding the supply decisions firms make as elasticity of demand is important for understanding the concept of demand. (You may find it helpful to review the portion of Chapter 2 dealing with demand elasticity.)

Industries differ in terms of their elasticities of supply. For industries with a high proportion of fixed costs, such as production agriculture, the quantity of total output changes slowly in response to changes in market prices. In the short run, even if prices fall sharply, firms in such industries may remain in full production. For example, farmers may, instead of changing their total output in response to a change in market prices, change from one crop to another, or change from one livestock enterprise to another. In contrast, most heavy-manufacturing firms, such as those in the steel and automobile industries, lay off workers and reduce production in bad times and increase employment and output in good times. These differing production responses are measured in terms of the price elasticity of supply.

Price **elasticity of supply** is the percentage change in the quantity of a good or service supplied in response to a given percentage change in its price, *ceteris paribus*. The coefficient of the elasticity of supply (E_s) can be calculated from the rate of change at any point on a supply curve, or from the relative percentage change in price and quantity supplied between any two points on the supply curve. The first is **point elasticity;** the second is **arc elasticity.** They are computed using the same formula as used in the calculation of the elasticity of demand:

$$E_s = \dfrac{\dfrac{\text{change in quantity supplied}}{\text{sum of two quantities}}}{2} \div \dfrac{\dfrac{\text{change in price}}{\text{sum of two prices}}}{2} = \dfrac{\dfrac{Q_2 - Q_1}{(Q_2 + Q_1)}}{2} \div \dfrac{\dfrac{P_2 - P_1}{(P_2 + P_1)}}{2}$$

Note that as in the case of demand elasticity the "2s" cancel algebraically, and therefore can be deleted.

POINT ELASTICITY

As in the calculation of demand elasticity, the formula for calculating elasticity at a point is $E_s = \Delta Q \div \Delta P \times P \div Q$.

There are three general cases for elasticity of supply. *If the percentage change in quantity is greater than the percentage change in price, the supply is elastic, $E_s > 1$. If the percentage change in price is greater than the percentage change in quantity, the supply is inelastic, $E_s < 1$. Finally, if these percentage changes are the same, the supply has unitary elasticity, $E_s = 1$.* The three cases of elasticity—elastic, unitary elastic, and inelastic—are presented in Table 4-6. The arc elasticity formula is used to calculate these elasticities.

Figure 4-11 shows how the elasticity of a supply curve changes at different price-quantity levels. It also shows that we can use the position and slope of a linear supply curve to determine whether the curve is elastic, inelastic, or unitary elastic. As shown, a point on a supply curve is elastic ($E_s > 1$) if a line tangent to the curve at that point intersects the price axis, inelastic ($E_s < 1$) if a tangent line intersects the quantity axis, and unitary elastic ($E_s = 1$) if the tangent passes through the origin. The unitary elasticity line has a constant elasticity ($E_s = 1$). Neither of the other two straight lines has a constant elasticity.

TABLE 4-6   CALCULATION OF SUPPLY ELASTICITY

Q	ΔQ	P	ΔP	$(Q_1 + Q_2)/2$	$(P_1 + P_2)/2$	$E = \dfrac{\Delta Q}{(Q_1+Q_2)/2} \div \dfrac{\Delta P}{(Q_1+Q_2)/2}$
2.5		95				
	0.5		5	2.75	97.50	$(0.5/2.75)/(5/97.5) = 3.55$
3.0		100				
	0.6		20	3.30	110.00	$(0.6/3.3)/(20/110) = 1.0$
3.6		120				
	0.2		20	3.70	130.00	$(0.2/3.7)/(20/130) = 0.35$
3.8		140				

Note: Hypothetical data.

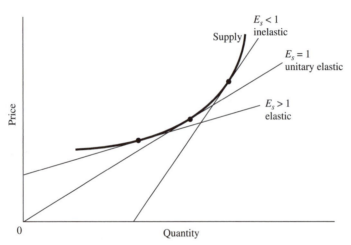

FIGURE 4-11   Elastic, unitary elastic, and inelastic supply curves.

Determinants of Elasticity of Supply

The elasticity of supply for a market for any good or service is a weighted average (the weights are based on the quantities supplied by the firms) of the elasticity of the supply curves of the firms that produce the good or service. Since it takes time for producing firms to react to a change in price for their output, time is the main factor in their planning horizon and in determining the elasticity of supply. Generally, the longer the producer has to plan and to adjust to an existing or an expected price change, the greater output response and therefore greater elasticity of supply. For example, if farmers observe or expect the price of wheat to increase and other prices to remain the same they may not be able to produce more wheat in the current growing season. However, given several years, farmers can produce more wheat. In the long run, supply is nearly perfectly elastic for some commodities because of the ability of producers to respond. However, total farm output is not perfectly elastic because the fixed quantity of land ultimately limits production.

Time Period

When analyzing the effect of time on the elasticity of supply, economists distinguish among three periods: the market period, the short run and the long run (as defined in Chapter 3). These concepts do not refer to a specific time period, as in number of days or years. They refer to the conditions of supply and to changes in the quantity supplied in response to a change in price of the commodity produced, *ceteris paribus*.

The Market Period   The **market period** refers to a period of time so short that producers cannot change the quantity in response to a change in demand or price. For example, if a farmer harvests the entire crop of vegetables and brings them to an outdoor market on a hot day, the vegetables must be sold whether the price is high or low because any unsold vegetables will have to be thrown away. Neither can the farmer offer more for sale, until another crop is produced. The supply curve of that farmer is perfectly inelastic. Figure 4-12 illustrates the farmer's perfectly inelastic supply curve in the market period. Regardless of a shift in the demand curve, the quantity supplied by farmers remains unchanged. A shift in demand from D_1 to D_2, for example, causes the market equilibrium price to fall from P_1 to P_2, while Q is unchanged.

A market period is not confined to any specific time or to any commodity. It may stretch over a complete production period. For example, the approximate quantity of seasonal merchandise in a clothing store is generally set well before the season begins: a change in price will not have much, if any, effect on the total quantity to be supplied. The quantity of livestock that will come to market in the near future tends to be largely set. In these cases, supply is inelastic and shifts in demand will bring much greater changes in price than in quantity supplied. If winter coats do not sell as well as predicted, you can count on large markdowns.

The Short Run   As explained in this chapter, output can be increased in the **short run** by increasing the variable inputs used (fixed inputs do not change in the short run). In Figure 4-13, a shift in demand from D_1 to D_2 causes the equilibrium price to rise

FIGURE 4-12   Perfectly inelastic supply.

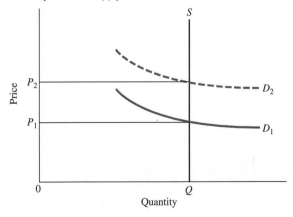

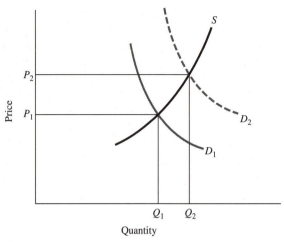

FIGURE 4-13   Short-run supply elasticity.

from P_1 to P_2 and the quantity to increase from Q_1 to Q_2. Farmers respond to higher grain prices by adding more fertilizer and other variable inputs. Many businesses purchase more raw materials, add labor, and perhaps add more shifts. The nature of the production process determines how long it takes to increase production by increasing variable inputs in the short run. Some firms can change the rate of output daily or hourly while for others it will require months or even years.

The Long Run   The **long-run supply curve** is always more elastic than the short-run supply curve because all inputs are variable. Firms can increase or decrease in size. The long-run supply curves may reflect increasing- and constant-cost industries. A long-run cost curve is upward sloping for an increasing-cost industry, as shown in Figure 4-14a, and flat and horizontal for a constant-cost industry, as shown in Figure 4-14b. Since all inputs are variable, whether the industry has decreasing, constant, or increasing cost depends on the prices of the inputs. If, as the demand for the product grows, the industry must pay higher prices to command enough resources to expand production, the industry's costs will rise. If the industry can command increasing resources at the same prices, it can expand with constant costs.

Elasticity of Supply of Inputs

The elasticity of supply for any output that the firm produces depends on the elasticity of the supply of the inputs used to produce it. Where resources, such as land or irrigation water, are relatively fixed in the short run, the elasticity of supply of the commodities that use those inputs is reduced. Many of the important inputs used in farm production are inelastic in supply in the short run, and this reduces the elasticity of supply of farm output in the short run.

In the long run, however, all inputs are variable resulting in a more elastic supply of output in the long run. For example, in the long run some land can be drained or cleared

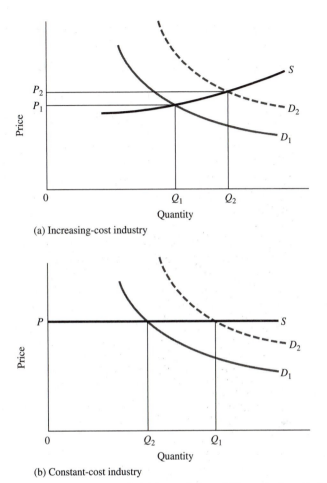

(a) Increasing-cost industry

(b) Constant-cost industry

FIGURE 4-14   Long-run supply curves in (a) increasing- and (b) constant-cost industries.

and brought into crop production. (There are millions of acres of land in the United States, and even larger areas of land in South America and Africa, that could be converted to crop production.)

Cross-Price Elasticity of Supply

Similar to the cross-price elasticity of demand, the **cross-price elasticity of supply** (E_{sxy}) refers to the effect of the price of good or service X on the quantity of good or service Y supplied. Commodities that are complementary in supply, such as soybean oil and soybean meal, have a positive cross-price elasticity. Commodities that compete for resources have a negative cross-price elasticity because as the price of one goes up the production of the other goes down, (*ceteris paribus*). Since many agricultural commodities are produced using the same resources, the cross-price elasticity of supply (E_{sxy}) for those agricultural products is negative and in many cases very elastic. For example, if the price of corn goes up (*ceteris paribus*), the production of soybeans will go down, and vice versa because corn and soybeans are produced on the same type of land. Cross elasticity would be high between corn and soybeans, but low between cattle and cotton.

SUMMARY

The economist's concept of costs is based on limited resources, alternative uses, and measuring values in terms of opportunity costs of forgone alternatives. Costs are either explicit or implicit. The short run includes both fixed and variable costs, which account for the typical U-shape of short-run cost curves. Average and marginal costs are derived from total costs. The MC curve cuts the AVC and ATC curves from below at the minimum point of each curve.

A firm can use either the total cost and return approach or the marginal cost and return approach to determine the most profitable level at which to produce. The portion of its marginal cost curve that lies above the average variable cost curve is the supply curve for a firm. Short-run equilibrium is at the level of output at which MR = MC.

In the long run, all production costs are variable. The U-shape of the long-run ATC curve is due to the technical and pecuniary economies and diseconomies of scale as a firm expands. Typically, the long-run ATC curve at first declines sharply as the scale of production increases and then tends to decline more slowly or level off. As the firm reaches a size at which technical or pecuniary diseconomies are encountered, ATC begins to rise.

Many farms and agribusinesses experience significant technical and pecuniary economies of large scale. The prevalence of the family farm suggests that most of the technical or internal economies are realized on family-operated farms. For most very large farms, pecuniary economies of large scale are more important than the technical economies.

IMPORTANT TERMS AND CONCEPTS

abnormal profits   90
arc elasticity   101
average fixed costs (AFC)   83
average revenue (AR)   90
average total costs (ATC)   83
average variable cost (AVC)   83
break-even point   93
close-down case   94
cross-price elasticity of supply (E_{sxy})   105
diseconomies of scale   100
economies of scale   99
excess profits   90
explicit costs   80
implicit cost   80
law of supply   96
long run   97
long-run average cost curve   98

long-run equilibrium   98
long-run supply curve   104
loss minimization   94
marginal cost (MC)   85
marginal revenue (MR)   90
market period   103
opportunity cost   79
pecuniary advantages of
   large scale   100
point elasticity   101
price elasticity of supply   100
profit   88
profit-maximizing output   88
short run   103
short-run supply curve   95
technical advantages of
   large scale   99

total costs (TC)   82   total variable costs (TVC)   81
total fixed costs (TFC)   81   windfall profits   90
total revenue   88

QUESTIONS AND EXERCISES

Name That Term

Read the following sentences carefully and fill in the missing term or terms.

1 An _____ cost is the value of an input that the firm or a family business owns or controls.
2 An _____ cost of using a resource (input) is the value of the contribution that resource could make in the highest priced alternative use.
3 _____ costs are the normal "out-of-pocket" costs of the inputs purchased for use in producing a product.
4 _____ is the change in total cost with one additional unit of output.
5 The costs of factors of production, such as property taxes, land, and machinery, are referred to as _____.
6 _____ is total revenue of a firm less all its economic costs.
7 _____ is the change in revenue for a one-unit increase in output.
8 A seller (or buyer) of a good or service who is unable to affect the price at which a commodity sells by changing the amount it sells (or buys) is called a _____.
9 _____ is the sum of total fixed costs and total variable costs.
10 _____ are the savings realized by purchasing at a lower price or selling at a premium, usually resulting from the size of the firm.

True/False

Read the following sentences, then decide whether each statement is true (T) or false (F) and mark it accordingly.

T F 1 Total costs increase as output increases because total variable costs increase while total fixed costs remain unchanged.
T F 2 A typical total variable cost curve starts at zero, increases rapidly at first, continues to increase but at a less rapid rate, then increases more rapidly again, and finally curves back toward the vertical axis when Stage III of the production function is reached.
T F 3 A profit-maximizing firm will continue to hire labor inputs up to the point where labor's marginal physical product equals its wage rate.
T F 4 Firms that cover variable costs but do not make economic profits in the short run should shut down in order to minimize losses.
T F 5 A profit-maximizing corn farmer should use the level of fertilizer that maximizes total output of corn per acre.
T F 6 A profit-maximizing farmer continues to increase the amount of a variable input used so long as the additional units of that input contribute to total profit.
T F 7 As output increases, average variable costs first decline, then reach a minimum, and then increase.
T F 8 The law of supply says that the price of output and the quantity produced are positively related.
T F 9 A firm that uses a single variable input and has a falling marginal product will have marginal costs that increase with output.

T   F   **10** Marginal costs intersect average variable costs at the level of output where AVC reach a minimum.

T   F   **11** Marginal costs intersect average total costs at the level of output where ATC reach a minimum.

T   F   **12** Marginal costs intersect average fixed costs at the level of output where AFC reach a minimum.

T   F   **13** For firms operating with some fixed costs, MC intersects AVC at a lower level of output than MC intersects ATC.

Multiple-Choice Questions

Circle the letter of the response that best answers the question or completes the statement.

1 Which of the following is most clearly an implicit cost for a firm?
   a rental payment on a fleet of delivery trucks
   b wages paid to unskilled labor
   c depreciation on company-owned computer equipment
   d payments to a consulting firm
   e none of the above

2 If a firm produces 9 units of output at an average total cost of $8 per unit and sells all 9 for $10 per unit, its profit is
   a $2.
   b $9.
   c $18.
   d $20.
   e none of the above.

3 For most firms, as output is increased
   a MC rises and then begins to decline.
   b TC rise and then begin to decline.
   c AFC decline and then begin to rise.
   d AVC decline and then begin to rise.
   e none of the above.

4 In the long run a firm
   a has no fixed inputs.
   b can enter or exit the industry.
   c can change the combination of products offered for sale.
   d has a more elastic supply curve than in the short run.
   e all of the above.

5 Which of the following describes the relation between cost curves?
   a AVC intersect MC at the minimum point on MC.
   b MC intersects AVC at the minimum point on AVC.
   c AFC intersect MC at the minimum point on MC.
   d MC intersects AFC at the minimum point on AFC.
   e None of the above.

6 The vertical distance between AVC and ATC gives
   a implicit costs.
   b AFC at a given level of output.
   c the law of diminishing returns.
   d the economies of scale.
   e none of the above.

7 If the total cost of producing 10 units is $70 and 11 units is $75 and fixed costs are $20.00, what is MC?

   a $5
   b $25
   c $15
   d $90
   e none of the above

8 If we compare the changes in TVC and TC as an additional unit of output is produced

   a the changes in TVC and TC are equal.
   b TVC change more than TC.
   c TVC change less than TC.
   d the changes in TVC and TC equal TC.
   e none of the above.

9 If a firm incurs a FC of $20 and a AVC of $10 when it produces 50 units of output, what are the firms TC?

   a $30
   b $1500
   c $500
   d $520
   e none of the above

10 Economies of scale are indicated by

   a the rising segment of the LRAC.
   b the declining segment of the LRAC.
   c the difference between TC and TVC.
   d a rising MC curve.
   e none of the above.

11 If a firm increases all of its inputs by 10 percent and its output increases by 8 percent, we can say

   a it is encountering constant returns to scale.
   b the marginal products of all inputs are falling.
   c it is encountering diseconomies of scale.
   d it is encountering economies of scale.
   e none of the above.

12 The price elasticity of supply measures how

   a easily inputs can be substituted one for another in the production process.
   b responsive the quantity supplied of X is to changes in the price of X.
   c responsive the quantity supplied of X is to changes in the price of Y.
   d responsive the quantity supplied of X is to changes in the income levels.
   e none of the above.

13 Supply curves tend to be

   a perfectly elastic in the long run because consumers have time to fully adjust to any change in amount supplied.
   b more elastic in the long run because there is time for firms to enter or leave the industry.
   c perfectly elastic because there is a maximum amount that can be produced from a given set of resources.
   d perfectly inelastic because consumers can switch their purchases to other goods.
   e none of the above.

14 Suppose the price of widgets rises by 20 percent and the quantity supplied increases by
18 percent, then the elasticity of supply is
  a negative and therefore widgets are inferior goods.
  b positive and therefore widgets are normal goods.
  c less than 1 and therefore widget supply elasticity is inelastic.
  d more than 1 and therefore supply is elastic.
  e none of the above.

Technical Training

1 Use the following illustrative cost curves to answer questions a–l.

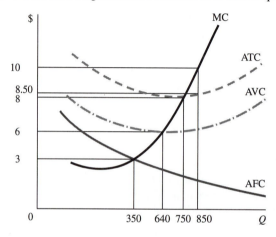

  a At $P = \$10$, how many units of Q does a profit-maximizing firm produce?
  b At $P = \$10$, what is the difference between average revenue and average total cost for
the firm?
  c At $P = \$10$, what is the maximum profit the firm can earn?
  d At $P = \$8$, how many units of Q does a profit-maximizing firm produce?
  e At $P = \$8$, what is the difference between average revenue and average total cost for
the firm?
  f At $P = \$8$, what is the maximum profit the firm can earn?
  g What is the break-even price and quantity for the firm?
  h What is the long-run shutdown price and quantity for the firm?
  i What are total fixed costs for the firm?
  j What is the firm's short-run supply curve? Shade your answer directly on the graph.
  k For prices between $6 and $8, does the firm earn positive or negative profits?
  l Why does a firm continue to produce and sell output in the short run when prices are
between $6 and $8?

Questions for Thought

1 Use the following information to answer questions a–g. Table 1 depicts the production
and cost information for your Hungarian restaurant (see questions and exercises in
Chapter 3). The relationship of the variable input (hours of labor) and the monthly output (chicken wings) is typical of most production functions. Assume that you sell the
wings for $3.50, the cost of the variable input P_l is $9/hour, and the restaurant incurs
monthly fixed costs of $200 for rent, insurance, and taxes.

TABLE 1   PRODUCTION AND COST INFORMATION FOR HUNGARIAN RESTAURANT

Input (hours)	Output (TPP)	Total fixed cost	Total variable cost	Total cost	Average fixed cost	Average variable cost	Average total cost	Marginal cost
0	0	$200	0	———	———	———	———	
								———
10	50	———	———	———	———	———	———	
								———
20	110	———	———	———	———	———	———	
								———
30	180	———	———	———	———	———	———	
								———
40	240	———	———	———	———	———	———	
								———
50	290	———	———	———	———	———	———	
								———
60	330	———	———	———	———	———	———	
								———
70	360	———	———	———	———	———	———	
								———
80	380	———	———	———	———	———	———	
								———
90	390	———	———	———	———	———	———	
								———

Note: AFC = TFC ÷ TPP; ATC = TC ÷ TPP; AVC = TVC ÷ TPP; Input = hours of labor; MC = change in TC ÷ change in output; TC = TFC + TVC; TFC = constant; TPP = chicken wings; TVC = inputs × cost per unit.

a Complete Table 1 by filling in the blanks, rounding to two decimal places.

b Graph the total cost (TC), total variable cost (TVC), total revenue (TR), and total fixed cost (TFC) on Graph 1. Remember to label your curves.

c Plot the average total cost (ATC), average variable cost (AVC), marginal cost (MC), and marginal revenue (MR) on Graph 2. (Note: MC and MR are plotted on the midpoint between quantities. In other words, plot the first MC and MR values where TPP = 25.)

d Assume that you can sell as many wings as you decide to produce at $3.50. How many wings will the restaurant produce? What is your profit or loss? Explain your answers, and shade in the profit/loss area on Graph 2.

e Referring to Table 1 and Graph 2, what is the lowest price you can receive for wings and still produce in the short run? Explain why you would halt production if the price fell below this value.

f Designate the firm's short-run supply curve.

g Give a specific example of an explicit cost incurred by the restaurant. Give an example of an implicit cost the restaurant might incur. How should the value of the implicit cost be determined?

2 Define price elasticity of supply. What is an elastic, inelastic, and unitary elastic supply?

3 What is the relationship between the marginal cost curve of individual firms and the elasticity of supply?

4 What is measured by the coefficient of elasticity of supply? Give an example.

5 How does time influence the elasticity of supply? Give examples of the elasticity of supply in the immediate market period, the short run, and the long run.

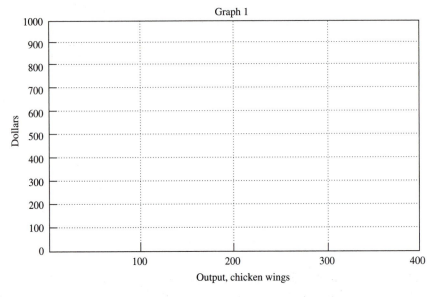

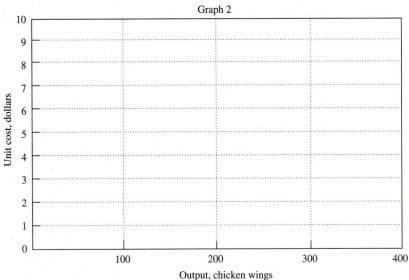

6 Within a given time period, how does the ease or difficulty of substituting among inputs and outputs affect the elasticity of supply?

7 Is the elasticity of aggregate agricultural supply more inelastic in the short run or in the long run? Why? What determines whether in the long run an industry is an increasing-, decreasing-, or constant-cost industry?

8 View yourself as producing a "college education" by taking courses, including this one. Make a list of your costs. Identify which are explicit and implicit. Which are fixed and which are variable?

9 Students pay sizeable fees for the right to attend college classes and work toward a degree, yet are often happy when individual class sessions are canceled. Use economic arguments to explain why this feeling might make sense.

10 Suppose you are managing a firm that is operating in that uncomfortable region between the close-down point and the break-even point. In the short run, you listen to your text and operate, lose money, but minimize your losses while you hope "things turn around." One way things can improve in the long run is for prices to go up and you, like every businessperson, hope the price or your output increases. What else can you do to improve the prospects for your firm's long-term survival?

11 Suppose things do not "turn around" for the firm described in problem 10. How long can you continue to follow your text's advice and operate at a (short-run minimum) loss? Assuming prices never fall below the close-down level, how do you know when it's time to say "enough is enough" and close down?

ANSWERS AND HINTS

Name That Term   1. implicit; **2.** opportunity; **3.** Explicit; **4.** Marginal cost; **5.** total fixed costs; **6.** Profit or loss; **7.** Marginal revenue; **8.** price taker; **9.** Total costs; **10.** Pecuniary advantages of large scale

True/False   1. T; **2.** T; **3.** F; **4.** F; **5.** F; **6.** T; **7.** T; **8.** T; **9.** T; **10.** T; **11.** T; **12.** F; **13.** T.

Multiple Choice   1. c; **2.** c; **3.** d; **4.** e; **5.** b; **6.** b; **7.** a; **8.** a; **9.** d; **10.** b; **11.** c; **12.** b; **13.** b; **14.** c

Technical Training   1. a. 850; **b.** $1.50; **c.** $1275; **d.** $750; **e.** $0; **f.** $0; **g.** $P = \$8$, $Q = 750$; **h.** $P = \$6$ and $Q = 640$; **i.** $1050; **j.** shade MC curve above AVC; **k.** negative profits or losses; **l.** it loses less than if it shuts down

THREE

MARKETS

5

THEORY OF MARKETS

CHAPTER OUTLINE

OVERVIEW

This chapter explains how supply and demand interact to determine market prices and quantities of goods and services. It applies these concepts to selected agricultural input and output markets. We begin by illustrating how the supply curves of individual firms and the demand curves of individual consumers and firms are summed to generate market supply and demand curves.

The basic economic factors that explain how a market works are considered beginning with a relatively simple case focusing on the basic principles of market operations. We remove the simplifying assumptions as we work through this and the following chapters. This allows us to consider more realistic cases. We discuss the perfectly competitive market at some length because of its importance in economic analysis, especially in the agricultural sector.

LEARNING OBJECTIVES

This chapter will help you learn:

- How supply and demand curves interact to determine the prices and quantities of goods and services produced and consumed
- About markets in time, space, and form
- The characteristics of a competitive market
- The determination of output in a competitive market

MARKET SUPPLY AND DEMAND CURVES

Supply

In Chapter 4 we explained that a profit-maximizing producer determines how much to produce by equating MR and MC. The producer operates in the short run if MR = MC at a price above the AVC curve. This is the reason the firm's supply curve is the portion of the firm's marginal cost curve above the AVC curve. The curve indicates that the higher the price for the good, the more the firm will produce and sell. If we know how much an individual firm offers for sale at any given price, it is rather easy to add up the quantities offered by these firms to determine how much they will offer collectively at each price. The total offered by all the firms in the market is the aggregate or market supply. **Market supply** is the various amounts of a good that producers are willing and able to produce and make available at each of a series of prices during a specified period in a given market.

The **supply curve** for a good in a market is the horizontal sum of the supply curves for the individual producers. In Figure 5-1, panels (a) and (b) show how much wheat two French farmers would produce for sale at various prices. These are supply curves for individual farmers. If they were the only French wheat farmers, we could determine how

FIGURE 5-1   Aggregation of supply curves.

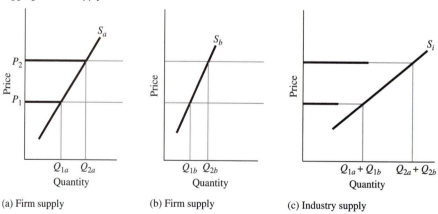

(a) Firm supply   (b) Firm supply   (c) Industry supply

much wheat French farmers would produce at various prices by adding horizontally the individual supply curves, that is, by summing the amount of wheat produced by each farmer at each price. This horizontal summation, shown in panel (c), would be the market supply curve for French wheat.

Demand

In Chapter 2 we indicated how the consumer, with a limited budget, makes choices among available goods to maximize utility. The result was that individuals purchase less of a good as its price increases. As in the case of supply, if we know how much individual consumers purchase, it is a simple matter of summing these individual demands to find the aggregate or market demand at any given price. **Market demand** is a schedule showing the amounts of a good consumers are willing and able to purchase in the market at a series of prices during a specified period in a given market. As with supply, this definition applies to all services, resources, commodities, and so on, including, of course, the input and output markets of agriculture.

As with market supply, the horizontal sum of the demand curves of all individuals in the market generates market demand. In Figure 5-2, panels (a) and (b) are demand curves for individuals. The demand curve in panel (c) is calculated by horizontally summing the quantities demanded by these two individuals at various prices.

In constructing supply and demand curves for a good, we assume that price is the only economic variable that changes and that it determines the quantity of a good supplied and demanded. The curves are drawn on the assumption that many important determinants other than price (technology, tastes and preferences, income, prices of other goods, etc.) do not change. We first explain how the supply and demand curves come together in a market to determine prices. Then we discuss how other economic factors result in *shifts* in supply and demand curves.

FIGURE 5-2   Aggregation of demand curves.

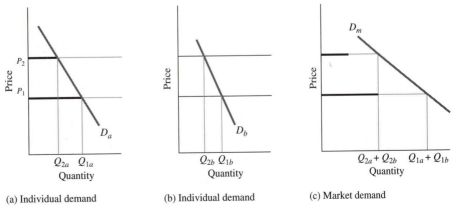

(a) Individual demand   (b) Individual demand   (c) Market demand

---

INDIVIDUAL AND MARKET DEMAND

To help you understand the relationship between quantity demanded and price, think of your favorite food. How much of that food do you purchase at its current price? How much would you purchase if its price were to double, or fall to half its current price? Your answers will provide three points on your individual demand curve for that food. Think about how much of your favorite food would be purchased by all the residents in your community at these three prices. These quantities are three observations on the demand curve in your local community market. Finally, if you were to estimate the amounts purchased worldwide at these prices, you would have an estimate of world demand for your favorite food.

MARKETS

Now we can bring the concepts of supply and demand together to see how buying and selling decisions interact to determine both the prices and the quantities of goods traded in a market. A **market** is an institution or an arrangement that brings buyers and sellers together. The **market price** is the mutually agreeable price at which willing buyers and willing sellers exchange a good.

MARKETS AND MARKETING

The word *market* has several definitions. It describes a food store (a supermarket), the act of selling a commodity (a farmer marketing strawberries), a commodity produced for sale (strawberries), and a segment of buyers (the teenage market). It is also a synonym for price (the market is up today), among other things. When we discuss the management of agribusinesses we use the term *marketing* in referring to the firm's sales efforts. These activities include decisions on pricing, advertising efforts of the sales staff, and so on.

There are innumerable markets. Examples include highly organized and formal financial and international commodity markets, wholesale and retail trading establishments, the neighborhood flea market, shadowy black markets, and casual transactions between individuals.

In every market there must be an agreement on the terms of exchange, most importantly the price. Often negotiation or bargaining on the spot occurs to determine price. Examples include major purchases such as land, houses, cars, and most barter transactions. In some developing countries, many markets are informal and bargaining is the predominant system of exchange. In other cases, the price is "posted" by one of the parties and the other must "take it or leave it." If sales are brisk, the merchant will likely raise the price. If sales lag and inventories accumulate, the merchant will likely lower the price to move the merchandise. Decisions such as these determine market prices.

MARKET EQUILIBRIUM

The market system of allocating goods brings together buyers and sellers. Sellers provide various quantities of a product at different prices, and purchasers will buy various quantities at different prices. At high prices the sellers are willing to sell large amounts but the buyers will take only small quantities. At very low prices, the buyers would be willing to purchase large quantities but the sellers will offer small amounts. There is one and only one price that equates the quantity of a good offered for sale with the quantity buyers are willing to buy. The price occurs at the market equilibrium and is the equilibrium price. *Market equilibrium occurs when the quantity of a good offered by sellers at a given price equals the quantity buyers are willing and able to purchase at that same price.*

A simple example is the pizza market around most university campuses. There are usually many businesses that sell pizza, and many students and faculty buy it with some regularity. Each of the businesses has a supply curve for pizza. Summing these supply curves generates a pizza market supply curve, as shown in Figure 5-1. Likewise, the pizza demand curve for the individuals in the campus community can be generated, as suggested in Figure 5-2. The supply and demand curves are combined on one graph, as in Figure 5-3, reflecting the market for pizza. At the point where the two curves cross, supply and demand are equal. At this price, the market for pizza is in equilibrium. At the price P_e, or market clearing price, the quantity of pizza demanded is equal to the quantity of pizza supplied, Q_e.

MARKET EQUILIBRIUM

There is an important difference between the optimal level of production in a firm and the equilibrium price and quantity in a market. The firm can calculate, given a certain set of prices, its optimal level of production. The market equilibrium, in contrast, is a constantly moving target. Both producers and consumers (or sellers and buyers) are constantly adjusting to new sets of circumstances. Each time a firm or a consumer makes a price or a quantity change, the market equilibrium changes. Thus, a market is seldom if ever in—but there is always pressure to move toward—equilibrium.

FIGURE 5-3   Market equilibrium.

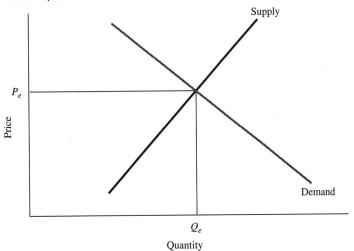

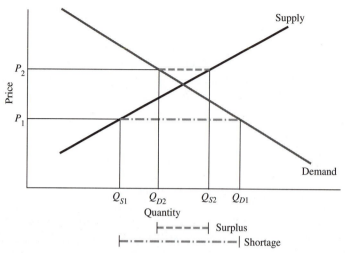

FIGURE 5-4   Disequilibrium prices, shortages, and surpluses.

In most markets a flow of goods is produced, and buyers come to the market repeat-edly. These markets are operating at an **equilibrium price** when, at the going price, sellers can sell the quantity of their products they wish to sell, and all buyers can buy as much as they want at that price.

Markets that are not in equilibrium have either **shortages** or **surpluses,** as shown in Figure 5-4. If the price is higher than the equilibrium price, the producers would be will-ing to sell more than the buyers are willing to purchase. (We will discuss the causes later in the chapter.) Then, surpluses develop ($Q_{S2} - Q_{D2}$ at P_2 in Figure 5-4). Merchandise accumulates as unsold stocks (inventories), and the merchants respond by lowering the price to encourage buyers to purchase more, and by producing less.

However, with *disequilibrium prices,* if prices are below the equilibrium, there will be shortages in the market. This is shown in Figure 5-4 as $Q_{D1} - Q_{S1}$ at P_1. In this case some willing buyers are turned away empty-handed. Sellers will raise prices and produce more, and the higher prices cause the consumers to demand less. Thus, in an unregulated market, if the market price is either too high or too low, economic forces come into play to move the price toward the equilibrium price. These economic forces are the result of decisions of individual producers and consumers doing what is best for themselves.

The tendency of markets to move toward an equilibrium price and quantity is a gen-eral phenomenon (sometimes called the invisible hand) that has many important impli-cations. Markets that are out of equilibrium provide price signals to producers and con-sumers. These signals result in changes in production and consumption levels. If there is a shortage, more resources are allocated to the production of this good. If there is a surplus, fewer resources are allocated.

The tendency of markets to move toward equilibrium also has important implica-tions for the performance of policies that seek to modify market outcomes. For in-stance, if a government implements a policy that results in all market sales of an agri-cultural commodity at a price above the free-market equilibrium level, there will be a market surplus. If the price is held above the equilibrium level, the government may have to (1) restrict production, (2) subsidize exports, (3) subsidize some consumers, or (4) destroy the surplus of the commodity. We will return to these questions in the chapter on trade and in the chapter on price and income policies in Part Five.

MARKET FORECASTING

Some agricultural economists devote their professional efforts to analyzing the outlook for agricultural prices at the regional, national, or world level. On the basis of estimates of expected crop and livestock production, expected demand, and the effects of government programs, these analysts project future prices for agricultural commodities. Farmers, agribusiness firms, commodity traders, and concerned consumers depend on such estimates in making their production and consumption plans.

Shifts versus Changes

Supply and demand curves reflect specific conditions, in a specific market, during a specified period. But from one period to another, supply and demand can change. This can happen in two different ways. First, if the market price changes, the equilibrium quantity moves from one point to another on the curve, *ceteris paribus*. In the case of a demand curve, if the price increases, the quantity demanded will decrease as the price line crosses the demand curve at a higher point. Thus the *quantity demanded* has changed, in this example decreased, but *demand* has not changed. In the case of a supply curve, if the price increases, the quantity supplied will increase as the price line crosses the supply curve at a higher point. Thus the *quantity supplied* has changed, in this example increased, but *supply* has not changed. This is called a change in the quantity demanded or in the quantity supplied.

Second, it is also possible for the demand or supply curve to shift from one time period to the next. For example, if population increases from one year to the next, the amount of food demanded at any price will increase, *ceteris paribus*. In this case the demand curve has shifted to the right. Supply curves can also shift. If the number of producers increases from one year to the next, more will be supplied at any price, *ceteris paribus*. Thus a *shift* of the supply or demand curve means that the curve has changed positions.

The difference between a change in the quantity demanded *without a change in demand* and a shift of the demand curve is shown in Figure 5-5. If the price changes from P_1 to P_2 without a shift in demand, the quantity demanded will change from A to B as the purchaser moves along a given demand curve, from Q_1 to Q_2 on the Q axis. *Changes in quantity demanded (or supplied) associated with movements along a demand (or supply) curve result from price changes for that product.* If the demand curve shifts outward, or to the right, buyers will buy more of the good at that price, as indicated by points A and C on demand curves D_1 to D_2 in Figure 5-5. *Changes in the quantity purchased resulting from factors other than changes in the price of that good or service itself are associated with a shift of the supply or demand curve itself.* Note that if the demand curve shifts from D_1 to D_2, D_1 no longer exists.

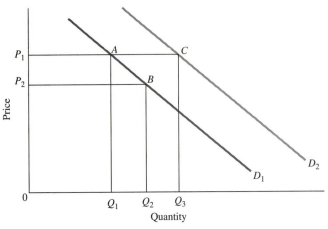

FIGURE 5-5   A shift and a change in demand.

Factors That Shift the Demand Curve

Most markets are constantly changing. Both supply and demand are constantly shifting resulting in price changes that, in turn, stimulate more changes. In the following, we briefly discuss some of the factors that cause demand curves to shift.

Population   The more buyers, all else being equal, the greater is demand. Population in the United States has been growing slowly. In some European countries population is declining. In many developing countries population is growing rapidly. Since food demand is directly related to population, in some European countries food demand is shifting in while in the United States it is shifting outward slowly. In some developing countries, food demand is shifting outward rapidly. World population is growing at a rate that causes many to be concerned about the long-term ability of the agricultural system to continue to meet the shifting demand without a considerable increase in prices.

Tastes   Consumers' tastes change for many reasons. Age, environment, and other geographic and cultural factors can change tastes. New information about how a product affects health may shift the demand curve for it. Advertising or changes in fashion might change demand. Technological change can make new products available, which shifts the demand curves for new and existing products.

Income   For most goods, an increase of consumer income shifts demand outward. When increases in income shift the demand curve outward, the good is called a **normal good** or **superior good.** However, for some goods, an increase in income causes a decline in consumption at all price levels; the demand curve shifts inward. For example, as your income rises after graduation, you will buy more of most goods and services but your demand for beans and hot dogs may fall dramatically (this would be measured by your income elasticity). In Japan, people consumed less rice as incomes rose. In these cases, beans, hot dogs and rice are **inferior goods.**

Price of Related Goods   Changes in prices of other goods can cause a demand curve to shift. The direction of change in demand depends on whether the other goods are *substitutes* or *complements*. A **substitute** is a good that can replace another good. For example, bacon and sausage are substitutes. When the price of a substitute rises, the demand for the other good increases as its demand curve shifts outward. A **complement** is a good generally used in combination with another good. For example, bacon and eggs are complements as are bread and butter. When the price of a complement rises, the demand curve for the other good shifts inward.

Expectations   Expectations about future prices, product availability, and income can affect demand. If higher prices are expected, consumers might buy more now, shifting the demand curve outward. If a future shortage is expected, consumers might increase purchases today so as not to be caught short. If higher incomes are expected, consumers might borrow to buy more now and cause the demand curve to shift outward.

Factors That Shift the Supply Curve

We now turn to some of the factors that can cause supply curves to shift.

Resource Prices   Resources are the inputs used to produce a good. If the price of a resource goes up, the cost of producing the good will go up and fewer will be made and sold. The supply curve for the good is shifted inward by an increase in resource prices.

Technology   Technological improvements allow producers to produce more of a good with the same amount of inputs, or the same quantity with fewer inputs. In either case, the effect is an outward shift of the supply curve.

Taxes and Subsidies   Most taxes are treated as costs and an increase in taxes causes an inward shift of the supply curve in the long run. Subsidies reduce costs and cause an outward shift of supply.

Prices of Other Goods   Changes in prices of other goods can cause the supply curve to shift. For example, some land can be used for either corn or soybeans. An increase in the price of soybeans would cause farmers to grow less corn, thus shifting the supply curve for corn inward.

Expectations   Expectations about prices in the future can affect supply today. Producers will change the amount they produce in the present in order to improve their expected future income. The supply curve for present production may be shifted inward or outward by the producer to take advantage of the expected prices in the future.

Number of Sellers   The supply curve shifts outward as new firms enter an industry and inward as firms leave the industry.

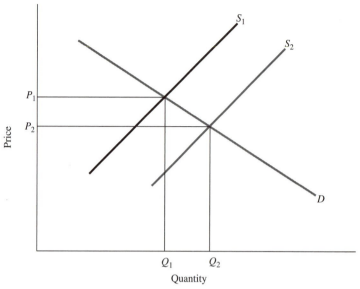

FIGURE 5-6   A shift right of a supply curve.

In sum, then, there are two explanations for changes in the quantity demanded or supplied of a good. If the supply or demand curve does not move but prices do, we refer to this as a *change in the quantity* supplied or demanded. We refer to a movement of the curve itself as a change or *shift* in supply or demand. Figure 5-6 shows the effect of an outward shift (to the right) of the supply curve. A shift of the supply curve to the right causes a *decrease* in the equilibrium price, a movement downward along the demand curve, and an *increase* in the equilibrium quantity demanded and supplied.

Figure 5-7 shows that a shift of the demand curve to the right causes an increase in both price and quantity demanded. A shift of the demand curve to the left causes a decrease in both price and quantity demanded. Thus, price and quantity supplied or demanded move in the same direction when the demand curve shifts. They move in opposite directions when the supply curve shifts, other things being equal.

Markets in Form, Space, and Time

A sheet of metal and a block of plastic in a factory in the Far East are worth little to you or me. If, however, they are made into a computer or a camera and placed on a retailer's shelf, we may by willing to pay a large amount to purchase that metal and plastic. An economic system exists to change material resources (metal and plastic) into goods and services that provide consumers utility. In that process, products are stored for future use, transported from place to place, and converted from basic inputs to consumption goods to create utility in time, space, and form. Markets direct and coordinate these activities. The classification system of form, space, and time is helpful in understanding many economic questions. We consider first the category of form.

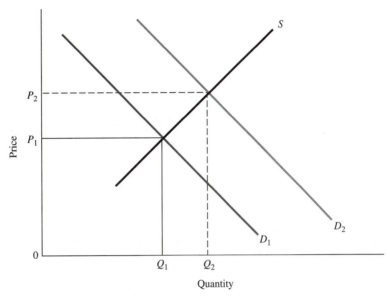

FIGURE 5-7   A shift right of a demand curve.

Form   Many natural resources do not provide utility in their natural state. A lump of coal, a barrel of crude oil, plant nutrients such as nitrogen, and ores are examples of resources that have little use unless they are transformed. Usually the resources go through many changes as they are converted into products and services valued by consumers. Markets establish the values of the raw materials, the intermediate products, and the final goods. Operating through these markets, the economic system provides signals that allocate these inputs to their highest valued uses. For example, steel producers sell finished steel to the manufacturers of automobiles, tractors, and computers. The quantity of steel used to produce these and many other products depends on the supply of steel and competing materials and the demands from the manufacturers.

Changes in form do not end with the production of goods and services. Waste products remain after a good is used. For example, power plants and automobiles generate air pollutants when they transform energy. The consumption of food produces large quantities of scrap material, animal wastes, and human wastes. Managing both the cost and the environmental damage associated with the waste presents challenging economic and political problems. We explore some of these problems in Part Five.

Space   In addition to being changed from one form to another to create products that provide utility, goods are transported to the locations where they are in demand. Most goods are not consumed where they are produced. For example, iron ore is mined where it is found. Most of it is transported considerable distances for refining and processing. Manufactured steel is shipped and delivered to manufacturers of the products used by consumers, and these products are transported to the markets where consumers buy the final products.

In some countries, a relatively underdeveloped transportation system is a major constraint on the performance of the economic system. Many underdeveloped nations do not have an adequate system of roads, bridges, trucks, railroads, harbors, barges, and airport facilities to move the products of agriculture from the producing areas to the cities, or to ports for export. In some countries, spoilage due to inadequate transportation and storage systems destroys a high percentage of the products. Likewise, it is difficult to move the industrial products used by farmers to support production on farms. For example, fertilizers may be unavailable or higher priced than in a country with a good transportation system. The development of a transportation system will usually involve both the public and the private sectors. The public sector usually develops the roads, bridges, railways, and similar major facilities. The private sector often develops the firms that own and operate the trucks, ships, and similar equipment. In many countries the government operates the transportation industry or regulates it as a public utility.

Time   The **time** dimension is also an important determinant of the amount of utility provided by a good. Utility is generated by storing goods and selling them at a later time. Grain-handling firms purchase and store grain for sale later to buyers, such as livestock feeders, millers, and exporters. These grain-handling firms generate their profits by anticipating future supply and demand conditions and making appropriate purchase, storage, and sale decisions.

Economists often refer to local, national, and world markets. In economic terms, areas are separate markets if the prices for a product are different between or among areas. If the market is efficient and trade is unrestricted, price differentials between areas will be close to the cost of transportation between areas. However, separate markets may exist for several reasons. First, not all markets are efficient in setting equilibrium prices. Separate markets can exist if a product cannot be stored or if the cost of storing a product is more than marketing firms or consumers are willing to pay. If market information does not move among market areas, price differences can exist. Although a lack of price information may occur in some undeveloped areas of the world, modern communication facilities have essentially eliminated the lack of market price information as a major factor in the creation of separate markets.

Second, trade may not be free or open. An "open market" exists when there is unlimited freedom to trade the product. Trade reduces the price differentials between two countries to the cost of transportation. Governments create separate markets if they enact barriers to trade. At the extreme, a government may prohibit the import or export of a product, creating a "closed market" in which prices can be significantly higher or lower than prices in other countries. The implications of open and closed markets are discussed at length in Chapter 8.

In this chapter we consider the basic economic factors that operate in both open and closed markets. We begin with a simple case to focus on basic market principles. We remove the simplifying assumptions as we move through this chapter and into the following chapters and consider more realistic cases.

FUTURES MARKETS

In a **futures market,** buyers and sellers agree to exchange a certain amount of a standardized commodity at a set point of time in the future for a set price. Since most traders use the futures market to protect themselves from risk or to profit from price changes, few contracts are actually fulfilled. Instead, the contracts are liquidated before the date of delivery arrives.

The futures markets perform an economic function for buyers and sellers of commodities and financial securities as well as for those who produce or own them. Market participants face less uncertainty associated with price changes because there is a price at which a product may be bought or sold for future delivery. These markets are particularly beneficial for agricultural products because price fluctuations are common.

There are several uses of the futures markets. People who own a product, or who will produce it, face a risk that the price will fall. The futures market allows an owner to sell a contract to someone else in order to lock in a price in the future. This is called hedging. The price protection is a result of the individual having offsetting positions in the cash and futures markets, owning the product in one and selling ("going short") in the other. If the price falls, the individual loses value in the cash market but gains an approximately equal amount in the futures market. Rather than waiting until the end of the growing season or until livestock are ready for market, the producer can buy a futures contract to sell the grain or livestock at the appropriate time. In some cases banks and other financial institutions are more willing to loan funds to producers who have used the futures market to reduce price risk.

Some individuals, called speculators, buy and sell contracts, attempting to realize a profit. (The structure of the futures market makes large profits, and losses, possible.) Many individuals use the prices of futures contracts to plan their future production and marketing decisions.

PERFECTLY COMPETITIVE MARKETS

At this point we turn to the organization of markets. We describe the major market types and their implications for how the economy allocates resources. There are four basic market structures: pure competition, monopolistic competition, oligopoly, and monopoly. In this chapter we discuss the **perfectly competitive market,** a benchmark against which other market forms may be compared. In Chapter 7 we describe the other forms of market competition.

Pure Competition

We begin with the case of **pure competition** because it is basic to the understanding of the economic system. Pure competition has the following market characteristics.

Many Buyers and Sellers   With many buyers and sellers no one, or small group of, market participants can significantly influence the market. All buyers and sellers have the option of trading with several others, often in a highly organized or structured market. Farm commodity markets are often cited as classic examples, with many farmers selling to many buyers. The stock and bond markets are also examples.

Homogenous Product   Competing firms in a market offer a uniform good or service for sale, or one standardized in terms of specific market grades or classes. Grain and livestock, for example, are traded in specific classes or grades, as are diamonds and crude oil.

Most farm commodities can be classified by class or grade, color, texture, or some other characteristic. Within a specific class or grade, buyers do not discriminate among sellers. Often neither the buyer nor the seller physically examines the commodity exchanged.

Freedom of Entry and Exit   New firms can enter, and existing firms are free to leave, the industry. There are no **barriers to entry.** Barriers include patents, licensing requirements, entry quotas, extremely high capital requirements or loyalties to the brands of other firms. Existing firms may shut down without getting approval from a government commission or another body. Farming usually meets this condition even though the capital requirements may be rather high for many individuals. Many agribusiness industries do have barriers to entry, and in some cases these barriers are quite substantial.

Perfect Information   **Perfect information** requires that all buyers and all sellers know the prices of all other buyers and sellers. Buyers and sellers are aware of the alternatives, particularly the prices charged by many others in the market. Because of the government's market information services, the extension service, and the farm press, most markets for agricultural commodities meet this condition. (In many markets there is less than perfect information, but sufficient information is available to market participants to allow them to make informed decisions.)

Economists have shown that if all markets met the conditions of a purely competitive market and the factors of production were mobile, goods would be allocated among market participants in a way that would maximize the productivity of the economic system. Each seller would face a horizontal demand curve, as shown in Figure 5-8. That is, the seller could sell all the product it could produce at the market price. Trying to charge more than the market price would result in nothing being sold because buyers would purchase from other sellers. There would be no reason to sell at a lower price because all output could be sold at the higher market price. This is the situation of a grain farmer who can sell all grain produced at the market price but none above the market price.

FIGURE 5-8   Individual firm demand in a competitive market.

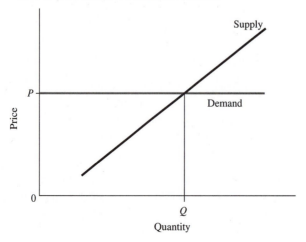

Long-Run Costs of Production

The above discussion shows that in the short run the market price determines the quantity of production as firms make decisions on the basis of the output price (MR) and MC. In the long run the opposite occurs. The **long-run cost of production** determines the price of a product or service. Why? In markets with competitive prices, if the price is high enough to generate excess profits, additional firms will enter the industry. The increased use of inputs by the entrants tends to increase the price of the inputs used in production (the price of land is especially responsive). The increased production supplied by the entrants drives down the market price for the product. This continues until the prices, costs, and production are at equilibrium levels. If, however, the market price is low enough to result in losses, firms leave the industry. This results in less demand for the inputs used in production (lower costs) and less output produced (higher prices). Eventually the industry will approach equilibrium conditions.

Some governments set prices for products. Prices of agricultural commodities are often set above the existing market price because of the political power of agricultural organizations. When a government sets the price of a product above equilibrium levels, that price will become the actual cost of production for the product in the long run. If the price is higher than the equilibrium market price, firms will enter the market and bid up the price of inputs until the cost of production is equal to the price support level. If the government was to set a price at lower than equilibrium levels, firms will exit the industry. Exit will continue until the cost of inputs falls, because of decline in demand for inputs, to the point that the cost of production is in line with the price established.

EXAMPLES OF ADJUSTMENT OF FARM PRICES

The real world is never in a stable economic equilibrium. Firms and industries are constantly adjusting toward a long-run equilibrium position. However, this position is also constantly changing as population, technology, tastes, and preferences change. Agriculture provides many illustrations of this principle. In the early 1970s high grain prices in the United States brought increased profits to many grain farmers. The efforts of farmers and other investors to buy more land to increase production caused land prices to rise very rapidly. The implicit opportunity costs of using land went up accordingly, and some explicit costs such as property taxes began to rise as well. The average total cost curves of grain farmers, including both explicit and implicit costs, shifted upward. This was followed by an increase in grain output and a significant reduction in export demand (the demand curve shifted) for U.S. agricultural products, due in large part to U.S. macroeconomic policies and European agricultural policy. This forced grain prices back down toward equality with average total cost. By the early 1980s, many farmers were claiming that the price of grain had fallen below their cost of production, meaning that the price was below their minimum average total cost. The farmers in this position were in either a loss-minimizing or a close-down situation. Many farmers were forced out of agriculture, and land prices fell precipitously. In the early 1990s land prices increased in response to higher grain prices but by the end of the 1990s, and into the 21st century, crop and land prices were again falling.

SUMMARY

You have learned how to aggregate the supply and demand curves of individuals to find market supply and demand curves. Supply and demand interact in a market to generate an equilibrium in which the market clears with all the product selling at a price. We noted that markets operate in time, space, and form. We indicated the distinction between a shift of a supply or demand curve and a change in the quantity supplied or demanded as prices change. The characteristics of a perfectly competitive market were discussed. Other forms of market competition will be discussed in Chapter 7.

IMPORTANT TERMS AND CONCEPTS

barriers to entry   130
complement   125
equilibrium price   122
factors that shift the demand curve   124
factors that shift the supply curve   125
form   127
freedom of entry and exit   130
futures market   129
homogenous product   129
inferior good   124
long-run costs of production   131
market   120
market demand   119

market equilibrium   120
market price   120
market supply   118
normal good   124
perfect information   130
perfectly competitive market   129
pure competition   129
shortage   122
substitute   125
superior good   124
supply curve   118
surplus   122
time   128

QUESTIONS AND EXERCISES

Name That Term

Read the following sentences carefully and fill in the missing term or terms.

1 A _____ curve shows the amounts of a good or service buyers purchase at various prices during some period of time, other things equal.
2 A _____ curve relates quantities producing firms bring to market at market price, other things equal.
3 A _____ is any institution or mechanism that brings together the buyers (demanders) and sellers (suppliers) of a good or service.
4 A _____ is the amount by which the quantity demanded of a product exceeds the quantity supplied at a given (below-equilibrium) price.
5 A _____ is the amount by which the quantity supplied of a product exceeds the quantity demanded at a given (above-equilibrium) price.
6 _____ products of competing firms are perfect substitutes for each other.
7 Patents, high start-up costs, and loyalty to the brands of other producers are examples of _____.
8 Market _____ is a schedule showing the amounts of the good consumers are willing and able to purchase in the market at a series of prices during a specified period in a given market.

True/False

Read the following sentences, then decide whether each statement is true (T) or false (F) and mark it accordingly.

T F **1** An increase in the price of a substitute good will cause the demand curve for widgets to shift downward and to the left, causing the price and quantity of widgets to decrease.

T F **2** An increase in consumer income causes the market clearing price and quantity of a normal good to increase.

T F **3** Population, tastes, income, and prices of related goods are all factors that shift the aggregate demand curve.

T F **4** Resource prices, technology, taxes and subsidies, prices of other goods, and number of sellers are all factors that shift the aggregate supply curve.

T F **5** Many buyers and sellers, homogenous product, freedom of entry and exit, and perfect information are all characteristics of a perfectly competitive market.

T F **6** Shortages put downward pressures on a good's price.

T F **7** Surpluses put upward pressure on a good's price.

T F **8** An increase in the price of fertilizer, other things equal, would increase the market price and quantity of corn.

T F **9** An increase in the tax on cigarettes, other things equal, would increase the price and decrease the quantity of cigarettes produced and consumed.

T F **10** An increase in the price of a complement, other things equal, causes the market price and quantity of a good to decrease.

Multiple-Choice Questions

Circle the letter of the response that best answers the question or completes the statement.

1 If gasoline is a normal good for Stan, what happens to Stan's consumption of gasoline as his income increases?

a consumption increases

b consumption remains unchanged

c consumption decreases

d consumption may increase or decrease

e none of the above

2 Which of the following will not affect the demand for motorcycles?

a change in the price of pickup trucks

b change in the price of gas

c change in the price of steel

d changes in the tastes and preferences for vehicles

e change in motorcycle helmet laws

Questions 3–8 refer to the market for widgets. Widgets are normal goods sold in a competitive market. If everything else is unchanged, what happens to the price and quantity of widgets sold if the following changes occur? For each question choose among:

a price increases and quantity increases

b price decreases and quantity increases

c price decreases and quantity decreases

d price increases and quantity decreases

e none of the above

3 Consumer income increases.

4 The price of a substitute for widgets goes down.

5 The price, a major input in widget production, goes up.
6 The sales tax on widgets is reduced.
7 A health warning causes consumers to like widgets less.
8 New widget producers enter the market.
9 In a competitive market, both the price and the quantity sold must increase over time if
   a supply increases and demand decreases.
   b supply decreases and demand increases.
   c demand increases and supply is unchanged.
   d supply decreases and demand is unchanged.
   e demand increases and supply increases.
10 If there is a shortage or a surplus in a competitive market this will
   a not happen in a competitive market.
   b cause the buyers and sellers to react in ways that will eliminate the surplus or shortage.
   c cause the supply curve to shift.
   d cause the demand curve to shift.
   e cause both the supply and the demand curves to shift.

Technical Training

For the following, suppose the market for iglets, a normal consumer good unless noted, is in equilibrium at P^* and Q^*. Draw supply and demand curves to show what happens to the price and quantity of iglets if the changes given in questions 1–8 occur and everything else remains unchanged.

1 Consumer income decreases.
2 The price of a substitute goes up.
3 The price of labor, a major input in iglet production, decreases.
4 A new advertising campaign causes people to like more iglets.
5 Technological improvements reduce the cost of producing iglets.
6 Income decreases and iglets are an inferior good.
7 The price of a complement to iglets goes down.
8 The U.S. Surgeon General announces that iglet consumption may reduce the incidence of AIDS.

Use the aggregate demand and supply curves shown in the following graph to answer questions 9–12.

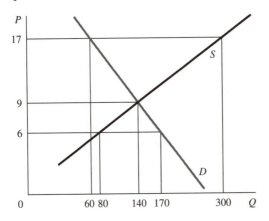

9 What is the market-clearing price and quantity?

10 What is the shortage at $P = 6$?

11 If price increases from 6 to 9, by how much does aggregate quantity supplied increase?

12 If price increases from 6 to 9, by how much does aggregate quantity demanded decrease?

Questions for Thought

1 The table below displays the yearly market supply and demand data for high-top basketball sneakers. On a separate sheet of graph paper plot the *market* supply and demand. Indicate the market quantity and price equilibrium.

MARKET SUPPLY AND DEMAND FOR
BASKETBALL SHOES (In Thousands)

Price ($ per pair)	Quantity demanded	Quantity supplied
60	2,300	1,100
70	2,100	1,200
80	1,900	1,300
90	1,700	1,400
100	1,500	1,500
110	1,300	1,600
120	1,100	1,700
130	900	1,800
140	700	1,900

2 Explain how the market supply and demand for basketball shoes would change if demand increased by 10 percent. Show the appropriate shifts on your graph.

3 In Chapter 1, you were asked why water, a commodity crucial to life, was inexpensive while diamonds, one of life's frills, were expensive. Use supply and demand curves to illustrate why this is true. What factors could change the relative prices of diamonds and water?

4 When asked about a surplus of their product, dairy farmers occasionally make the following argument "We produce so much because the price of milk is so low we need to produce a lot so we can sell enough to cover all our fixed costs. If the price of milk were only higher we wouldn't need to produce so much to make a living." What's wrong with this argument?

5 Lines are a sign of a possible shortage of a good or service. Some people are inclined to join lines in anticipation of buying something (tickets for a concert or merchandise on sale, for example) that may be for sale at below market-clearing price. Under what conditions is this a good strategy?

6 The flip side of shortages is surplus. Some people do not purchase items that appear to be in surplus for fear of paying above market-clearing price. Under what conditions is this a good strategy?

ANSWERS AND HINTS

Name That Term   **1.** demand; **2.** supply; **3.** market; **4.** shortage; **5.** surplus; **6.** Homogenous; **7.** barriers to entry; **8.** demand

True/False   **1.** F; **2.** T; **3.** T; **4.** T; **5.** T; **6.** F; **7.** F; **8.** F; **9.** T; **10.** T

Multiple Choice   **1.** a; **2.** c; **3.** a; **4.** c; **5.** d; **6.** b; **7.** c; **8.** b; **9.** c; **10.** b

Technical Training   Answers for questions 1–8 should look like graph a, b, c, or d.

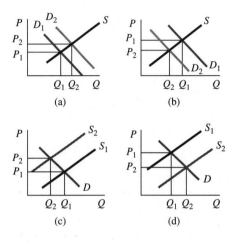

1. Graph b; **2.** Graph a; **3.** Graph d; **4.** Graph a; **5.** Graph d; **6.** Graph a; **7.** Graph a; **8.** Graph a; **9.** $P = 9$, $Q = 140$; **10.** 90; **11.** 60; **12.** 30

6

MULTIPLE INPUTS AND OUTPUTS

OVERVIEW

This chapter expands on the concepts developed in Chapters 2 through 5. In Chapter 2 we showed that the quantity of a product the consumer demands is negatively related to the price of that product. In Chapters 3 and 4 we used the example of a firm using one variable input to produce one product to show that the firm's supply curve is positively related to price. In Chapter 5 we showed how supply and demand curves are used to determine market prices and quantities. In this chapter we consider the case of firms using several inputs to produce several outputs. We then show the primary determinants of the demand for inputs and the supply of outputs.

LEARNING OBJECTIVES

This chapter will help you learn:

• How a firm determines the cost-minimizing combination of inputs to use in the production process

- What influences the firm's demand for inputs
- How a firm decides how much of several products to produce
- About the factors that influence whether a firm specializes or diversifies
- What influences the quantities supplied by firms

PRODUCTION WITH MULTIPLE VARIABLE INPUTS

In practice, a producer must make many decisions about the combinations of several variable inputs to use with one or more fixed inputs. For example, at the beginning of a growing season a grain producer has a fixed amount of land and machinery. The producer decides how much seed, fertilizer, pesticide, labor, fuel, and so on to use in producing the crop. A retailer decides which lines to stock and how much of each to buy. The agribusiness manager decides whether to use a labor-intensive operation or invest in more labor-saving machinery.

The economically rational producer seeks to combine inputs in the proportions that result in the highest profit. The profit-maximizing combination of two variable inputs can be determined using an isoquant. The **isoquant** is derived from, or is another way of presenting, the production function. It shows all of the combinations of two inputs that can be used to produce a given quantity of an output. The isoquant is analogous to the consumer's indifference curve.

In Chapter 3 the concept of production efficiency was presented. A firm that produces as much output as is technically possible will be operating "on" the production function. A firm that produces less than the maximum amount of output is inefficient and is operating "below" the production function. In the case of an isoquant, a firm that produces the maximum amount possible from the inputs used will be "on" the isoquant. An inefficient firm will use more than the necessary amount of the inputs and will operate at a point "above" the isoquant.

Isoquants are shown in Figure 6-1. The several isoquants taken together, as shown in Figure 6-1, are called an **isoquant map.**

The isocost line is analogous to the consumer's budget line. An **isocost line** indicates the combinations of two inputs that can be purchased with a given amount of money. The prices of the inputs determine the slope of the isocost line. The slope is equal to the negative inverse ratio of the prices. In Figure 6-1a, it is the negative of the price of buns divided by the price of hot dogs. In Figure 6-1b, it is the negative of the price of equipment divided by the price of labor.

Just as a consumer maximizes utility by operating where an indifference curve is tangent to the budget line, a firm minimizes costs by operating where an isoquant is tangent to an isocost line. Figure 6-1a shows an isoquant representing two inputs used in fixed proportions that yield a given output. (This is similar to an indifference curve for a consumer who purchased in fixed proportions of two goods.) A hot dog stand operator will buy the same number of hot dogs as buns. An operator may order 20 dozen of each if sales are expected to be poor but 30 dozen of each if sales are expected to be strong. Note that even if the prices were to change, which would change the slope of the isocost line, the operator would still buy equal amounts of hot dogs and hot dog buns. Similarly,

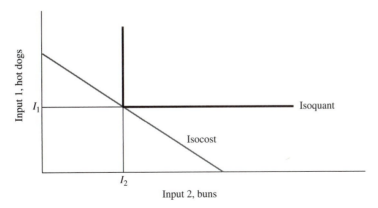

(a) Fixed proportions

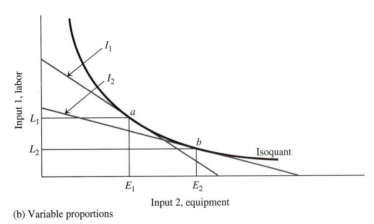

(b) Variable proportions

FIGURE 6-1   Isoquant maps with inputs used in fixed and variable proportions.

on a farm, each tractor requires only one operator, unless operators work in shifts. And one driver is needed for each truck a transportation firm has on the road.

In many cases inputs used in the production process are good substitutes. Figure 6-1b suggests such a case for labor and equipment. With the prices resulting in isocost I_1, the firm uses L_1 units of labor and E_1 units of equipment. With substitutable inputs, differing input prices result in different combinations of inputs. For example, consider the situation if the price of labor is higher than in isocost I_1 relative to the price of equipment, as shown in isocost I_2. In this case, the firm will use only L_2 units of labor but will increase equipment use to E_2 units, as suggested by the tangency of the isoquant and isocost line at that point. In each case the firm would be producing the same quantity of output. The point is that when inputs are good substitutes, their proportionate use is strongly influenced by the prices of the inputs but when inputs are not substitutes their relative prices do not change the proportions used.

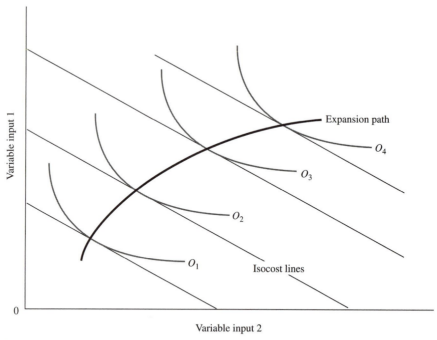

FIGURE 6-2   Expansion path of a firm using two inputs.

Expansion Path

As you saw above, the tangency of an isoquant and an isocost curve gives the least-cost combination of inputs to produce a given level of output. Figure 6-2 shows an **expansion path** which gives the least-cost combinations of inputs at various levels of output. The several isoquants indicate different technically efficient combinations of inputs that can be used to produce output levels O_1 to O_4. As in the previous discussion, each of the isoquants shows the various combinations of two inputs that can be used to produce a given quantity of output. The isocost lines show the cost of purchasing the inputs at the existing prices. The tangencies of the isoquant curves and isocost lines determine the cost-minimizing combinations of two inputs used to produce levels of output O_1, O_2, O_3, and O_4. The line connecting these points is the expansion path. It gives the least-cost combinations of two inputs used by a firm as it expands output from O_1 to O_4.

In the hypothetical case shown in Figure 6-2, the firm uses a higher proportion of variable input two at high levels of output than at low levels of output. In an industry, generally small firms use relatively more labor and less equipment than large firms do, all else being equal. In some cases, equipment is too large for use in a small firm or on a small farm. You wouldn't purchase a new computer to accomplish one new project, but if you were operating an office that routinely did such projects, you wouldn't consider using an existing, slow computer.

Demand for Inputs

In Chapter 2 we examined the economic concepts determining the consumer's demand for goods. In this section we study a firm's demand for inputs. When applying the law of demand to inputs, we stress this most crucial point: *The **demand** for inputs is a **derived demand;** it is derived from the demand for the finished goods and services that are produced with these inputs.*

Most resources used as inputs in producing other goods do not directly satisfy consumers' wants. No one directly consumes a pound of fertilizer, a barrel of crude oil, or a tractor. Production processes transform such resources into food, and other goods and services to meet the demands of consumers. The demand for an input depends on (1) its contribution to production (its marginal physical product) and (2) the price of the good produced. That is, the demand for an input reflects the value it adds, or its **marginal revenue product (MRP).** *The value of the marginal physical product, the marginal revenue product, is the increase in total revenue resulting from the use of an additional unit of an input.*

$$MRP = MPP \times P$$

The marginal resource cost (MRC) is the amount that each additional unit of an input adds to the firm's total cost. It is *not* the same as marginal cost (MC), which was discussed in Chapter 3. MC is calculated on the basis of a change in *output* whereas MRC is calculated on the basis of a change in the use of an *input*. **Marginal resource cost (MRC)** is the increase in total cost resulting from the use of an additional unit of an input. (MRC is sometimes called marginal factor cost, MFC.) In perfect input markets, where the action of the buyer does not affect the market price, the MRC of an input is the same as its price.

To maximize profits, or to minimize losses, a firm purchases and uses additional units of each variable input at the point where the last unit adds the same amount to the firm's total revenue as to its total cost, or MRP = MRC. This relationship, called the **MRP = MRC rule,** is very similar to the MR = MC rule presented earlier. The rationale of the two rules is the same, but for MRP = MRC the calculation is based on a change in the *use of inputs, not a change in the production of outputs.* That is, the decision maker determines whether using one more unit of an input will cost more or less (MRC) than the revenue added (MRP) by the resulting increase in output.

The rational manager will buy and use an amount of each input just up to the point where the MRC of using each input is equal to its MRP. Clearly the input price is a major factor in determining how much the firm purchases for its use. The MRP curve, which can be used to find the number of units of a variable input demanded at various prices, is created by plotting the amount of the input purchased at a series of prices of the input.

We can use watermelon production as illustrated in Table 6-1 as an example of the MRP = MRC rule. The manager can use the rule to determine how much irrigation water to use in producing watermelon. In this example, watermelon is produced using from 1 to 10 units of water. Without irrigation water, production is zero. Each of the first 3 units of water added results in an increasing MPP. Additional water results in a diminishing marginal physical product.

TABLE 6-1   TOTAL PHYSICAL PRODUCT, MARGINAL PHYSICAL PRODUCT, AND MARGINAL
REVENUE PRODUCT

Input water (#)	MRC ($)	TPP watermelon (#)	MPP watermelon (#)	MRP watermelon ($)
0	25	0		
			10.0	25
1	25	10.0		
			12.4	31
2	25	22.4		
			14.0	35
3	25	36.4		
			15.2	38
4	25	51.6		
			14.4	36
5	25	66.0		
			12.8	32
6	25	78.8		
			10.0	25
7	25	88.8		
			6.8	17
8	25	95.6		
			2.4	6
9	25	98.0		
			−6.0	−15
10	25	92.0		

Suppose that a unit of irrigation water costs $25.00, and the price of a watermelon is $2.50. The first unit of water, which costs $25.00, used on this watermelon patch generates 10 watermelons and thus $25.00 of revenue. The second $25.00 unit of water generates an additional 12.4 watermelons and $31.00 more of revenue. The manager will apply the second, third, fourth, fifth, sixth, and seventh units of water, because each returns as much or more additional income as the $25.00 spent for each unit of water. The manager will not add the eighth unit of water because cost increases another $25.00 but revenue increases only $17.00. Note that if the farmer did use the eighth unit of water, total output and total revenue would increase. However, the increase in revenue would be less than the increase in cost of production. This relationship clearly shows that maximizing output, in this case watermelons per patch, (more generally, yield per acre) does not maximize profits. Although nine units of water maximizes yields, seven units of water maximizes profits.

Figure 6-3 shows the relationship of MRP to MRC for the hypothetical watermelon production example in Table 6-1. The MRP curve can be used to find the number of units of the input used at varying prices. At prices of water higher than $25.00 the producer would use less. For example, at a water price of $38.00 the firm would use only four units of water. At $10.00 for water, the firm would use eight units of water and produce 95.6 watermelons. The ninth unit of water would not be used even at this lower price because it only returns $6.00 of additional revenue.

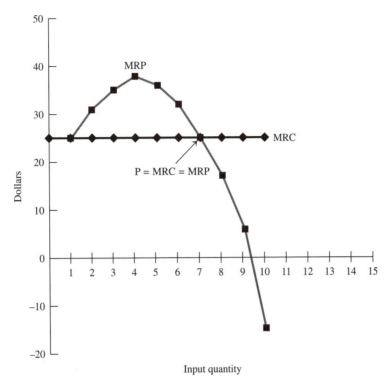

FIGURE 6-3   Determining the level of input usage using the MRP = MRC rule.

Remember, in this illustration all other inputs are fixed. The costs of the fixed inputs do not have to be considered because they do not change. We would need to know the costs of these fixed inputs to determine whether the firm is realizing a profit or a loss. Thus, *the MRP schedule is a firm's demand curve for an input, ceteris paribus.* It gives the profit-maximizing quantities of an input purchased by a firm at various input prices.

The Firm's Demand for Several Inputs   Generalizing from above, a firm maximizes profits by employing each resource at the point where MRP = MRC. Again, if we assume perfect markets, where the action of the buyer does not affect market prices, the MRC of an input is equal to the price of the input. Then the profit-maximizing amount of each input is the quantity at which the price of the input (its MRC) is equal to its MRP. Or more simply:

$$P_1 = \text{MRP}_1 \quad \text{and} \quad P_2 = \text{MRP}_2 \quad \text{and} \quad \ldots P_n = \text{MRP}_n$$

This rule is sometimes expressed as:

$$\text{MRP}_1 \div P_1 = \text{MRP}_2 \div P_2 = \text{MRP}_n \div P_n$$

The resulting combination of inputs is identical to the cost-minimizing combinations of inputs found when using the isoquant and isocost curves of the firm. The profit-maximizing producer must consider both the productivity of the resource, as reflected in diminishing marginal physical productivity, and the prices (marginal resource costs) of the inputs. For example, a relatively small amount of a resource (such as fertilizer, water, labor, or machinery) is employed if its price ($P_r = \mathrm{MRC}_r$) is very high. Thus the firm operates at a point high on the MRP$_r$ curve. Conversely, if $P_r(P_r = \mathrm{MRC}_r)$ is low, much more of the resource is used to reach a point low on the MRP$_r$ curve. Thus for any resource, a high input price results in the use of less (a low price, the use of more) of that resource by firms as they maximize profits.

The producer's choice is also analogous to the consumer's choice. Finding the combination of goods that maximizes utility for a given budget, the consumer reaches a decision based on preferences and the prices of the various products. Thus, the MRP $\div P$ rule is analogous to the $\mathrm{MU}_i \div P_i = \mathrm{MU}_j \div P_j$ result presented in the discussion of consumer demand.

Determinants of the Demand for Inputs

Next we turn to the **determinants of the demand for an input** and the factors that cause it to shift. The MRP curve is the demand curve for an input used in the production process. The factors that determine the location of this curve fall into three categories: (1) the demand for the output produced, (2) the productivity of the input, and (3) the prices of other inputs in relation to their productivities.

1 Demand for the Output Produced   An increase in the price of a product shifts the demand outward for the inputs used in its production. For example, if the price of peanuts increases, farmers increase production by using more of the variable inputs such as fertilizer. If prices are expected to remain high, farmers will shift land from other crops, say cotton, into peanut production because peanuts have become more profitable. This will further increase the demand for peanut seed, farm chemicals used on peanuts, and peanut-processing facilities. Shifting land out of cotton production reduces the demand for cotton seed and fertilizers, herbicides, and insecticides used on cotton. Thus, in addition to the price of the inputs, the demand curves for inputs are changed by the price of the commodity produced and the prices of competing commodities.

2 Productivity of the Input   The demand for an input depends on its productivity and the productivity of the other inputs used. For example, the demand for a higher-yielding seed will be greater than the demand for a lower-yielding seed. Or, the demand for well-educated managers is strong because education and training make them more productive than less well educated employees.

Anything that changes the productivity of an input will shift the demand for that input. For instance, a new set of locks and dams on the Illinois River would increase efficiency, reduce the costs of barge shipping of grain, coal, and other bulk products and thus increase the demand for barge transportation, *ceteris paribus.* At the same time the demand for the inputs used in barge operations—fuel, river captains, and even barges—would increase.

3 Prices of Other Inputs   Just as the demand for an output shifts with changes in the prices of other outputs, so the demand for an input shifts with changes in the prices of other inputs. The existence of substitutes and complements affects the relationship, as discussed below. Some inputs are **substitutes** for other inputs, such as plastic for steel, herbicides for cultivators, trucks for railroad cars, cotton for nylon, and so on. A decrease in the price of an input reduces the demand for inputs that are close substitutes, *ceteris paribus.* For example, a decline in the price of diesel fuel relative to gasoline prices reduces the demand for gasoline, because of the substitution of diesel engines for gasoline engines.

Complements are inputs that are used together and are jointly demanded. In the case of complements, a decrease in the price of one input increases the demand for the other, *ceteris paribus.* If the price of fertilizer falls relative to all other input prices, this increases the quantity of fertilizer demanded and the demand for labor and machinery needed to handle the larger crop.

PRODUCT DECISIONS

Specialization or Diversification

Up to this point, we have discussed firms that produce one product. There are many firms that do this because they have found it more profitable to specialize. Most firms, however, diversify and produce several products. Examples of **specialized** firms are wheat farms in Kansas, sugarcane plantations in Indonesia, fabric producers in India, and vineyards in France. A farm in the U.S. cornbelt, which produces only corn and soybeans, is nearly as specialized as these single-crop operations. Specialized agribusiness firms operate primarily at the local level and provide services such as trucking, chemical application, and management. Specialized retail firms offer only one type of product such as up-scale women's fashions, video rentals, or seafood.

Most major corporations produce and sell many products; they are highly **diversified.** Most agribusinesses and all the major agribusiness corporations produce and market several products (product "lines") or agricultural services. A diversified farm can take several forms. It might combine crop and livestock enterprises or grow different crops such as cotton, soybeans, and alfalfa. It might produce several animal or poultry products such as milk, pork, wool, and broilers. Alternatively, it might produce several crops for family consumption and sale, as is common in developing countries.

What are the advantages and disadvantages of specialization and diversification? Specialization has the advantage of increasing the volume produced and concentrating management expertise on a smaller number of tasks. With a limited number of managers in the firm, as on a family farm, the manager can become more proficient by concentrating on one or two enterprises. A farm specializing in one enterprise may be large enough to buy feed, fertilizer, and some other inputs at discounted prices. Many specialized farmers receive a higher price by selling in large volume, or by selling a product that conforms to the buyers' standards. Also, they may become more proficient at selling in a volatile market.

Specialization, however, usually exposes farmers to greater risks of price uncertainty and crop failure due to weather or pest problems, and this makes income more uncertain. Also, specializing in one or two crops places peak-period burdens, and other times of little activity, on farm labor each year. Specializing in a livestock enterprise, such as dairy or hog production, ties the family to daily chores, unless the operation is large enough to justify hiring a labor force.

Technological advances in farming and improvements in markets and agricultural policy have resulted in cost advantages for U.S. farms that specialized and expanded. Indeed, as producers concentrate on their most profitable product or products, specialization occurs and agriculture develops. Also, as labor requirements in production agriculture have declined with greater specialization, many agricultural workers have pursued off-farm employment. In the United States this has helped to fuel the development of the rest of the economy.

Production Possibilities Frontier

If a firm is producing two products, how does it decide how much of each to produce? Suppose a food processor has a certain amount of land, labor, and capital to use to produce two goods—frozen dinners and canned foods. The manager must decide how much of each product to produce with the available labor and management, and how to integrate these operations with the marketing plan. Available resources, such as labor and capital, limit the amount of frozen dinners that can be produced without giving up some of the canned goods production. The decision depends in part on the degree of supplementarity, complementarity, or competition between the canned and frozen goods production enterprises.

Figure 6-4 shows how the manager of a firm can use the isorevenue line and the **production possibilities frontier (PPF)** to choose the most profitable combination of two products to produce. The **isorevenue line (IR)** shows the amount of revenue the firm will receive from the sale of various combinations of the two products. Assuming the firm is in perfectly competitive markets, the market determines product prices. The slope of the IR is the negative ratio of the prices of the two products. Thus, it is similar

FIGURE 6-4   Production possibilities frontier and isorevenue line.

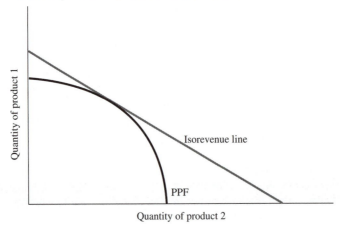

Quantity of product 2

to the consumer's budget line and the firm's isocost line. A change in the price of either product changes the combination of products produced. An increase in the price of one product, relative to the other, results in the firm producing more of that product and less of the other.

The PPF shows the maximum amount of various combinations of the two products that can be produced by the firm. It is called a frontier because it is the maximum the firm can produce. A firm could produce less but it would be operating at a point inside the curve which would be inefficient. The shape of the curve shows that producing more of one product means producing less of the other. For example, given a certain amount of meat and vegetables, increasing the production of frozen dinners means the firm's canned goods production must be reduced. Similarly the equipment used by a breakfast cereal manufacturer to produce cornflakes cannot simultaneously be used to produce oat-bran flakes. The manager must choose the most profitable combination. The profit-maximizing combination of products occurs where the PPF and IR lines are tangent.

The PPF is curved as a result of the law of diminishing returns. As discussed in Chapter 3, because of diminishing marginal returns, devoting more of a firm's inputs to the production of one product eventually makes the inputs less productive. There are several practical reasons for this. For example, a farm that produces several crops can plant each crop on the land where the soil is best suited for that crop. Producing only one crop means that some of the crop will be produced on less suitable land. In addition, by producing several crops, the work of planting and harvesting can be spread out to use inputs more efficiently. Producing only one crop may mean planting or harvesting part of the crop at a non-optimal time. Finally, in many cases, producing crops in rotation results in lower input costs for pesticides (less insect carryover reduces the need for insecticides) or fertilizer (legume crops produced one year provide nitrogen for the next year's crop).

Competition between the activities producing two products occurs when increasing the output of one product reduces the output of the other. Usually competition between producing activities occurs because they use the same inputs. Corn and soybean enterprises in the midwestern states of the United States provide a good example. Both can be grown on the same land, and the tillage and planting activities tend to occur at about the same time of the year. Examples from agribusiness include the competition for the use of plant and labor inputs to process several foods harvested at the same time of the year. In a manufacturing plant, an assembly line and workers on it can be used to produce only one line of cars or trucks at any given time. In a retail grocery store there is intense competition among the firms that supply the store for shelf space to display their products. Allocating more space to one product means that less space is available for another product. In each of these examples, increasing the output of one product decreases the output of another.

Enterprises are **complementary** if increasing the production of one increases the output of another. Soybean oil and meal production are complementary activities in a soybean processing plant since both are produced in the crushing operation. When legumes such as alfalfa or sweet clover are grown in rotation with corn or small grains, the legumes are complementary. They improve the fertility of the soil by adding nitrogen, which increases the yield of the crop that follows. Many agribusinesses develop complementary products (a product line) that result in full use of the production equipment and that benefit from a coordinated marketing program. In the case of complementary products, the curved PPF does not apply to the production decision made by the firm.

Technological change can alter the relationships among commodities produced and thus the shape of the PPF. For example, sheep compete for feed with cattle. Therefore, changes in production technology that favor cattle may be disadvantageous to sheep production and vice versa. The technology of synthetic fiber production has produced many products that compete with wool and are therefore disadvantageous to sheep production. These technological changes have affected costs as well as relative market prices and have resulted in a sharp drop in world sheep population.

Enterprises are **supplementary** to each other when they use different resources, or the same resources at different times. For example, in some food processing plants several different commodities are processed at different times of the year. In some locations, livestock can be grazed on land that is unsuitable for crop production. In this situation, crop and livestock enterprises may be supplementary (they use different land resources). They may also be supplementary to a degree in their use of labor. Cattle may graze on pasture during the summer when labor needs for crops are great and be fed in confinement during the winter when labor is not needed for cropping operations.

AGRIBUSINESS DECISIONS

Corporations such as agribusinesses seek to develop a line of products that are *supplementary* (the production and sales forces will be utilized at all times), or complementary (efforts to promote each product help to promote other products), but not *competitive* (no activity of the firm adversely affects another). For example, it would be surprising if a cereal business allowed the advertising campaign for one of its brands to criticize another of its brands.

Determinants of Supply

In Chapter 4 we demonstrated that the firm's marginal cost curve is its supply curve. We indicated that as the price received for an output increases, the firm produces more of the product, *ceteris paribus*. Since marginal cost and supply are synonymous, any change in the marginal cost curve shifts the supply curve. The basic factors that determine the firm's supply curve are:

- production technology,
- prices of the inputs used in production,
- taxes and subsidies related to the good,
- prices of other goods produced using the same resources,
- uncontrollables,
- expectations about prices, and
- the number of sellers in a market.

A change in one of these determinants causes the supply curve for a good to shift to the right or the left. Let us consider the effect of each of these determinants on supply.

- **Production Technologies**   The effect of technological advances on the production function was discussed in Chapter 3. With technological advances a firm can produce more output at the same cost. This is equivalent to a shift downward and to the right of the upward-sloping portions of the AVC and ATC curves as well as the marginal cost curve, the firm's supply curve. The shift of the marginal cost curve to the right means that the firm will produce (supply) more at any output price above the minimum of AVC (ATC in the long run).

- **Prices of Inputs Used in Production**   Declines in the prices of inputs used in production reduce marginal costs and therefore increase supply. A reduction in the price of fertilizer, for example, shifts the supply curve for corn to the right; a reduction in the price of steel increases the supply of automobiles, *ceteris paribus*. Conversely, increases in the prices of inputs shift the supply curve to the left. An increase in the price of machinery shifts to the left the supply of products produced using that machinery. Note that rising input prices can offset the positive effects of advances in technologies. For example, if the prices of computers had risen over the last decades instead of falling dramatically, the extent of their use and the impact on the supply curve of the goods and service produced with them would have been dramatically less.

- **Taxes and Subsidies on Variable Inputs**   A tax or subsidy on a variable input alters the input's effective price to the purchaser. This changes the variable and marginal costs and therefore the supply. For example, a tax on fertilizer or pesticides raises the effective price to the producer. As a result, the marginal cost of producing every unit of grain increases, and the supply curve shifts inward.
  A tax or subsidy on a fixed-cost input, such as an increase in the farm property tax, does not in the short run change the supply because it does not affect marginal cost.

- **Prices of Other Products Produced**   Changes in the prices of other products produced from the same set of inputs shift the supply curve for a product because the opportunity cost of the inputs change. Where corn and soybeans can be grown on the same land, an increase in the price of soybeans would cause some farmers to increase the production of soybeans and reduce the production of corn. If the prices of frozen food products increase, it is likely that the food processors will reduce the production of canned food products.

- **Uncontrollables**   Factors that are beyond the control of the operator—"acts of God"—can influence the supply curve. In agriculture, changes in weather can dramatically change the production function and the supply of product placed on the market. Bad weather reduces the amount of the crop produced, shifting the supply curve to the left, and good weather can push yields well above normal, shifting the curve to the right. In an agribusiness firm mechanical failures or a work stoppage can shift the supply curve inward.

• **Expectations about Future Prices**   Because the production of most goods such as farm and food products takes time, expectations about the future price of a product affect the current supply. Here, however, the effect is uncertain. For example, the expectation of a higher price for corn would induce some farmers to withhold part of their current harvest from the market, *ceteris paribus,* thus reducing the current supply. It might also induce some livestock ranchers and cattle feeders to reduce their herds, thus reducing the magnitude of the price increase. However, the expectation of a higher price for corn might cause farmers to plant more corn which would also lower future prices. Reactions such as these set up cycles of increases and decreases in the supply of agricultural commodities. These cycles vary in length and magnitude, depending on production technologies and other economic variables.

Expectations about future prices are important to all producers because production takes place with some uncertainty about the price that will be received for the product. For example, retailers purchase seasonal goods months before the goods are offered for sale. If retailers anticipate a stronger consumer demand in the coming season, they order larger quantities. If so, the retailers' demand curve shifts to the right in anticipation that consumer demand will also shift to the right. During the sales season, when the goods are on hand and the supply is fixed, retailers adjust the price to change the quantity demanded. If, for example, the actual consumer demand has not shifted to the right, retailers will reduce prices (have sales) to increase the quantity purchased by consumers. Farmers face uncertainties as to both prices and yields, and therefore tend to use several strategies to reduce uncertainty. These strategies include diversification, hedging in the futures market, contracting production at set prices, purchasing crop insurance, and keeping open-ended lines of credit.

Price expectations also play a major role in the supply decisions of owners of natural resources. If mine owners expect to receive a much higher price in the future for the mineral they are mining they might store the mineral for sale in the future or they might reduce or cease current production. If they expect prices to be lower in the future, they may tend to produce as rapidly as feasible so that they can sell more of their output at the current higher price. Thus, their supply curves can shift with a change in expected prices.

• **Number of Sellers**   If existing firms are earning an excess profit, additional firms will enter the industry and the industry supply curve will shift to the right. Conversely, continuing losses cause firms to leave an industry, thus decreasing supply. While the previous six determinants apply to the decisions of an individual firm, and thus the quantity the firm will supply, this determinant relates to the number of firms in the industry and thus to the industry supply curve.

SUMMARY

In this chapter we considered the decisions of a firm using multiple inputs and producing multiple outputs. We found that the relative prices of inputs and the productivities of these inputs work together to determine the amounts of the several inputs used. Similarly, in firms producing multiple products, as most do, the relative prices of the products produced and the shape of the production possibility frontier determine the combination of products produced. We also noted the determinants of the demand for inputs and the determinants of the supply of outputs by firms. The determinants of the demand for inputs are the demand for the output produced, the productivity of the input, and the prices of other inputs. The determinants of supply are the production technologies, input prices, taxes and subsidies on inputs, uncontrollables, and expectations about future prices.

IMPORTANT TERMS AND CONCEPTS

competition   147
complementary   147
demand for inputs   141
derived demand   141
determinants of supply   148
determinants of the demand for inputs   144
diversified   145
expansion path   140
isocost line   138

isoquant   138
isoquant map   138
isorevenue line   146
marginal resource cost   141
marginal revenue product   141
MRP = MRC rule   141
production possibilities frontier   146
specialized   145
supplementary   148

QUESTIONS AND EXERCISES

Name That Term

Read the following sentences carefully and fill in the missing term or terms.

1 A map of the alternative combinations of two inputs that can be used to produce a given quantity of output is called an _____.

2 A curve representing combinations of two inputs that can be purchased with a given amount of money is called an _____ line.

3 An _____ represents the least-cost combinations of two inputs that can be used to produce increasing quantities of a product.

4 The demand for inputs by a firm is _____ from the demand for the finished goods that these inputs combine to produce.

5 If a firm uses one more unit of an input and holds all other inputs constant, then the resulting change in total revenue is called the _____.

6 If a decrease in the price of one input increases the demand for another input, other things being equal, then the two inputs are called _____.

7 If an increase in the price of one input increases the demand for another input, other things being equal, then the two inputs are called _____.

8 Firms that produce numerous different outputs are said to be _____.

9 Firms that concentrate on one output are said to be _____.

10 The _____ line shows the different combinations of two outputs that yield the same amount of total revenue for the firm.

11 The _____ curve shows the combinations of two products that a firm can produce with a given level of inputs and technology.

True/False

Read the following sentences, then decide whether each statement is true (T) or false (F) and mark it accordingly.

T   F   **1** A profit-maximizing firm should continue to hire units of an input so long as the next unit hired costs less than the revenue added by the resulting increase in output.

T   F   **2** A profit-maximizing firm should hire inputs so long as its MRC is more than its MRP.

T   F   **3** Other things equal, an increase in the price of output will cause a profit-maximizing firm to employ more units of each variable input.

T   F   **4** Other things being equal, a decrease in the productivity of an input will require a profit-maximizing firm to hire more units of that input to make up for the loss in productivity.

T   F   **5** Other things being equal, an improvement in technology will reduce costs, increase supply, and lower output prices.

T   F   **6** An increase in the price of an already expensive input used in producing a good will lower costs as firms substitute away from the now more expensive input, increase supply, and lower output prices.

T   F   **7** One way to get firms to use less of an input that has undesirable externalities is to impose a tax on that input.

T   F   **8** Lousy weather means reduced supply and higher prices for agricultural products.

T   F   **9** If you expect the price you receive for your product to fall in the future you should plan to increase output to make up for the revenue you are likely to lose.

T   F   **10** If you are making a lot of money selling a new product you are happy to see other firms entering your market.

Multiple-Choice Questions

Circle the letter of the response that best answers the question or completes the statement. The figure below shows that the least-cost combination of labor (L) and capital (K) to make 100 units of output (Q) is $L = 15$ and $K = 4$. Questions 1–4 are based on this figure. All questions assume "other things equal."

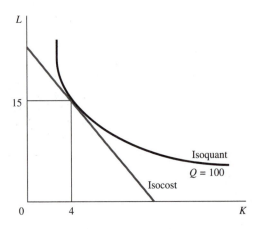

1 If the firm wants to make less than 100 units of output, the amount of labor it needs to hire
   a increases.
   b stays the same.
   c decreases.
   d is uncertain.

2 If labor gets a raise and the price of capital remains constant, then the cost-minimizing level of labor
   a increases.
   b stays the same.
   c decreases.
   d is uncertain.

3 If the price of capital and labor both increase 10 percent, then the cost-minimizing level of labor
   a increases.
   b stays the same.
   c decreases.
   d is uncertain.

4 If the price of capital and labor both increase 10 percent, then the total variable cost of capital and labor to the firm for making 100 units of output
   a increases.
   b stays the same.
   c decreases.
   d is uncertain.

The figure below shows the level of labor (L) a profit-maximizing firm should achieve if the price per unit of labor is \$22. Questions 5–7 are based on this figure. All questions assume "other things equal."

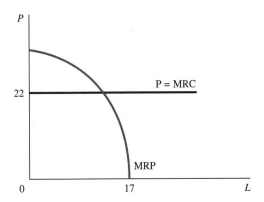

5 If the price of labor increases, the amount of labor the profit-maximizing firm should hire
   a increases.
   b stays the same.

    c decreases.

    d is uncertain.

  6 If the price of output the firm produces increases, the amount of labor the profit-maximizing firm should hire

    a increases.

    b stays the same.

    c decreases.

    d is uncertain.

  7 If labor becomes more productive, the amount of labor the profit-maximizing firm should hire

    a increases.

    b stays the same.

    c decreases.

    d is uncertain.

The figure below shows the level of two different outputs (X and Y) that a diversified firm should produce to maximize profit given its production possibility curve and the relative prices of X and Y. Questions 8–10 are based on this figure. All questions assume "other things equal."

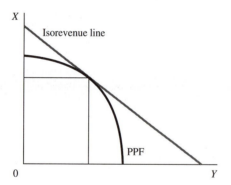

  8 If the price of X increases, the firm's profit-maximizing level of X

    a increases.

    b stays the same.

    c decreases.

    d is uncertain.

  9 If the price of both X and Y increase by 10 percent, the firms revenue

    a increases.

    b stays the same.

    c decreases.

    d is uncertain.

  10 If the price of both X and Y increase by 10 percent, the firm's profit-maximizing level of X

    a increases.

    b stays the same.

c decreases.
d is uncertain.

Technical Training

The figure below shows the least-cost combinations of labor (L) and capital (K) to make 100 units of output (Q) under several combinations of prices for labor and capital. Points a and b are cost minimizing tangencies of the isocost lines and the isoquant. Questions 1–4 are based on this figure. All questions assume "other things being equal." Let w = wage rate = price per unit of labor and r = rental rate = price per unit of capital.

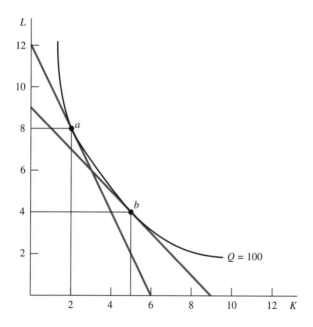

1 If $w = \$10$ and $r = \$20$, what are the cost-minimizing quantities of K and L?
2 If $w = \$10$ and $r = \$20$, how much must the firm spend on K and L in order to make 100 units of output?
3 If $w = \$50$ and $r = \$50$, what is the cost-minimizing level of K and L?
4 If $w = \$50$ and $r = \$50$, how much must the firm spend on K and L in order to make 100 units of output?

The figure on the next page shows the least-cost combinations of labor (L) and capital (K) to make 100, 200, and 300 units of output (Q). Points A, B, and C are cost-minimizing tangencies of the isocost lines and the isoquant and, thus, are points on the expansion path. Questions 5–8 are based on this figure. All questions assume "other things being equal." Let w = wage rate = price per unit of labor, and r = rental rate = price per unit of capital.

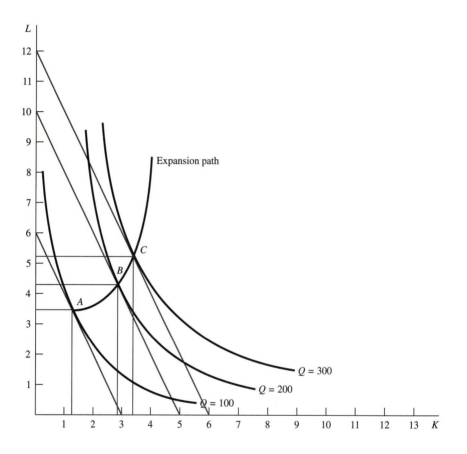

5 If the w = $24, what must r equal in order for the isocost lines to be valid?
6 At the prices in question 5, approximately how much must the firm pay for capital and labor to produce 200 units of output?
7 Is the firm's average cost increasing or decreasing as output is increased?
8 Does this firm use relatively more capital or labor as it expands output?

The figure on the next page shows the level of two different outputs (X and Y) that a diversified firm should produce to maximize profit given its productions possibility curve and two different relative prices of X and Y. Questions 9–11 are based on this figure. All questions assume "other things equal." Let P_x and P_y be the price per unit of outputs X and Y, respectively.

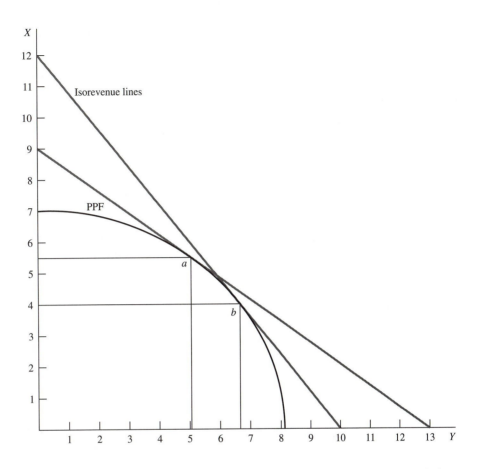

9 If $P_x = \$5$ and $P_y = \$6$, what is the revenue-maximizing combination of X and Y?

10 If $P_x = \$5$ and $P_y = \$6$, how much revenue does the firm earn?

Questions for Thought

1 In the questions for Chapter 4, the table on page 111 depicts the production and cost data for Hungarian Restaurant. Using those data, calculate and graph the marginal revenue product (MRP) and the marginal resource cost of labor (MRC). Indicate on the graph how many hours of labor maximize profits or minimize losses. Explain.

2 In relation to an input, what does the MRP curve represent?

3 What rule should firms follow when determining the optimum allocation of multiple inputs? Explain.

4 What does an expansion path represent? When constructing the expansion path, which values are kept constant? Which are varied?

5 What is meant by derived demand?

6 Briefly discuss the three determinants of the demand for inputs.

7 Give an example of a nonagricultural firm that is diversified in its production. What would be some of the advantages and disadvantages to this type of arrangement?

8 Given a production possibilities frontier and an isorevenue line, how would you determine the quantities that maximize your revenue? Explain why this combination is optimum.

9 Give a nonagricultural example of enterprises that are complementary in production. Explain why they are complementary.

10 Discuss the seven determinants of supply. Give three or four specific examples, not used in the text, of circumstances that affect supply.

11 Laborers have been known to tell overzealous workers, "Don't work so hard—we'll run out of work!" How does this sentiment square with the economic notion that increased labor productivity, other things equal, means more units of labor get hired?

12 Farmers often lament improvements in technology. "If we adopt a new technology, output will expand, prices of output will fall, and our profits will decrease." Despite this feeling, successful farmers nevertheless adopt new technologies. Why?

13 Consider yourself a firm producing output to sell the world. What products do you offer? Are you specialized or diversified?

ANSWERS AND HINTS

Name That Term   **1.** isoquant map; **2.** isocost; **3.** expansion path; **4.** derived; **5.** marginal revenue product (MRP); **6.** complementary; **7.** substitutes; **8.** diversified; **9.** specialized; **10.** isorevenue line; **11.** production frontier or production possibility curve

True/False   **1.** T; **2.** F; **3.** T; **4.** F; **5.** T; **6.** F; **7.** T; **8.** T; **9.** F; **10.** F

Multiple Choice   **1.** c; **2.** c; **3.** b; **4.** a; **5.** c; **6.** a; **7.** a; **8.** a; **9.** a; **10.** a

Technical Training   **1.** $K = 2, L = 8$; **2.** \$120.00; **3.** $K = 5, L = 4$; **4.** \$450.00; **5.** \$48; **6.** \$236; **7.** decreasing; **8.** labor; **9.** $X = 4.0, Y = 6.75$; **10.** \$60.50

7

FORMS OF MARKET COMPETITION

CHAPTER OUTLINE

OVERVIEW

This is the second of four chapters on the theory and operation of markets. In this chapter, we present the forms of competition, indicate how firms operating in such industries make decisions, and discuss the social implications of the several forms of competition. The perfectly competitive model is the standard against which the performance of other market structures is compared. Chapter 8 covers international trade.

LEARNING OBJECTIVES

This chapter will help you learn:

- How the firms operating in different competitive environments determine prices and outputs

- About the economic problems associated with each form of competition
- About the social costs and benefits of each form of competition

ALTERNATIVE ECONOMIC SYSTEMS

As you begin this chapter, be aware that various societies have different goals and systems for handling agricultural inputs and products as well as other products. These societies use different systems to manage their resources to pursue social goals. Generally speaking, there are two philosophies for answering the basic questions of what, how, and for whom to produce. At one extreme is pure **capitalism,** which relies extensively on private ownership of resources, freedom of individuals and economic units to make the choices in their best interests, and a marketing system with flexible prices as a coordinating mechanism. The possibility of a wide range of income levels and wealth is not only recognized but also considered desirable because it provides incentives for creativity and productivity. At the other extreme is **communism,** which emphasizes public ownership of most resources, central economic planning and greater exercise of the authority of the state. Communist governments place more emphasis on having less divergence among individuals in incomes and wealth.

Between these two extremes of capitalism and communism, there are many other possibilities of resource ownership, government planning, and degree of economic freedom. One such alternative is liberal, or democratic, **socialism.** However, no economy is purely capitalist, socialist, or communist. For example, although the United States, Canada, Australia, and some other countries are capitalist, many of their basic resources are publicly owned or controlled by the state, and there are various degrees of central planning. Austria, Sweden, and some other countries in western Europe are examples of a blended socialist economy, or the countries "in between." A significant portion of the economy is in public hands.

During a major part of the twentieth century, the Soviet Union, China, Cuba, and some countries of central Europe operated near the other extreme. However even in these communist countries, organized under a central planning system, there are areas of private enterprise. Not everything is covered by a central plan, and the authority of the state cannot practically be extended to control every economic decision. Thus, the market model has applications to some economic activities in all economic systems. At the turn of the century, the countries created after the collapse of the communist economic system in the USSR and several central European countries adopted many of the characteristics of capitalism.

In this chapter we consider several forms of market organization that exist in the private sector of an economy.

ALTERNATIVE MARKET STRUCTURES

There are eight general forms of market competition or *market structures,* four characterizing buyers and four characterizing sellers, as depicted in Figure 7-1. With four basic models on both the selling and the buying side of a market, there are 16 (4 × 4) possible combinations for any given market. Since individuals and firms participate in

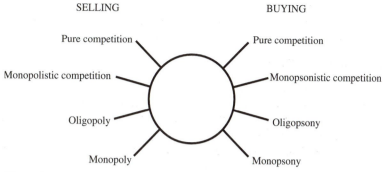

FIGURE 7-1   Alternative competitive relationships in markets.

many markets, they are likely to encounter most or all these forms of competition. The primary purpose of this chapter is to make it clear that the combinations experienced affect the economic performance of the market, especially the prices and output levels.

The sellers in an industry are characterized by the several types of competition as summarized in Table 7-1. A similar set of conditions characterizes the buyers in an industry. The economists' ideal occurs when both buyers and sellers are perfectly competitive because the optimal quantities are produced and sold at prices that provide the appropriate signals to producers and consumers.

The characteristics of a **perfectly competitive market** were described in Chapter 5. These markets are characterized by large numbers of buyers and sellers, homogenous products, freedom from barriers of entry to and exit from the industry, and perfect information. When these conditions are met (which seldom occurs) or even approximated the profit- and utility-maximizing decisions of sellers and buyers generate an optimal set of market prices and quantities. If these conditions were achieved in all markets, the transformation of resources into goods and services demanded by consumers would be optimal. The distribution of the benefits from this economic activity would depend on the allocation of property rights. Many agricultural markets approximate these conditions. Most other markets are not perfectly competitive.

TABLE 7-1   SPECTRUM OF MARKET STRUCTURES FOR SELLERS

Item	Perfect competition	Monopolistic competition	Oligopoly	Pure monopoly
Number of sellers	Numerous	Many	Few	One
Ease of entry/exit	Unrestricted	Unrestricted	Partially restricted	Absolutely restricted
Ability to set price	None	Some	Some	Absolute
Product differentiated	No	Yes	Yes or no	Unique product
LR excess π possible	No	No	Yes	Yes
Examples	Grain and livestock producers	Feed manufacturers Cereal manufacturers	Equipment producers	Power company

Many markets operate under imperfectly competitive conditions. When either buyers or sellers (or both) are in imperfectly competitive markets, it is impossible to specify the exact nature of the result. Market strategies involving advertising, sales techniques, pricing, and sometimes bargaining will determine the outcome. When one side of the market is characterized by competitive, or near competitive, conditions and the other side is not, the result will generally favor the participants with market power. Since the markets for consumer goods usually have many customers and fewer suppliers, the possibility of higher prices and reduced output exists. In agriculture, since there are many farmers and some industries that sell to and buy from the farmers are characterized as imperfectly competitive, farmers may be at a disadvantage. Over the years this perception of market inequality has been used to justify many agricultural policies, such as governmental price-support programs.

IMPERFECT COMPETITION

The conditions necessary for perfect competition are seldom met in the real world. Imperfect market competition exists over a large part of the economy, including many agribusiness industries and many markets in which we all make transactions in our professional and private lives. **Imperfect competition** *exists whenever the individual or firm buying or selling a product or service has some control over the prices.* Imperfect competition occurs when a seller faces a downward-sloping demand curve for the product or service, or when a buyer faces an upward-sloping supply curve for an input.

The key phrase in the definition is a firm's *control over price.* The control is seldom, if ever, absolute because most firms cannot ignore their competitors and set any price they choose. But sellers can raise or lower their prices within limits, and thereby reduce or expand their sales as they move along the demand curves they face. Also, through advertising and sales promotion, they can shift both the firm and the market demand curves. In addition they may see their own demand curves shift as a result of their competitors actions.

Also, in an imperfectly competitive market a firm's buying practices can influence the market price of the inputs they purchase. That is, the firm can shift the market demand for its inputs. For example, a large meatpacking firm can influence the market price for cattle, at least temporarily, through its buying practices. This, in turn, affects the supply available to other buyers and the prices they pay. A purely competitive firm, in contrast, cannot affect the market price because its share of the market is too small.

How much an imperfectly competitive firm can influence the market price depends on its position in the market. Important variables are the firm's size, its share of the market, and its strategies for buying and selling. A firm's influence also depends on the market itself, the actions of competitors, the firm's customers or clients, and the general strength of market conditions.

Imperfectly competitive industries can arise in many ways. A firm that holds a patent on a new product may be successful in establishing such dominance that when the patent expires, other firms find it difficult to enter the market. In some cases, one or more aggressive firms take advantage of economies of scale and grow until they control a sufficient market share to have some control over price. Imperfect competition may also result if the government develops and operates an enterprise on an exclusive basis.

The Firm in an Imperfectly Competitive Market

Next we consider the decision making in a hypothetical imperfectly competitive industry. This is followed by a discussion of the characteristics of several forms of imperfect competition among firms selling a good or service and of imperfect competition among firms buying a good (a resource or a product).

We illustrate output and price determination using a hypothetical case of a tractor division of a farm machinery manufacturer operating in an imperfectly competitive market. To simplify the case, we assume one firm operating without concern about the reactions of competitors. The firm manufactures and sells 7100 to 9900 tractors per year at prices from $44,000 to $52,000. The prices, quantities sold, and total revenues are shown in the first three columns of Table 7-2.

These first three columns are familiar. Total revenue equals price charged times quantity sold, or $(1) \times (2) = (3)$. Quantity sold in column (2) depends on the demand curve and the price charged by the dealer. Column (1) is the price received for a tractor, and the total revenue in column (3) corresponds to a point on the demand curve for tractors. In this example, total revenue increases with each $2000 per tractor price reduction and 700 more tractors are sold. Marginal revenue in column (5) is the addition to total revenue received as more tractors are sold with each $2000 price reduction. The marginal revenue per tractor in column (6) shows how much additional revenue is received by selling each of the additional 700 tractors. The equilibrium price per tractor is the price at which the marginal revenue (MR) and marginal cost (MC) curves intersect. A comparison of total revenue and total cost, or average revenue (AR) and average total cost (ATC), indicates whether the firm is breaking even or realizing a profit or a loss.

In Figure 7-2, the demand and marginal revenue curves are drawn as continuous lines. This could occur if the tractor prices changed in increments smaller than the $2000 used in the example. Given the practice of negotiating for expensive items such as tractors and automobiles, the smooth line is reasonably realistic. In industries where the price of the product does change in increments, the demand curve and the marginal revenue curve have a stair-step appearance.

An important difference between perfect and imperfect competition is the shape of the **average revenue** and **marginal revenue** (AR and MR) curves. Both the AR and

TABLE 7-2   SALES AND REVENUE DATA FOR TRACTOR DEALERS IN A HYPOTHETICAL IMPERFECT MARKET

(1) Price charged ($)	(2) Tractors sold	(3) TR (mil. $)	(4) Additional tractors sold	(5) ATR (mil. $)	(6) MR ($)
52,000	7100	369.2			
			700		29,714
50,000	7800	390.0		20.8	
			700		25,714
48,000	8500	408.0		18.0	
			700		21,714
46,000	9200	423.2		15.2	
			700		17,714
44,000	9900	435.6		12.4	

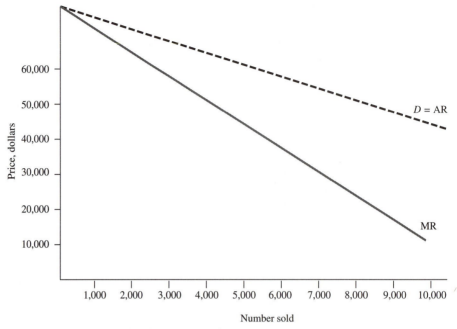

FIGURE 7-2   Demand (average revenue) and marginal revenue curves faced by a hypothetical machinery firm in an imperfectly competitive market.

MR curves are horizontal lines for a seller in a perfectly competitive market. The firm can sell as much of the product as it wishes to produce at the market price. In imperfectly competitive markets however, a firm must reduce its price to sell more of a product. Since the revenue from each additional unit is less than each previous unit, the average revenue must be decreasing and MR must be less than AR. In other words, the slope of MR is greater than the slope of AR; the MR curve is steeper than the AR curve. For a linear AR curve, the MR curve is also linear. The MR line intersects the vertical axis at the same point as the AR line and intersects the horizontal axis at a point halfway between the origin and the AR curve.

Determination of Output   Suppose this firm has the capacity to produce 8000 tractors per year, and for each level of output from 5000 to 8000, the costs are as shown in Table 7-3. What will be the equilibrium levels of sales, selling price, cost, and profit under these imperfectly competitive market conditions?

As in pure competition, the equilibrium output is at the point where the MC curve crosses the MR curve. As indicated in Figure 7-3 this occurs with the sale of about 5800 tractors. If 5800 tractors are produced, MC = MR = $22,500 and ATC is $45,000.

For a firm in an imperfectly competitive market we use the demand curve to find the market price. A vertical line through the point where MC = MR is used to determine the costs and returns. Point a, on the demand curve, shows that the firm can sell the 5800 tractors at $48,000 per tractor. Point b, on the ATC curve, shows that the average total cost of each tractor is $45,000. Hence, the firm realizes a profit of $3,000 on each tractor. The total profit before taxes is $3000 × 5800 = $17.4 million (Figure 7-3).

TABLE 7-3   TOTAL, AVERAGE, AND MARGINAL COSTS FACED BY A MANUFACTURER OF TRACTORS

Number	TFC[1]	TVC[1]	TC[1]	AFC[2]	AVC[2]	ATC[2]	MC[2]
5000	125	128	253	25	25.60	50.60	15.50
6000	125	141	266	20.83	23.50	44.33	27.00
7000	125	180	305	17.86	25.71	43.57	50.50
8000	125	237	362	15.63	29.63	45.25	72.00

[1]millions of dollars
[2]thousands of dollars
hypothetical data

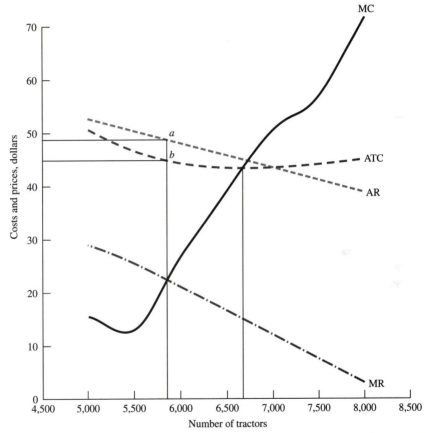

FIGURE 7-3   Costs, revenues, and equilibrium production of a firm in an imperfectly competitive market.

To confirm that this is the profit-maximizing output, consider how profits would change if the firm produced a different number of tractors. If more tractors were produced, the marginal cost would be more than $22,500, as indicated by the rising marginal cost curve. Because the firm is in an imperfectly competitive market, it would have to reduce the price of all tractors to sell the larger number manufactured. Therefore the marginal revenue, the change in revenue associated with an additional tractor, is less than $22,500. Since the additional costs are greater than the additional revenue, profit would fall. Or, if the firm did not produce the last unit, it would be forgoing some of its profit.

If fewer tractors are manufactured, marginal cost falls and marginal revenue rises. Profits fall because there are fewer tractors sold. The profit on each tractor produced would be higher, but the number sold would fall so that the total profit would be reduced.

Short-Run Profits and Losses and Long-Run Equilibrium   Figure 7-4 presents examples of short-run profits, short-run losses, and long-run equilibrium of firms in imperfectly competitive industries. In the short run, a firm may experience profits or losses, or break even, as in pure competition. However, in an imperfectly competitive industry the downward-sloping demand curve causes the firm in all three cases to reach equilibrium at a lower level of output and a higher price than in pure competition. This

FIGURE 7-4   Short-run profits and losses and long-run equilibrium in an imperfectly competitive industry.

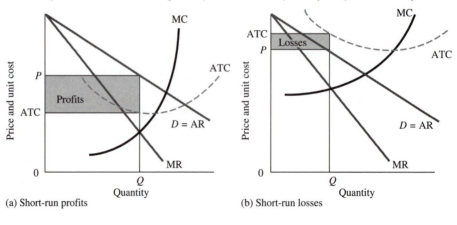

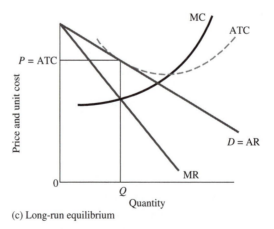

is a social cost of imperfect competition in that consumers purchase less of the product and pay more per unit than in a competitive market. If there are short-run profits, as in Figure 7-4a, there is an incentive for firms to enter the industry. This causes profits to be reduced or perhaps competed away, because of rising costs and reduced demand for each firm's products. Losses, in Figure 7-4b, create a tendency for firms to leave the industry until a normal profit occurs, because of increasing demand or falling costs.

Thus, profits and losses also work in imperfectly competitive markets to change the industry. Profits and losses in each case depend on several factors: the ease of entry, capital costs, economies of scale (both technical and pecuniary), the elasticities of supply and demand, tax rates, regulations, and so on. Profits also stimulate growth, and losses lead to contraction in competitive and imperfect markets. Profits are not necessarily higher among firms in imperfectly competitive markets than among firms in competitive markets. However, in general, because of the characteristics of imperfect markets, barriers to entry (and product differentiation particularly) lead to higher levels of profits for a longer time than would be expected in a perfectly competitive market.

Social Cost of Imperfect Competition

Because of restrictions on entry either nationally or in a local market, firms in some imperfectly competitive industries can earn excess profits for a very long time. Such entry restrictions can be related to technical factors, location, or pecuniary factors. Pecuniary factors relate to the large costs of getting established in the market. Alternatively, they can result from licenses or a favored position of an established firm in a given market. Such situations are important in agribusiness.

In general, firms in an imperfectly competitive industry produce less and charge higher prices than firms in a perfectly competitive industry. In a perfectly competitive industry, firms face a perfectly elastic demand at the point where MC intersects AR. In a perfectly competitive industry AR equals MR. If the price were $45,500 per tractor, the equilibrium output would be 6700 tractors. ATC is $43,000, with profits before income taxes of about $2500 per tractor. (In a competitive market new firms would enter, increasing output, and the price of inputs would be bid up, eliminating excess profits.) The short-run total profit is $2500 × 6700 = $16.75 million, or $0.65 million less than under imperfect competition. This difference, which is an added cost to buyers of the product, is the **social cost of imperfect competition** (the shaded area in Figure 7-5). Besides the difference in profit, note that under perfect competition the price of tractors would be lower, more tractors would be produced and sold, and more resources would be devoted to tractor production.

Does this mean that, for example, several hundred, or thousand, automobile firms operating in nearly perfect competition would sell better automobiles at a lower price than would a few oligopolistic firms? Not necessarily, because the economies of scale and the research and development efforts in large firms may offset the social costs of imperfect competition. The smaller firms under perfect competition would likely have higher average variable costs and higher marginal costs. They may not be able to support an aggressive research and development effort to improve the product.

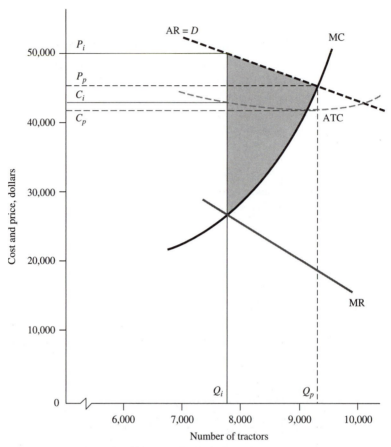

FIGURE 7-5 Comparative equilibrium positions of firms in purely competitive and imperfectly competitive markets and the social cost of imperfect competition.

Distribution of the Social Costs and Benefits of Imperfect Competition

Who pays the social costs of imperfect competition? The actual distribution of money income or profits does not capture completely the effects of imperfect competition on economic welfare. Producing 1500 fewer tractors each year reduces aggregate farm output, and reduces the total inputs used in agricultural production. Most of these costs are transferred forward to consumers as higher prices and reduced production of desired goods and services than would be expected under perfect competition. How gross and net farm incomes are affected depends on how much the supply is restricted because of imperfect competition and by the elasticity of demand for farm commodities. Imperfect competition in agricultural industries also affects other parts of the economy. In the example above, selling fewer tractors means that less steel and other physical resources, as well as labor, will be used to produce tractors, thereby reducing the total demand for those resources in the economy.

The added profits resulting from imperfect competition can be paid to employees, returned to stockholders, kept by the firm for capital investment in development and growth, used for acquisitions, or paid as higher corporate and personal income taxes. Some argue that excess profits earned because of imperfect competition enable investments in research and development that would not be possible in a perfectly competitive industry. Also, there is no assurance in the long run that average rates of profit will be higher among imperfectly competitive firms than among small firms in more perfect competition. However, as indicated, the prices would be higher and the output lower in an imperfectly competitive industry.

In the case of agriculture, the farm level is nearly perfectly competitive while the firms that provide inputs and purchase farm output are often imperfectly competitive. Although some social costs of imperfect competition within agricultural industries may be passed on to consumers through higher prices, the effects of these costs on farm inputs supplied by agribusinesses are generally felt most directly by farmers. Consequently, farmers have opposed the concentration of economic power within agribusiness supply and marketing industries. In most countries, farm organizations have supported government regulation of railroads, power companies, meatpacking companies, warehouses, and other agribusinesses; and in most democratic nations, they have led the fight for regulations or the breaking up of firms in industries where there is a high concentration of market power. In doing so, whether farm organizations have done more for farmers, for consumers, or for themselves is an unanswered question.

IMPERFECT COMPETITION AMONG SELLERS

There are three categories of imperfect competition in selling: (1) monopolistic competition, (2) oligopoly, and (3) monopoly. Remember that in perfect competition there are many buyers and many sellers, undifferentiated products, no barriers to entry or exit, and perfect information. Let us briefly consider each type of imperfect competition.

Monopolistic Competition

Monopolistic competition differs from pure competition in the following ways:

Differentiated Products   Product differentiation includes real differences such as among makes of automobile, brand names, and other advertised claims of superiority. Examples include breakfast cereal, seed corn, and designer labels. The differences among products may be real or imaginary. In monopolistic competition product differentiation is the main departure from pure competition.

The advertising industry is a central actor in product differentiation. Advertising provides information to buyers about quality and physical differences. It also attempts to create the perception of differences in the minds of consumers to instill brand loyalty and make it possible to charge a higher price. The important point is that in the minds of the buyers, differences exist, whether they are created by advertising or are physical differences among products.

Some farm commodities first sold in essentially purely competitive markets move through markets characterized by monopolistic competition. Corn, wheat, soybeans, peanuts, poultry, sugar, and many other agricultural commodities are processed into products identified by company or brand name.

PERFECT AND IMPERFECT COMPETITION IN AN INDUSTRY

Fruit and vegetable markets at the wholesale level approximate perfect competition. Buyers make their decisions without discriminating among growers or marketing firms. However, retail pricing may conform to monopolistic competition or oligopoly (a small number of sellers) as supermarket chains differentiate their services from those of their competitors. Also, an increasing amount of fruit is sold with company labels and firms in the poultry industry have developed brand name products.

Because products are differentiated, competition among firms is often vigorous on attributes other than price. These attributes include the conditions of sale, advertising, and real or perceived differences in product quality. For example, firms that supply supermarkets compete for preferred locations in the store and more shelf space because, by so doing, they can display their product in a larger or more highly visible area. Retail firms spend considerable amounts advertising their products to convince consumers that their products are superior to their rival's products. The next time you go to a grocery store, notice the brand names of various items, the different prices, and some differences among them. Ask yourself whether the price differences reflect actual differences among products or imagined differences based on advertising.

Many Firms, but Fewer Than Under Perfect Competition   Depending on the size and organization of the industry and economy, there may be many firms or just a few. Frequently producers do not know other producers of competing products. Each firm has only a small share of the national market, but there may be only three or four producers competing in a local market. Food processing and food marketing firms are examples.

Limited Control over Price   Firms in monopolistic competition have limited control over the prices of their products. There may be meaningful differences among the products, but a firm cannot increase prices significantly or consumers will shift to another brand or product. How far prices may diverge is an issue that economists study when analyzing the performance of such markets.

Limited or Restricted Entry   Although barriers to entry are usually not substantial, entry into a monopolistically competitive industry is generally more difficult than entry into a purely competitive industry. Product differentiation is a major reason, but also these firms are often large, and the capital requirements for starting up and gaining a profitable market share are substantial. Considerable spending for advertising may be needed to establish a new brand name.

The costs of advertising and sales promotion, which are passed on in higher prices that consumers pay, are an important factor in evaluating the performance of such markets. However, advertising also plays a role in providing information to consumers and to competing firms. In fact, most modern markets could not operate without advertising. Monopolistic competition provides variety in food and other agricultural commodities, and the emphasis on brand names generally ensures a high degree of quality in processing and manufacturing. How to retain these desirable features along with greater market efficiency is one of the central economic problems of agribusiness and the food industry.

Oligopoly

Oligopoly means "few sellers." Some agribusinesses in farm services and supplies—such as the major farm machinery companies, farm chemical companies, and others—are oligopolies. Some processing and marketing firms also tend to behave as oligopolists: the major meat packers, the large dairy manufacturing firms, and the major supermarket food chains. The most distinguishing characteristic of the oligopolistic firms is their interdependence in pricing and marketing practices. Interdependence exists when the firm's decisions about prices and other market variables depend in part on what other firms in the industry may do in response.

Oligopolistic industries can result from (1) more extreme product differentiation than under monopolistic competition, (2) the higher entry costs associated with plant and equipment, (3) the greater dependence on costly and long-term research and development efforts, (4) easier access to major financial markets, and (5) aggressive merger and acquisitions strategies. For several years, there has been significant growth of oligopolistic firms on an international scale. Most major corporations operate internationally. The following are characteristics of oligopolies:

Few Sellers Oligopoly exists when a few firms dominate the market for a product. Globally a few major firms dominate the automobile, heavy machinery, and oil industries. A few firms in the United States, for example, dominate domestic fertilizer and farm chemicals industries, and in some instances they have a major effect on prices in other countries. There are numerous food marketing firms, but in many cities four or five companies have 50 percent or more of the total retail food market. Each of the major firms has a large enough share of the market for its actions and policies to affect its rivals. The major firms are mutually interdependent. How they make decisions regarding price and quantity is of major concern to all participants, farmers and consumers as well as other marketing firms.

Some Control over Prices The degree of a firm's control over price is limited by the mutual interdependence of firms. Oligopolists usually try to avoid aggressive price competition because this can escalate into a price war. Thus, the prices charged by oligopolists tend to be rigid, or "sticky." This has important repercussions in the agricultural and food industries. For example, farmers often complain about the inflexible prices for machinery and farm chemicals and about the fact that the prices for their commodities do not increase when consumer prices rise. Consumers complain that grocery prices remain high even when farm prices fall. These markets would be more competitive if the few large firms were broken into two or three times as many firms. However, other factors might offset the benefits of increased numbers of firms. These factors include international competitiveness, control of funds for research and development, and access to major financial markets.

High Barriers to Entry   Because the obstacles to entry are effective in maintaining an oligopoly, an industry characterized by oligopoly tends to remain that way. Consider, for example, the difficulty of a small firm trying to enter the national or international market for farm machinery and competing with giant firms such as Deere and Company, Case International, and Kubota. Only the large, well-financed international firms have been able to enter the farm tractor industry in the United States, and fewer still have been able to enter the Japanese market. The oligopolistic nature of competition in the tractor market has resulted in only a few profitable firms internationally.

In some other industries, the possibility for entry and growth of small firms limits the prices an oligopolist can charge. For example, although it is very difficult for a new entrant to become a major factor in the food market at the national level, small local grocery stores can be a factor in a local market. Ethnic food stores and "health" food stores are examples of such local grocery stores.

Differentiated or Undifferentiated Products   An oligopolist may produce either undifferentiated or highly differentiated products. Those marketing raw materials generally offer a product that is highly standardized by class and grade, while more of those appealing directly to the consumer differentiate their product. The degree of differentiation affects the prices and market practices of the competing firms. Expenses for advertising and sales promotion are higher among oligopolists who sell a differentiated product than among those who sell a standardized commodity. Automobiles, coffee, cigarettes, and soft drinks, as well as many food products, are good examples of advertising and sales promotion by oligopolists in processing and marketing. The heavy advertising of farm chemicals and farm machinery in the major farming areas of developed countries suggests firms are oligopolistic.

Pricing Behavior in Oligopoly

The **mutual interdependence** among a small number of firms in an oligopolistic industry introduces several complexities into the decision-making processes of a firm. The essence of the problem can be appreciated with a simple example from game theory. **Game theory** *is the science of strategy between or among competitors.* Suppose there are two florists in the community, Rose's Buds and Violet's Blossoms. Further suppose that they compete for the floral business according to the outcomes in Table 7-4. The

TABLE 7-4   PAYOFF UNDER ALTERNATIVE PRICING STRATEGIES
IN A TWO-FLORIST INDUSTRY

Pricing strategies	Rose high and Violet high	Rose low and Violet high	Rose high and Violet low	Rose low and Violet low
	Payoff			
Rose	$100	$125	$25	$50
Violet	$100	$25	$125	$50

Note: Hypothetical data in thousands of dollars.

decision made by Rose affects Violet's profits and vise-versa. They each realize the highest income if they charge a low price and the other charges a high price. But if they both charge a low price, the profit is much lower than if they both charge a high price.

In this example Violet will realize $125,000 in profit if she charges low prices and Rose charges high prices (and the opposite if Rose is low and Violet is high). This is because most of the customers will purchase from Violet's Blossom. If both Violet and Rose follow the low price strategy, they will both receive $50,000 in profits, far less than if their competitor charges a high price. The third possibility is that they both choose the high price strategy, in which case they both earn $100,000 in profits.

Clearly is cases such as this the choices make a significant difference. If Violet and Rose can settle on the high price strategy, both do quite well. However, there is great temptation for them to lower prices to enhance their bottom line. If that happens, it is likely the other will also lower prices in order to be competitive and the lower profit outcome will be realized. This type of interaction among firms occurs with various outcomes depending on many factors related to the particular industry.

The point of the flower shop example is that firms must anticipate the reaction of other firms in the industry to their pricing decisions. If a firm raises the price of its product, other firms in the industry may not raise prices to the same extent. As a result, the firm raising its price may lose market share. But if the firm lowers its price, other firms are likely to lower their prices too to avoid losing market share. The result is that each firm in an oligopolistic industry faces a **kinked demand curve.** As shown in Figure 7-6, the kink in the curve is at the current price and quantity. For higher prices, the firm's demand curve is relatively elastic; for lower prices, it tends to be inelastic. The change in the slope of the demand curve creates a gap in the marginal revenue curve at the current quantity.

Next we turn to four types of pricing behavior in oligopolistic markets. They are (1) *noncollusive oligopoly,* (2) *collusive oligopoly,* (3) *price leadership,* and (4) *cost-plus pricing or cost-plus contracting.*

Noncollusive Oligopoly   In a **noncollusive oligopoly,** firms operate with a kinked demand curve.

PRICE CHANGES UNDER NONCOLLUSIVE OLIGOPOLY

A firm in an oligopolistic industry may increase prices if it determines that its product is good enough, or the loyalty of the consumers is strong enough, for it to do so without losing market share. Alternatively, a firm may cut prices in an attempt to drive rivals out of the market or to weaken them enough to make them good targets for takeover. Such behavior has occurred in the airline industry, and among agribusinesses in farm machinery, fertilizer, and pesticides.

A kinked demand curve is illustrated in Figure 7-6. If rival firms match price cuts but do not match price increases, the result is a kink in the demand curve and a break in the MR curve. In this situation prices tend to be stable. If MC "crosses" MR in the break, the firm will not change prices even if technological advances reduce costs as shown by the shift of marginal cost from MC_1 to MC_2. Instead the firm captures the benefits from the reduced costs.

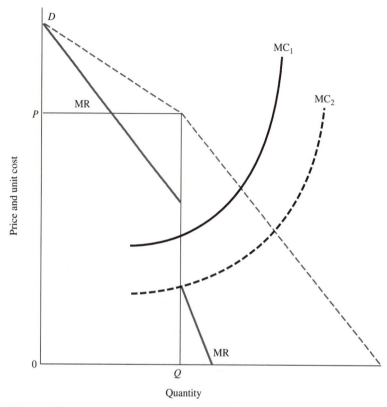

FIGURE 7-6   Price stability under oligopoly.

Collusive Oligopoly   Collusion exists if all firms in an industry reach an explicit or tacit (unspoken or unwritten) agreement to fix prices. Again the demand curve for that industry will be kinked. The firms will compete for market share on the basis of non-price factors such as service or advertising. The intent of the 1890 Sherman Act, and several subsequent acts, was to limit agreements in restraint of trade. These laws are difficult to enforce. They do, however, tend to limit collusive oligopoly.

Returning to Rose and Violet, we can see the incentive to collude. Suppose they are both charging low prices. It is quite likely that both will realize that if either or both firms charged higher prices, profits will increase. However, neither florist can take the initiative alone to raise prices because it will mean a reduction in profits for that firm and a big increase in profits for the competitor. However, if they can agree to both raise their prices, they can both double their income. Thus there is a significant incentive to collude and raise prices.

Cheating is another factor in collusive pricing, as can be seen in Table 7-4. If Rose and Violet are both charging high prices, there is considerable incentive to cheat on the agreement. That is, if either florist charges lower prices, she will increase profits at the expense of her competitor.

Finally, if a "discount" firm, willing to sell at low prices, should enter the market, Rose and Violet would then face the difficult decision of matching the low prices and sharing the lower profits with the discount firm or attempting to maintain their clientele on some basis other than price.

Price Leadership   There is a recognized price leader in some industries. In some cases, there is little apparent explanation for why a particular firm is the price leader. (Rose may follow Violet's lead on prices simply because it "has always been that way" in this community.) In other cases, the price leader is a dominant firm in the market. American Telephone & Telegraph (AT&T) achieved and held this position until the U.S. Supreme Court declared that it violated the nation's antitrust laws. Price leadership does, however, appear to operate in many other industries. In agribusiness, examples can be found in farm machinery, some pesticides, and retail grocery stores in some regions.

Cost-Plus Pricing or Cost-Plus Contracting   It is common practice in some industries, especially those selling unique products to state and federal governments, to sell products at cost plus a given percentage markup.

Monopoly

Strictly speaking, a pure **monopoly**—a one-firm industry—is rare, but there are many situations where a monopoly is the expected result of unregulated competition. These situations include railroads, especially in long-distance hauling of bulky heavy raw materials, such as coal, grain, and lumber. Local monopolies occur in utilities—such as gas, electric, water, and telephone service—resulting from the high cost of the distribution system. In the agribusiness sector, a monopoly can occur when there is just one local elevator company, when one firm holds patents on pesticides or other products, or when a fertilizer company gains a dominant position in mining phosphorus.

In some instances, regulations create monopolies. For example, U.S. patent laws grant the rights of monopoly for seventeen years to the inventor or firm that creates a new product. The prospect of monopoly profits for this period is intended to spur firms to do research and development to create new products.

Monopolies sometimes arise because this is the most efficient way to organize production and marketing activities. Monopolies often occur if there are significant economies of scale as a firm expands to a size that enables it to control the entire market. For example, a public utility has high initial costs to build a plant and distribution system, but the cost of serving an additional customer is quite small. Also, the cost of developing software was high for a firm such as Microsoft but the cost of producing another copy is small. It is common for these types of firms to be regulated because of the imbalance of power between the single seller and the many buyers. It is difficult or impossible for a competing firm to enter a market served by an existing monopolist. Instead of breaking up large and efficient firms, it has been common policy in the United States to regulate such monopolies in terms of rates of production and pricing. The measures taken to regulate or control monopolies are sensitive issues in economic analysis and public policy. The basic characteristics of monopolies are the following.

Single Seller   The concept of pure monopoly is important theoretically; and for agriculture the condition of monopoly, or near monopoly, is important in many areas such as transportation, public utilities, and communication. Farmers who have access to only one railroad on which to ship their grain or livestock may be faced with a monopoly unless there are economically viable alternatives, such as trucks and barges. In most cases urban businesses have numerous alternative suppliers and customers.

Unique Product   The monopolist's product is unique—there are no close substitutes. Close is not always easy to define. The test of substitutability is based on whether the consumers are willing to shift to some other product as the price increases. A firm that enjoys a monopoly will engage in advertising and sales promotion if the market situation indicates that this will shift the demand curve outward. Sometimes a monopolist may advertise for public relations purposes or to enhance employee morale.

Although the seller imposes monopoly, it does not affect all buyers in the same way. A farmer may have no real alternative to the local grain dealer; hence, to this farmer the grain dealer is a monopolist. Another farmer might be located near several dealers or have access to a river grain terminal, and so there is no monopoly for that farmer. The monopoly-like condition depends on both the seller and the buyer.

Ability to Set Prices   The monopolist is the price maker, although in some cases the price is regulated by a government agency. The monopolist controls the amount offered for sale and the conditions of the sale. The closer the market situation approaches a monopoly-like condition, the more need there is to regulate the firm to control the social costs of higher prices and lower output. For example, whether a railroad should be regulated is an important issue in agriculture. In countries where railroads are the only means of shipping farm commodities, the case for regulation of rates and practices is strong; but some studies have suggested that where good highways have been constructed there is less need for regulation.

Discriminating monopolists charge different classes of customers different prices for the product, thereby increasing their profits. For example, utility firms charge residential users higher rates than industrial customers.

Barrier to Entry   A monopoly must have an economic, legal, or technical barrier to the entry of other firms. A company may decline to enter a market in which another firm is already established. Patent protection prevents firms from entering some markets. In some cases, existing companies have a substantial advantage in knowledge and experience in producing a given product, which makes it difficult for competitors to enter.

The U.S. government has established many agencies to protect buyers from the adverse consequences of monopolies. The **Interstate Commerce Commission** regulates transportation; the **Federal Communications Commission** regulates telephone rates, television, radio, and so on; and state and local public power commissions regulate local gas and electrical power companies. The regulated firm need not be concerned about the reaction of rivals or potential rivals, although it is concerned about the actions of the regulatory agency. Rapid changes in communication technology have resulted in changes in the regulations and their effect on the industry.

The U.S. Sherman (antitrust) Act of 1890 and the Clayton Act of 1914 were the first of several acts with the goal of preventing collusive pricing and restrictive trade practices by oligopolists. Laws such as these, and more recent laws, are enforced to a varying extent by the successive presidential administrations, depending on the philosophy of each administration.

IMPERFECT COMPETITION AMONG BUYERS

The terms, *monopolistic competition, oligopoly,* or *monopoly* refer to the sellers of a product. Similar market structures, called *monopsonistic competition, oligopsony,* and *monopsony,* exist on the buyer's side. Because these forms of competition are analogous to the forms of competition on the seller's side, we only briefly comment on their market characteristics.

Monopsonistic competition designates the existence of a modest number of buyers facing a large number of sellers, such as might exist in a local auction market for livestock, some fruits and vegetables, or other commodities.

Oligopsony refers to a market with just a few buyers. Many central livestock markets, for example, link a large number of purely competitive farmers and ranchers with a few large meatpacking companies. Livestock farmers and ranchers sometimes complain that this market structure does not result in a "fair" price because the buyers drive down prices. Producers allege that buyers time bids to take advantage of weak market conditions and tacitly agree not to be aggressive in seeking a larger market share.

Monopsony refers to a situation in which there is a single buyer, as might be the case if a single packing plant is located a long distance from other plants or if there is only one buyer for fresh produce operating in a given area. One of the functions of farmers' cooperatives is to establish more bargaining power for farmers to grapple with the concentrated market power on the buyers' side of the market.

SUMMARY

This chapter shows how a firm operating in an imperfect market determines optimum output and price levels. Firms use the same MR = MC decision rule as in competitive markets, but because they have some control over prices in imperfect markets, profit maximization results in reduced output and higher prices than under perfect competition. As a result, resource use is lower and consumer prices are higher in imperfectly competitive industries.

The several categories of imperfect competition among both buyers (monopolistic competition, oligopoly, and monopoly) and sellers (monopsonistic competition, oligopsony, and monopsony) are described.

IMPORTANT TERMS AND CONCEPTS

average revenue   163
barrier to entry   172
capitalism   160
collusive oligopoly   174
communism   160
cost-plus pricing   175
differentiated products   169
discriminating monopolist   176

Federal Communications Commission   176
game theory   172
imperfect competition   162
Interstate Commerce Commission   176
kinked demand curve   173
marginal revenue   163
market structures   160
monopolistic competition   169

QUESTIONS AND EXERCISES

Name That Term

Read the following sentences carefully and fill in the missing term or terms.

1 The economic system in which most of the economy's resources are privately owned and managed is called _____.

2 In an _____, a few firms sell either a standardized or differentiated product, entry is difficult, a firm's control over the price at which it sells its product is limited by mutual interdependence, and there typically is a great deal of nonprice competition.

3 A _____ is a market in which there is only one buyer of a good, service, or resource.

4 A market in which there is only one seller of a good, service, or resource is called a _____.

5 An _____ is a market in which there are few buyers.

6 _____ competition refers to a market in which many firms sell a differentiated product, entry is relatively easy, the firm has some control over the price at which the product it produces is sold, and there is considerable nonprice competition.

7 Patents, high start-up costs, or loyalty to the brands of other producers are all examples of _____.

8 A _____ product is a product such that, if prices are the same, buyers are indifferent to the sellers from whom they purchase.

9 _____, the science of strategy between and among competitors, can be applied to oligopolistic firms.

10 A noncollusive oligopolist faces a _____ demand curve for its output based on the assumption that rivals will follow a price decrease but not a price increase.

11 In an oligopoly, sometimes the _____ will announce a change in price and the other firms quickly announce similar changes.

True/False

Read the following sentences, then decide whether each statement is true (T) or false (F) and mark it accordingly.

T F   1 Compared to a competitively organized industry, a monopoly sells more output at a higher price and thus earns more profit.

T F   2 When one side of a market is characterized by competitive or near-competitive, conditions and the other is not, the result will generally favor those on the competitive side of the market.

T F   3 Economists like competitive markets because they believe individual buyers and sellers should have some control over the prices they pay.

T F   4 Firms operating in a market that is imperfect competitively strive to make their market more competitive in order to improve their profits.

T F   5 The equilibrium quantity for a firm to produce is the level of output where marginal revenue equals marginal cost of the output.

T F   6 For a firm in an imperfectly competitive market, the marginal revenue associated with selling an additional unit of output is usually greater than the average revenue from selling output.

T F   7 After finding its profit-maximizing level of output by equating marginal revenue and marginal costs of output, the firm sets its price by "going up to the demand curve" and "charging what the market will bear."

T F   8 Having a monopoly guarantees that a firm will make a great deal of economic profit.

T F   9 Because products are differentiated, firms in noncompetitive markets often compete vigorously in areas other than price.

T F   10 Cost-plus pricing is a procedure used by oligopolistic firms to determine the price they will charge for a product by adding a percentage markup to the estimated average cost of producing the product.

Multiple-Choice Questions

Circle the letter of the response that best answers the question or completes the statement.

1 All of the following are characteristics of monopoly except
  a no entry of new firms.
  b small number of sellers.
  c a unique product.
  d ability to control its prices.
  e operates where marginal cost equals marginal revenue.

2 In the short run a pure monopolist's profits
  a will be maximized when price equals average variable cost.
  b may be positive, negative, or zero.
  c are always positive.
  d are zero.
  e will be maximized when price equals average total cost.

3 Which of the following is not a characteristic of a perfect market?
  a small number of buyers and sellers
  b availability of perfect information
  c a homogenous product
  d freedom of entry and exit
  e none of the above

4 Which of the following forms of market competition incur social costs?
  a monopoly
  b perfect competition
  c monopolistic competition
  d oligopoly
  e all except b

5 A discriminating monopolist
  a earns profits equal to a nondiscriminating monopolist.
  b charges the same price to all of its customers.
  c chooses to produce where marginal cost is equal to marginal revenue.
  d has costs that are lower than a nondiscriminating monopolist.
  e none of the above.

6 A firm exhibits some or all of the following characteristics of a monopsony if
  a it is the only buyer in the market.
  b it is a price taker.

  c it can make marginal cost equal marginal revenue by adjusting its price and output.

  d its selling price is lower than its marginal revenue.

  e it faces a kinked demand curve.

7 Imperfect competition within the agricultural industry

  a is generally limited to the firms that supply inputs to farms.

  b can't exist in the long run because all costs are variable.

  c may be found in the farm machinery and farm chemical industries.

  d does not exist in the grain firms because grain is openly traded on the commodity futures market.

8 Agribusiness firms in monopolistically competitive industries are similar to those in oligopoly only in that

  a non-price competition is common to both.

  b the kinked demand curve is experienced in both.

  c the number of firms is about the same.

  d mutual interdependence is common in both.

  e none of the above.

9 A kinked demand curve is characteristic of

  a pure competition.

  b oligopoly.

  c monopolistic competition.

  d monopoly

  e monopsony.

10 Capitalism, socialism, and communism are words that

  a are used to characterize different forms of economic and political systems.

  b allow for many alternative possibilities of resource ownership, governmental activity, and individual freedom.

  c represent different means of achieving a countries' economic goals.

  d describe systems that place different weights on concerns about equity and efficiency.

  e all of the above.

Technical Training

The graph below shows the demand curve (*D*), marginal revenue curve (MR), marginal cost (MC), and average total cost curve (ATC) for a monopolistic firm. Use this diagram to answer questions 1–11. Price per unit in dollar (*P*) and units of output (*Q*) are on the two axes.

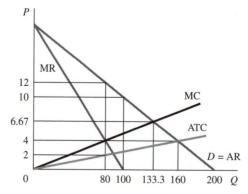

1 What is the marginal cost to the firm when $Q = 80$?
2 What is the marginal revenue to the firm when $Q = 80$?
3 How many units of output should the firm produce in order to maximize profits?
4 What price per unit should the monopolist charge?
5 What is the total average cost per unit if the monopolist produces 80 units of output?
6 What is the maximum possible profit the monopolist can earn?
7 What is the firm's total fixed cost?
8 How many units of output should a monopolist interested in maximizing revenue rather than profits produce?
9 What is the maximum obtainable total revenue for the firm?
10 If the monopolist produced the level of output that maximized revenue, would the result be a profit or a loss?
11 At what level of output would the monopolist break even?

Questions for Thought

Use the following information to answer questions 1–5:

Consider a local T-shirt industry that can be characterized as imperfectly competitive. Because of unique designs, the T-shirts that each firm produces are highly differentiated from the competition's shirts. That is, in contrast to a purely competitive industry where a large number of manufacturers produce a homogenous product and all sell at the same price, the producers of T-shirts may control the price they charge. The table below gives the cost and demand data for a T-shirt factory in this industry.

COST AND DEMAND DATA FOR T-SHIRT FACTORY

Shirts sold Q(s)	Price charged P(s)	Total revenue	Marginal revenue	Total fixed cost	Total variable cost	Total cost	Average variable cost	Average total cost	Marginal cost
700	5.00	———		———	2400	———	———	———	
			———						———
800	4.75	———		———	2500	———	———	———	
			———						———
900	4.50	———		———	2575	———	———	———	
			———						———
1000	4.25	———		———	2675	———	———	———	
			———						———
1100	4.00	———		———	2825	———	———	———	
			———						———
1200	3.75	———		———	3030	———	———	———	
			———						———
1300	3.50	———		———	3340	———	———	———	
			———						———
1400	3.25	———		———	3800	———	———	———	

Note: ATC = TC ÷ output; AVC = TVC ÷ output; MC = change in TR ÷ change in output; MR = change in TR ÷ change in output; P(s) = price of shirts; Q(s) = quantity of shirts sold; TC = TFC + TVC; TFC = fixed; TR = P(s) × Q(s); TVC = given.

Notice that the price of shirts $P(s)$ is inversely related to the quantity of shirts demanded $Q(s)$.

1 Complete the table on page 181 assuming that the firm's fixed cost is $1000.
2 Plot the demand curve ($D = AR$), marginal revenue (MR), average total cost (ATC), average variable cost (AVC), and marginal cost (MC) on the graph below. Remember to label the axes and all entries on your graph.

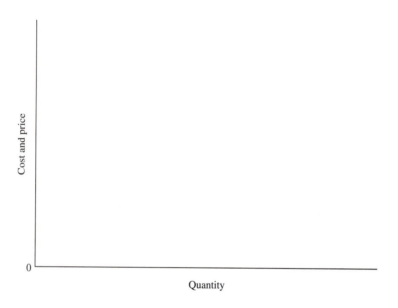

3 At what quantity and price would the shirt factory operate? Calculate the firm's profit or loss (approximate quantities and prices from the graph). On the graph, shade the area representing profits.
4 Which form of imperfect competition (monopolistic competition, oligopoly, or monopoly) does this represent? Explain.
5 Suppose consumers no longer cared about T-shirt design, and therefore the firm could not set prices based on product differentiation but instead was forced to sell T-shirts in a perfectly competitive market. Discuss the effect on the quantity of T-shirts produced and the market price. In which case (perfect or imperfect competition) would society be better off? Explain.
6 What are the characteristics of an oligopolistic firm? Give an example of a nonagricultural industry that can be described as an oligopoly.
7 Give three ways in which the United States departs from a purely capitalistic economy. Discuss why economies are not purely capitalist, socialist, or communist.
8 Do economists and businesspeople mean the same thing by "competition"?
9 Farmers, as price takers, are often said to operate in competitive markets. Yet farmers tend to be willing to help their neighboring (and "competing") farmers when trouble arises. Is this helpful behavior consistent with economic notions of competition? Is such helpful behavior common among firms operating in markets characterized by monopolistic competition?

10 A common farmer's lament goes like this. "When we go to sell our products, the buyers tell us the price. When we go to buy inputs, the sellers tell us the price. When do we get to set prices?" Based on this lament, which do you think is the most comfortable position to be in; price taker in a competitive market, price taker dealing with an oligopoly, or price-setting imperfect competitor?

11 Are monopolies always bad? Would drug companies invest millions of dollars inventing new drugs if they couldn't patent their findings and enjoy several years as a monopolistic producer?

12 Trade associations (plumbers, electricians, and, for that matter, college professors) sometimes are accused of erecting unnecessarily high standards for training and competence for people aspiring to these jobs. Why do people presently working in these jobs have a vested interested in "maintaining high standards"?

ANSWERS AND HINTS

Name That Term   **1.** capitalism; **2.** oligopoly; **3.** monopsony; **4.** monopoly; **5.** oligopsony; **6.** Monopolistic; **7.** barriers to entry; **8.** homogenous; **9.** Game theory; **10.** kinked; **11.** price leader

True/False   **1.** F; **2.** F; **3.** F; **4.** F; **5.** T; **6.** F; **7.** T; **8.** F; **9.** T; **10.** T

Multiple Choice   **1.** b; **2.** b; **3.** a; **4.** e; **5.** c; **6.** a; **7.** c; **8.** a; **9.** b; **10.** e

Technical Training   **1.** \$4; **2.** \$4; **3.** 80; **4.** \$12; **5.** \$2; **6.** \$800; **7.** \$0; **8.** 100; **9.** \$1000; **10.** profit; **11.** 160

8

INTERNATIONAL TRADE

CHAPTER OUTLINE

OVERVIEW

International trade is vital to the U.S. economy and is a driving force behind the fortunes of U.S. agriculture. It is no coincidence that periods with rapid growth in agricultural exports—the 1910s, the 1970s, and the early 1990s—have accompanied prosperity for farmers and the farm sector. When world demand for U.S. agricultural products declines, as happened during the Great Depression of the 1930s, in the late 1990s, and again in the early 2000s, agriculture suffers.

Why do nations trade? This is the fundamental question addressed in this chapter. A key to the answer is the consequences for the domestic economy. A related question is, Why do nations erect barriers to trade? The answer to this question also lies in the effects of trade on the domestic economy.

We start with a discussion of how international trade affects the domestic market for a commodity. When profit-maximizing producers and utility-maximizing consumers can buy and sell in international as well as domestic markets, they are likely to change their behavior. Their responses determine how much is imported or exported, and what happens to domestic prices.

As we look beyond a single market and examine a nation's trade flows, we see how a nation's resource endowment and available technologies determine its comparative advantage and thereby its exports and imports. We show how a nation gains from trade and point out that while a nation as a whole gains, some members of the economy lose from trade, at least in the short run. If those adversely affected have enough political power, they can block moves to freer trade that benefit the majority. We examine three common measures that governments have used to restrict trade, and how their use affects equilibrium in the market. Finally, we end with a brief discussion of the history of negotiations to liberalize international trade under the auspices of the World Trade Organization (WTO).

LEARNING OBJECTIVES

This chapter will help you learn:
- How international trade affects the determination of domestic prices, production, and consumption
- How resource endowments and production technology determine a nation's comparative advantage and thereby its exports and imports
- How international trade policies can be used to alter the domestic market equilibrium
- About the history of U.S. agricultural trade and the role of agriculture in the GATT and WTO negotiations

INTERNATIONAL TRADE

International trade is the sale and purchase of goods across national boundaries. If international trade is not restricted, buyers and sellers in one country may purchase goods (and services) from any other country. In essence, each buyer and seller has the option to make a transaction either in the domestic market or abroad. For example, a Washington

State apple grower can sell apples in the United States or sell them to buyers in Japan, South Korea, or in any other country. The choice of where to sell the apples depends on from whom the grower can get the highest price.

Comparing prices at home and abroad is complicated by the fact that prices in other countries are denominated in their home currency—Japanese Yen, Korean Won, and so on. To compare foreign prices with the domestic price, the foreign-currency-based price is converted into its domestic currency equivalent using the **exchange rate,** *the price of the foreign currency in terms of the domestic currency.* For example, if the Japanese price of a bushel of apples is ¥540 and the exchange rate is 130¥/$, the dollar equivalent of the Yen price is $4.15 (540 ÷ 130). Or, a bushel of wheat worth $3.00 in the United States costs ¥390 ($3.00 ÷ 0.00769). (Note that $1 ÷ 130 = 0.00769$ and $1 ÷ 0.00769 = 130$). In the section on macroeconomics we discuss some of the determinants of the exchange rate. For the remainder of this chapter, we assume that all prices are expressed in dollars (which is often the case in international transactions). Also, instead of examining prices from all potential foreign buyers or sellers, we assume that there is a **world market** made up of the combined markets of all countries in the world. In the world market, products produced at home can be sold and products produced abroad can be purchased at the world price. We also ignore transportation costs to and from the world market so that we can examine the essential features of international trade.

The Small-Country Assumption

For almost all commodities, the volume in the world market, total world production, is much larger than the production or consumption of any one country. An important consequence is that for most goods and services, *the trade of any single country has little effect on the world price.* A country that cannot change world prices by altering its exports or imports is called a small country. A large country, in this context, is one that can significantly influence the world market for a commodity. Examples of large countries include South Africa in the diamond market, Brazil in the coffee market, and the United States in advanced military hardware.

For commodities where the small-country assumption holds, the world market can buy as much as the economy can produce, or sell as much as the economy will purchase, at a given **world price** (P_w). Another way of saying this is that at the world price domestic consumers can buy as much as they want from the world market at P_w and domestic producers can sell as much as they want in the world market at P_w. This means that if the government does not intervene in international trade, *the equilibrium domestic price is equal to the world price.* Since the value of American trade, including agricultural trade, is by far the largest of any nation it may appear that the United States can dominate the world market. However, if we compare U.S. production and trade with that of the rest of the world (ROW) for several important agricultural commodities we find this is not the case.

The United States is the largest producer of oilseeds, coarse grains, poultry, and beef and in each case produced between 25 and 30 percent of total world production. China is the largest producer of foodgrains (rice and wheat) and hogs with approximately 25 and 50 percent of world production, respectively.

In terms of trade, the U.S. exports represent about 6 percent of world wheat production, 10 percent of world corn production and approximately 20 percent of world soybean production. For all other agricultural commodities U.S. exports are less than 1 percent of world production, and in some instances the United States is a net importer. This suggests that in the short run, before producers and consumers in other countries can respond (perhaps a year or two), a change in U.S. trade policy can have a significant effect on world prices. This is especially true for corn and soybeans. In the longer run, however, any change in U.S. production and exports would lead to offsetting changes in production and exports in the ROW, and the effect on the world price would not be large.

Since even the United States is a *small country* for most agricultural commodities, for the remainder of this discussion we make the **small-country assumption,** which is that shifts in domestic demand or supply and trade policy changes do not change world prices.

Equilibrium Price with Trade

Given the small-country assumption, we turn to the operation of the international market and the impact on production, consumption, and prices in a trading country. We use the graphs of supply and demand introduced in Chapter 4.

Exports   With no international market, and with government policy prohibiting imports and exports, the equilibrium price (and quantity) is found at the intersection of the domestic demand and supply curves. This is sometimes called the closed-economy case. Suppose the economy in Figure 8-1 is opened to international trade, with the world price above the intersection of the domestic demand and supply curves. Producers can

FIGURE 8-1   Exports with the world price above the domestic price.

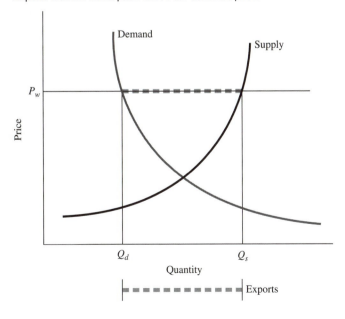

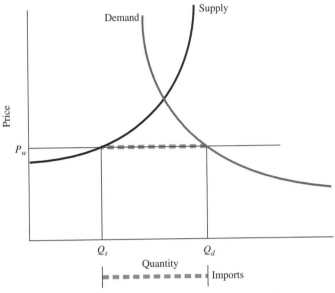

FIGURE 8-2   Imports occur when the world price is below the domestic price.

sell all they can produce on the world market at the higher world price: P_w. Producers will sell to domestic consumers only at this higher price. Since the world price is higher than in the closed economy, producers expand production and consumers reduce consumption. The quantity that would be surplus in a closed economy is exported. Although domestic demand is not equal to domestic supply, the market is in equilibrium. Producers can sell as much as they want, and consumers can buy as much as they want at the market price. The world price is the equilibrium price.

Imports   Suppose the world price is below the intersection of the domestic demand and supply curves, as in Figure 8-2. In this case consumers can purchase as much of the commodity as they want at the lower world price. Domestic producers must lower their prices to compete. Since the world price is lower than the domestic price would be if trade were prohibited, consumers expand consumption and producers reduce production. The would-be shortage in a closed economy is imported. Again, although domestic demand is not equal to domestic supply, the market is in equilibrium. Producers can sell as much as they want, and consumers can buy as much as they want, at the market price.

A Shift in the Demand Curve in an Open Economy

As the previous discussion shows, price determination in an open economy differs from that in an economy closed to international trade. The effects of shifts in demand and supply curves are also very different.

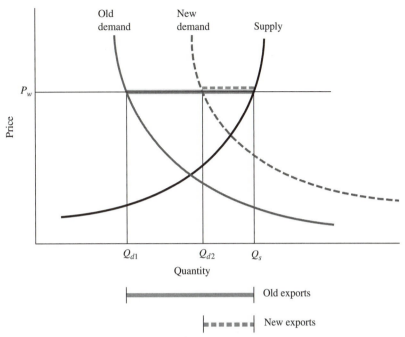

FIGURE 8-3   A demand shift in an open economy reduces exports.

In Figure 8-3 we show what happens to the quantity produced, consumed, and traded in an open economy when the economy's demand curve shifts outward because of, say, income growth in the country. In a closed economy, the outward shift of the demand curve causes a movement along the supply curve, increasing the domestic price and production. The higher domestic price reduces the increase in domestic consumption. In an **open economy,** the world price and the domestic price are the same at P_w. Changes in domestic demand don't change the price because this country's exports are not large enough to alter the world price. Since the domestic price doesn't change, production remains the same at Q_s. However, consumption increases from Q_{d1} to Q_{d2} as the demand curve shifts outward. The result is that the quantity of exports falls as indicated. Note that if income growth continues and there is no shift in the supply curve, the country will change from an exporter to an importer when the demand curve shifts beyond the supply curve. Also, note that the effect of a shift in the world demand resulting from world population growth would be quite different. World demand would grow, causing the world price to increase. This would stimulate production in all countries that are open to world trade. (You may want to try your hand (brain!) at graphs of a shift of the supply curve or of an economy with imports.)

So far, we have examined how the market outcome in an open economy differs from that in a closed economy. Producers and consumers will import or export if it is to their personal advantage to do so. But what determines the kinds of goods and services that a nation exports and imports? We address this issue next.

WHY NATIONS TRADE: THE ROLE OF COMPARATIVE ADVANTAGE

Bill Gates is a master programmer but likely didn't write a line of code during the years he directed the Microsoft Corporation. Microsoft employs many programmers who are less proficient than he. If asked to justify why he does not spend his time programming, Gates may observe that if he spent time programming he would save the time of the programmer, but the firm would lose the revenues he generates from management. Since his value as an executive is much greater than the value of the time of the programmer that would be saved, paying the less proficient programmer is a good choice for the company. Mr. Gates and the programmer have different resource endowments. Mr. Gates's "resources" or abilities are superior for both management and programming. The programmer does not have equal management skills, but is a good programmer. The firm gains by taking advantage of these differences. (Interestingly, in 2000 Mr. Gates stated his intention to retire to his first love—programming.)

For nations as well as individuals, differences in productive capacities determine who or which nation specializes in producing various goods and thus international trade flows. To understand how these differences translate into trade flows, we begin by examining the benefits from international trade. Exporting is beneficial in that jobs and profits are generated. The income produced can be used to buy foreign goods for consumption. But why do consumers want to consume foreign goods? There are two basic reasons. First, it might not be possible to produce the foreign good at home—it is impossible to mine diamonds if they do not exist. Second, it might be cheaper to purchase goods from another country. The second reason is the focus of this section.

Absolute and Comparative Advantage

The economic basis for trade among nations is embodied in the principle of comparative advantage. To understand this principle, we begin with the concept of absolute advantage and return to the example of Bill Gates and the programmer. Mr. Gates has the absolute advantage in both management and programming. He is better in absolute terms than the programmer described above. A country has an **absolute advantage** in the production of a good if it can produce more of that good with a given set of resources than another country. If one country has the absolute advantage in the production of rice and the other has an absolute advantage in beans, the trade flows are obvious. But, offsetting absolute advantage is not necessary for trade to be in the best interests of both nations. In our example Mr. Gates had an absolute advantage in both tasks but had a comparative advantage in management. Thus he concentrates on management and hires programmers, who have a comparative but not absolute advantage in programming, to concentrate on writing computer code.

A country has a **comparative advantage** *in those goods for which it has the lowest opportunity cost compared to its trading partners.* A crucial question is raised by this definition. What is meant by "lowest opportunity cost"? If countries export those goods and services for which they have a comparative advantage and import goods for which they have a comparative disadvantage, trade is mutually beneficial to all nations.

If all nations were the same, there would be no reason to trade. However, there are three important differences that explain why countries trade: resource endowments, production technologies, and tastes and preferences. In this chapter, we focus on the first two. We begin with a simple, hypothetical example of international trade. In this example, there are only two countries in the world, Richland and Poorland, and each country produces only two goods, apples and oranges, with only one resource—labor. Furthermore, Richlanders and Poorlanders have similar preferences for apples and oranges. This means that if a citizen of each country had the same budget and faced the same market prices for apples and oranges, the quantities of the two goods consumed would be the same.

To emphasize the importance of *comparative* advantage, we assume that Richland has an *absolute* advantage over Poorland in both apples and oranges. The situation in each country without trade is shown in Figure 8-4a. Both countries have 100 workers. With its 100 workers, Richland can produce 100 apples (1 apple per worker), or 80 oranges (0.8 oranges per worker), or any linear combination of them. The line connecting these two points gives a simple **production possibilities frontier (PPF).** In this example there are no diminishing marginal returns so the PPF is a straight line (a more realistic PPF would be curved). With its 100 workers, Poorland can produce only 50 apples (0.5 apples per worker), or 70 oranges (0.7 oranges per worker), or any combination. Comparing the two PPFs, Richland can produce more of both apples and oranges than can Poorland with the same amount of resources.

FIGURE 8-4 Richland and Poorland production and consumption with and without trade.

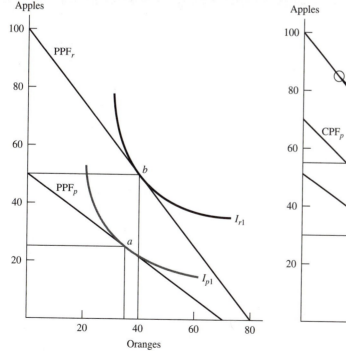

(a) Without trade

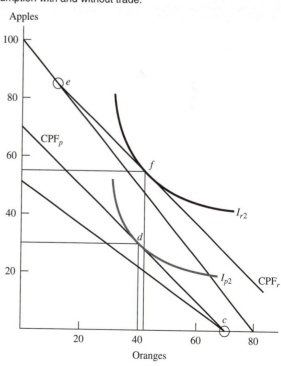

(b) With trade

TABLE 8-1 PRODUCTION AND CONSUMPTION LEVELS IN RICHLAND
AND POORLAND WITHOUT TRADE

	Poorland		Richland	
	Apples	Oranges	Apples	Oranges
Production	25	35	50	40
Consumption	25	35	50	40
Number of workers	100		100	

Note: Hypothetical data.

Figure 8-4 also gives the **social indifference curves** for both countries. The social indifference curve is analogous to the indifference curve for an individual presented in Chapter 2. *A social indifference curve reflects the collective preferences of all individuals in a country.* Without trade, domestic production and consumption must be the same as shown in points *a* and *b*. The equilibrium is where the indifference curve is tangent to the PPF. Table 8-1 summarizes the production and consumption levels in both countries without trade.

Does the fact that Richland has an absolute advantage in both apples and oranges mean that Richland would not gain in trade with Poorland? No, both would gain. We need to look at opportunity cost to see how this happens. To produce an additional apple in Richland, it would have to give up 0.8 oranges. To produce an additional apple, Poorland would give up 1.4 oranges. Clearly, the opportunity cost of apples is lower in Richland than in Poorland. This suggests that Poorland should concentrate on oranges and Richland on apples.

Suppose Poorland offers to trade oranges for apples with Richland on a one-for-one basis. This rate of exchange, known as the terms of trade, means that for every orange sent to Richland, Poorland receives an apple. Suppose Poorland offers 30 oranges, expecting 30 apples in return. Should Richland accept this offer? Absolutely! Richland can shift its production to 85 apples and 12 oranges, accept the trade, and consume 55 apples and 42 oranges. Table 8-2 summarizes the "without" and the "with" trade production and consumption levels. The most important result is that consumers in both countries consume more apples and more oranges than without trade. They both achieve a higher social indifference curve.

Figure 8-4 will help you to understand why this occurs. The PPFs without trade are reproduced for each country. Each country must choose to operate at some point on its PPF. We see in Figure 8-4a the choices without trade. Poorland will choose to produce and consume at point *a* (25 apples and 35 oranges) and Richland will produce and consume at point *b* (50 apples and 40 oranges).

In Figure 8-4b we add a **consumption possibilities frontier (CPF)** showing *the consumption possibilities with trade.* With trade, both countries are able to reach a higher social indifference curve. This is simplest to explain for Poorland. Poorland can improve its situation by producing 70 oranges and no apples, as shown at point *c* in Figure 8.4b. Poorland can then offer to trade some (or all) of its oranges to Richland on a one-for-one

TABLE 8-2 PRODUCTION AND CONSUMPTION LEVELS IN RICHLAND AND POORLAND BEFORE AND AFTER TRADE

	Poorland		Richland	
	Apples	Oranges	Apples	Oranges
Before trade				
Production	25	35	50	40
Consumption	25	35	50	40
After trade				
Production	0	70	85	12
Consumption	30	40	55	42

Note: Hypothetical data.

basis. By trading, Poorland moves along its CPF, which connects 70 oranges and zero apples and zero oranges and 70 apples and is above its PPF. Thus, by moving along the higher CPF Poorland can reach a higher social indifference curve.

Suppose Poorland offers to trade 30 oranges to Richland for 30 apples. This places Poorland at point *d*, which is clearly better for its consumers than are points *a* or *c*. To take advantage of this offer, suppose Richland chooses to produce 85 apples and 12 oranges, as shown at point *e*. Richland's consumption possibility frontier starts at point *e*. With trade at a one-for-one basis, Richland can trade apples for oranges along its CPF. If Richland accepts the 30/30 trade offered from Poorland, it moves along its CPF to point *f* where it is consuming 55 apples and 42 oranges. Both countries are able to consume more apples and more oranges with trade than without. Both countries are on a higher indifference curve, as shown in Figure 8-4b.

Of course, this all gets more complicated when a more realistic curved production possibilities frontier is considered, when prices have to be adjusted for an exchange rate between currencies, when numerous countries are involved, and when there are trade polices restricting the trade flow. But this example shows the basic concepts of comparative advantage and why it is in the interest of both countries to trade.

Terms of Trade

In the discussion above, we assumed terms of trade of one apple for one orange. How much each nation gains from trade depends on the **terms of trade,** *the rate at which units of one product can be exchanged for units of another product.* We can determine the minimum terms of trade each nation must receive to gain from trade. The terms of trade must be better than the tradeoff in domestic production for the country to benefit. In the above example, Richland would not be willing to sell apples for oranges if the Poorland trader offered less than 0.8 pounds of oranges per pound of apples. Poorland will sell oranges as long as the terms of trade are 1.4 pounds of oranges per pound of apples or less. Hence, the terms of trade in this example must be between 0.8 and 1.4 pounds of oranges per pound of apples. The actual value depends on resource endowments, productivity, and tastes and preferences. It cannot be determined without further information.

Limits to Specialization

In the apples and oranges example, opening the countries to trade resulted in Poorland specializing in one commodity. In today's world we do see specialization in some products (for example, computer chips, jet aircraft, and disposable diapers are made in only a few countries), but many other products are made around the world. In our example, specialization occurred because we assumed that production technologies or resource endowments led to different productivity levels. Economies of scale can lead to specialization as well. If there are diminishing returns or if one country is much larger than the other, then the gains from moving to free trade generally will be exhausted before complete specialization occurs in either or both countries.

Comparative Advantage and Gains from Trade

In the previous example, we indicated how and why two countries each gain from trade. To make the model clear enough to bring out the basic principles, we assumed that the production technology had constant returns and that consumer tastes and preferences were the same in both countries. In this section, we focus on the gains available to a single small country in the transition from a closed economy (no trade) to an open economy (free trade). We again assume only two goods, but the production technology results in increasing marginal costs as more and more resources are devoted to the production of one product. This assumption, which makes the example more realistic, results in a curved PPF.

Figure 8-5 depicts a **closed economy** producing two goods—industrial machines (on the Y axis) and agricultural products (on the X axis)—with a fixed amount of resources. Consumption must take place on or inside the PPF; that is, the PPF and the consumption possibilities frontier are the same if trade is not allowed. The social indifference curve farthest from the origin and tangent to the PPF determines where production and consumption take place (point *a*). The tangency between the social indifference curve and the PPF also determines domestic prices (indicated by the straight line tangent to both

FIGURE 8-5   Production, consumption, and domestic prices in a closed economy.

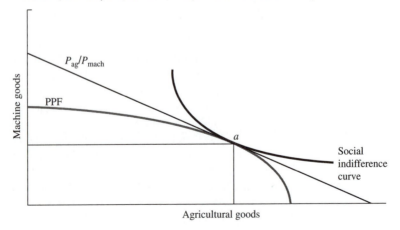

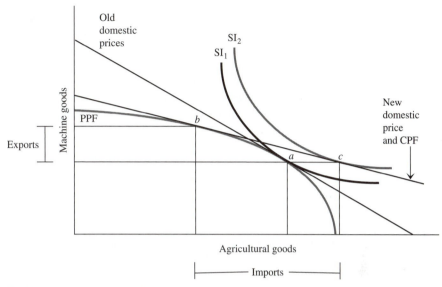

FIGURE 8-6   Production, consumption, and domestic prices in a open economy.

the PPF and the indifference curve at point *a*). (Exactly why this determines prices is a long story about general equilibrium that we defer to a more advanced course.)

In countries that are open to trade, domestic prices tend to adjust until they are the same as world prices. This is illustrated in Figure 8-6. Before trade, suppose that in some country there is a relationship between machinery prices and agricultural prices represented by the line labeled "old domestic prices." With a social indifference curve, such as SI_1, and a production possibility frontier, such as PPF, the equilibrium is at point *a*.

Then if trade is allowed, the prices for agricultural commodities and machines shift to the line labeled "new domestic prices." Because of the higher machinery prices and lower agricultural prices, fewer agricultural goods and more industrial machines are produced, indicated by point *b*. With trade, the world price line also represents the consumption possibilities frontier (CPF), because the country can reach any point on this line by trading in the world market. That is, consumption can happen at any point along the world price line. With the new domestic prices and the possibility of trade, *consumption happens outside the PPF,* where a social indifference curve (higher than the previous one) is tangent to the world price line (point *c*). This is possible because the country imports agricultural goods and exports machines in the amounts indicated on the figure.

REASONS FOR OPPOSITION TO FREE TRADE

In the examples presented above, trade made consumers better off, as indicated by the moves to higher social indifference curves. It remains a fact, however, that no country allows completely free trade, and many countries have policies that severely restrict trade. If trade is so clearly beneficial, why do policies that restrict trade exist? We discuss three reasons why a country might decide to intervene in its international trade flows.

Strategic Goods

Strategic Goods are goods that play an important role in the survival or well-being of a nation, for example, materials used in national defense or staple food crops in developing countries. Many countries are unwilling to depend on imports for substantial portions of such goods. Since trade can make production of some commodities less profitable and lead to declining production at home, the country might decide to implement a trade-restricting policy to reduce imports and encourage more domestic production than would occur with free trade.

Adjustment Costs

When a country trades a larger proportion of its production, as would occur with a shift from a closed economy to a free-trading economy, some industries decline and others expand. The costs associated with these changes are **adjustment costs.** Look back to Table 8-2 and ask yourself how you would react to trade if you were an apple producer in Poorland or an orange producer in Richland. Some workers in the declining industries must change jobs, move to new locations, and learn new skills to work in the industries that expand production for export. Also, owners of the declining firms will see the value of their assets decline. The costs of this transition can be very large—it is expensive to convert an apple orchard to orange production. If the workers and owners have to bear these costs, they may vehemently oppose moves to freer trade.

"Unfair Competition"

In many developed countries, those opposed to free trade argue that it permits *"unfair competition"* from developing countries where wages are only a fraction of wages in the developed countries. A similar argument is made in many developing countries. Opponents of trade say that imports of industrial goods from developed countries are "unfair competition" to "infant" firms in less-developed countries (LDCs) trying to expand domestic production of industrial goods. Also, some in these countries argue that agricultural products produced on large and mechanized farms represent unfair competition to the small and labor-intensive farms in developing counties.

To determine the validity of these arguments, we examine how trade affects incomes. One of the conclusions of trade theory is that opening a country to trade reduces the incomes received by owners of the scarcest resources. The reason is that trade allows the country to import goods whose production requires heavy use of the scarcest resource, reducing demand for that resource at home. (The country's exports will be produced with resources that are relatively abundant domestically and therefore relatively cheap, which increases the incomes of owners of these resources.)

No two countries have the same resource endowments. However, if we compare the developed countries as a group with LDCs, we find that developed countries have abundant capital, both as equipment and as human skills, or human capital. The LDCs, on the other hand, generally have a relative abundance of unskilled labor but a scarcity of capital and equipment. Therefore, developed countries (DCs) generally export goods produced with relatively more capital and skilled labor. Exports from LDCs embody more low-paid unskilled labor but less capital. Since demand for unskilled labor is relatively low in DCs, many of the imports into those countries are produced with large quantities of unskilled labor.

As DC imports of labor-intensive products grow, the demand for unskilled labor in the exporting countries grows, increasing the income of unskilled workers in those countries. In the importing countries, the labor-intensive imports compete with domestic products, pushing their prices down and reducing the demand for (and incomes of) unskilled workers in the DCs. Therefore, complaints from workers in DCs about "cheap foreign labor" have economic justification. However, if an importing DC reacts with trade barriers (such as on textiles from China), it increases prices of these goods, penalizes its own consumers, and decreases workers' incomes in China. In LDCs, capital in the form of high-technology equipment and skilled labor is relatively scarce, and since LDC imports use relatively more capital and skilled labor, free trade may result in lower incomes in LDCs for the owners of these resources. For example, a developed country such as Europe or Japan with relatively lower cost of capital and skilled labor has an advantage over a LDC in automobile manufacturing. Cars manufactured in a labor-rich developing country such as Indonesia don't have those advantages. The LDC owners of expensive capital invested in car manufacturing facilities in their country will lose if imports of cars from Europe or Japan are allowed.

Does this mean that it would be sound economic policy for the DCs to restrict trade to increase the incomes of unskilled workers, or for LDCs to restrict trade to increase the incomes of skilled workers and capital owners? No. While a trade-reducing policy in a DC protects the incomes of affected nonskilled workers, it also increases the consumer costs of products whose prices increase. In fact, as we demonstrated in our previous example using oranges and apples, trade increases consumption in both countries. Trade theory demonstrates that, in principle, it is possible for everyone to be better off with trade. A more appropriate public policy response is to help those adversely affected by freer trade to adjust to the new economic environment. For example, the government of a DC might increase funding for education, training, and relocation programs for unskilled workers. Governments of LDCs might support the development of new technology in industries with a good potential for growth. Governments in both DCs and LDCs might support international marketing and extension programs, and implement the transition to freer trade gradually to allow firms and workers more time to adjust.

THE COSTS OF RESTRAINING TRADE: AN EXAMPLE

The United States negotiated "voluntary" export restraints by the Japanese in order to reduce the imports of automobiles during the 1980s. According to a government study, these implicit import quotas resulted in about 30,000 more jobs in the automobile industry in 1983 and 1984. They also raised the average cost of a car by about $500. Since 9.5 million cars were bought annually during this period, consumers paid an additional $4.75 billion each year, or more than $150,000 for each job created. Clearly the U.S. economy would have been better off if that money had been used to retrain those workers for other jobs or to support other forms of adjustment assistance.

Source: "Has Trade Protection Revitalized Domestic Industries?" Congressional Budget Office, November 1986.

THE EFFECTS OF TRADE POLICIES

Government **trade policies** (and other policies) are implemented to change the performance of the economy—to cause market outcomes different from those that would occur with no policy. Whether the goals are appropriate is a normative question (see Chapter 1) that is not our concern here. In this section, we focus on the effects of three commonly used policy instruments (the mechanisms used to implement policy goals) affecting agricultural trade: tariffs, export subsidies, and import quotas. We illustrate this by examining the effects of tariffs, subsidies, and quotas on five variables:

1 Domestic price
2 Domestic production
3 Domestic consumption
4 Exports or imports
5 Government tax revenues and expenditures

A trade policy might also affect the world price in the short run. However, we ignore the effects on the world price because these effects are moderated over time.

Effects of a Tariff

A **tariff** *is a tax on imports.* There are two basic types of tariffs. An **ad valorem tariff** is an import tax that is *assessed as a percentage of the value of the imports;* for example, a tax of 5 percent of the value (price) of an import shipment is an ad valorem tariff of 5 percent. A **specific tariff** is an import tax that *is a fixed amount per unit of the good imported.* For example, a tax of $100 per cubic meter of plywood imports, regardless of the price, is a specific tax.

TARIFFS AND SELF-SUFFICIENCY

In some countries self-sufficiency is a goal that influences trade policy. One way to achieve this goal is to raise the tariff to a level such that the world price plus the tariff is equal to the price in the closed economy, that is the price that sets domestic demand equal to domestic supply. Imports would be eliminated as domestic production increased and domestic consumption fell.

Figure 8-7 demonstrates the effect of a tariff on the equilibrium price, production, and consumption in the domestic market. Without the tariff, the domestic and world prices are the same (P_w). Hence, domestic production is S_1, domestic consumption is D_1, and there are imports equal to $D_1 - S_1$.

With either an ad valorem or a specific tariff on imports, the price of imports increases by the tariff to $P_w + T$. Given a choice, consumers will not buy imports if domestic goods are available at a lower price. But this will lead to a shortage of domestic goods and an increase in their prices. Domestic producers will continue to raise their prices until the prices are equal to the world price plus the tariff. When the domestic

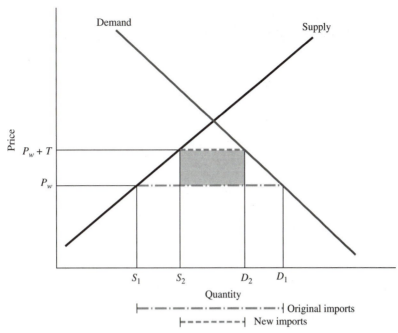

FIGURE 8-7   The effects of a tariff.

price is equal to the world price plus the tariff, domestic production will be S_2, domestic consumption will be D_2, and imports will be $D_2 - S_2$. At this price, consumers can buy all they want, producers can sell all they want, and the market is in equilibrium.

What happens to the five variables affected by this trade policy?

Domestic production	⇑	from S_1 to S_2, because the domestic price is up
Domestic consumption	⇓	D_1 to D_2, because the domestic price is up
Imports	⇓	because domestic production is up and consumption is down
Domestic price	⇑	from P_w to $P_w + T$, because the tariff raised the price of imports
Government revenue	⇑	from the tariff. The total government revenue from the tariff, indicated in Figure 8-7 by the shaded area, is equal to the tariff per unit of imports $(P_w + T - P_w)$ times $(S_2 - D_2)$, the quantity of imports.

Effects of an Export Subsidy

An **export subsidy** is a payment to exporters from the government. Just as there are ad valorem and specific tariffs, there are also both ad valorem and specific export subsidies. An ad valorem export subsidy is paid as a percentage of export value; a specific export subsidy is a set number of dollars paid per quantity exported.

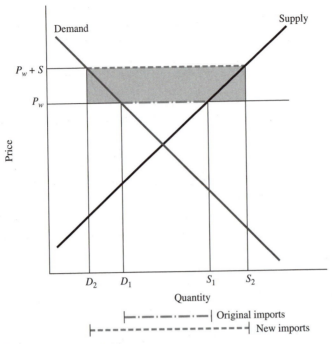

FIGURE 8-8   Effect of an export subsidy.

Figure 8-8 demonstrates the effect of an export subsidy. Initially the domestic market is in equilibrium, with the domestic price equal to the world price. When a subsidy is implemented, producers initially receive more for their goods on the export market (world price plus the subsidy) than in the domestic market. As a result, they increase the amount exported creating a domestic shortage. The shortage leads to higher domestic prices. At equilibrium the domestic price equals the world price plus the subsidy, and the exporting firms realize an increase in revenue.

What happens to the five variables affected by trade policy?

Domestic production	⇑	from S_1 to S_2, because the domestic price is up
Domestic consumption	⇓	D_1 to D_2, because the domestic price is up
Exports	⇑	because domestic production is up and consumption is down
Domestic price	⇑	from P_w to $P_w + S$, because the export subsidy forces consumers to pay more to compete with the export subsidy
Government revenue	⇓	because the export subsidy must be paid. The cost of the subsidy to the government, indicated in Figure 8-7 by the shaded area, is equal to the subsidy per unit times $(S_2 - D_2)$, the quantity of exports.

U.S. Export Subsidies

Since the late 1980s, the United States has used two export subsidy programs for selected agricultural commodities: the Export Enhancement Program and marketing loans. The goal of these programs was to augment the U.S. market share for exports of selected commodities.

Under the **Export Enhancement Program (EEP)** certain importing countries were targeted as important markets. Exporters to those countries negotiate the terms of sale with a buyer, and the government "enhanced" the transaction by augmenting the quantity shipped from government reserves. The effect was to provide a subsidy equal to the value of the commodity supplied by the government. The highest level of supports occurred in 1994 when they were over $1 billion. By 1997 they had diminished to $121 million, mostly on dairy products and have since dropped to zero.

The European Union provides substantially more export subsidies than any other nation. They reached nearly $15 billion in the late 1980s but had fallen to $7 billion by 1997, less than $1 billion by 2000 and are expected to continue to fall under pressure from other nations in trade negotiations.

The **marketing loan program** allows U.S. exporters of rice and cotton to borrow from the government to finance exports. The subsidy takes place in the repayment provisions. That is, the amount of the loan to exporters is determined by the U.S. domestic support price of the commodity, but the repayment is based on the international price. For example, suppose the support price of rice was $4 per kilogram and the international price was $3. A rice exporter can borrow $4 but has to repay only $3. In this example, the exporter would get a subsidy of $1 for every kilogram of rice exported.

Effects of an Import Quota

The previous two trade policies influence the quantity of a good traded by altering the relationship between the domestic and world price. An **import quota** is a limit placed on the quantity of a good that can be imported, regardless of the price. Whereas the previous two policies modified the domestic/world price relationship, the import quota severs it.

Figure 8-9 demonstrates the effect of an import quota. When the quota is instituted, there is an initial shortage. The shortage generates higher domestic prices for both imports and domestic goods. Eventually, prices reach equilibrium at P_d, the price at which domestic consumption is just equal to production plus imports under the quota.

What has happened to the five variables affected by trade policy?

Domestic production	⇑	because the domestic price is up
Domestic consumption	⇓	because the domestic price is up
Imports	⇓	because of the import quota
Domestic price	⇑	because the import quota limited the amount of the good available in the domestic market
Government revenue	0	no significant change (administrative and enforcement costs only)

The effects of an import quota are similar to those of a tariff, with one important exception. With a tariff, the government collects tariff revenues equal to the difference

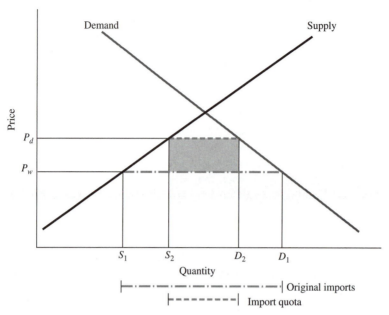

FIGURE 8-9   The effect of an import quota.

between the domestic price and the world price (the per unit tariff) times the quantity of imports. However, with an import quota the government does not receive the revenues, equal to the dotted rectangle in Figure 8-8. These revenues accrue to the individuals or firms who are given the right to import the quota amount.

A common mechanism for implementing a quota is for the government to issue certificates giving the right to sell a specific amount of the imported good in the domestic market. The holder purchases the good abroad at the world price and sells it at the higher domestic price, earning the revenue identified by the dotted area in Figure 8-9. This revenue, the difference between the domestic price and the world price times the quota amount, is called the **quota rent.** It is revenue that is received because of the quota and will be earned as long as the quota is in effect.

Effects of the U.S. Sugar Quota Policy

The United States had an import quota on sugar until the Uruguay Round of world trade negotiations, when it was converted to a tariff-rate quota (discussed below). Let's examine the effects of the sugar quota, before the change. The goal of sugar policy was to keep the U.S. domestic sugar price at about 20 cents per pound. Each year government officials would examine domestic demand and supply and choose a quota amount so that the quantity supplied at 20 cents plus the quota equaled demand at 20 cents.

The world sugar market is volatile; but for many years, the world price was about 10 cents per pound. With transport costs of about 2 cents per pound, imported sugar would cost about 12 cents per pound. With a quota of 2 billion pounds the quota rent is

$(0.20 - 0.12) \times 2,000,000,000 = \$160$ million. The U.S. sugar quota (and the associated quota rent) is given to governments in selected sugar-producing countries who pass that quota on to their exporters. It is important to note that the cost to U.S. consumers and the value of this protection to U.S. sugar producers is much larger than the quota rent because the producers receive the higher domestic price (instead of the world price) on all their production.

To conform to world trade rules from the Uruguay Round of world trade negotiations, the United States converted its sugar quota to a tariff-rate-quota system. In place of banning above-quota sugar imports, a very high tariff was imposed on imports over the quota. This tariff was set so high that it is unprofitable to import more sugar than the quota. However, under the agreement, the tariff will be gradually reduced. At some point it will become profitable to import above-quota sugar. In addition, the North American Free Trade Agreement (NAFTA) allowed Mexican sugar imports at a tariff rate lower than for other countries. This lower tariff has resulted in Mexican sugar exports to the United States. Under these agreements, the U.S. sugar industry will receive less protection from international producers. The future of support for the sugar industry is uncertain.

U.S. AGRICULTURAL TRADE AND ECONOMIC DEVELOPMENT

The last half of the twentieth century saw rapid growth in international trade in agricultural commodities, driven by technical change in agriculture and economic growth in the developing economies. The range of comparative advantage in agricultural commodities of some countries has been expanded, and this increases the potential benefits that can be achieved through trade. Economic growth combined with rapid population growth augmented the demand for goods and services including food and feed.

International trade and trade policy have had an important role in the shaping of U.S. agriculture. And, agriculture has been central in the performance of the U.S. economy in international markets. Early in the eighteenth century, U.S. foreign trade began a rise that has continued to this day. For much of the nineteenth century, agricultural products dominated U.S. exports. Cotton dominated international trade until the late 1800s; and cotton, tobacco, wheat, and wheat flour typically made up two-thirds or more of U.S. exports. It was not until 1890 that exports of manufactured products came to assume an important role. In the so-called golden age of agriculture, from the mid-1890s to the beginning of World War I in 1914, the steady growth in industrial employment and in exports of U.S. farm products undergirded the prosperity of agriculture. During the war and for about a year after that, exports increased even more.

Beginning in the 1920s and compounded by the Great Depression of the 1930s, U.S. (and world) agriculture suffered from the collapse of export demand. Conditions were reversed during World War II and the Korean conflict, with high demand for American food exports until about 1948. The farm sector suffered again from weak export demand during the 1950s and 1960s.

Another period of boom and bust in agricultural trade came in the period from 1970 to the early years of the 21st century. Several factors combined in the 1970s to produce record growth of U.S. exports. Population and incomes continued to grow in developing countries, generating an increase in demand. At the same time there were crop failures in several foreign countries, especially the Soviet Union. And, there was a substantial drop in the value of the dollar, making U.S. exports less expensive to foreign purchasers. The resulting rise in the real value of agricultural exports between 1972 and 1977 was about 75 percent.

The experience of the 1980s, however, was different. Economic growth slowed in the developing countries. A substantial increase in the value of the U.S. dollar made U.S. agricultural exports more expensive, and U.S. price supports tended to overprice grain in most export markets. As a result, U.S. agricultural exports declined sharply in the early 1980s before recovering again in the late 1980s.

In the 1990s, agriculture contributed about 20 percent of U.S. exports, and agricultural exports have used about 30 percent of U.S. farm production. Although agriculture no longer dominates U.S. trade as it did in the 1800s, it is still a major component. And as discussed above, the agricultural export market is crucial for many agricultural producers.

Agricultural Trade, Government Policy, and Development

A major determinant of world economic growth, and therefore U.S. agricultural performance, is the extent to which trade producers and consumers are allowed to adjust to exploit comparative advantage through trade. Yet we observe a curious phenomenon if we compare the structure of policies affecting agricultural trade in the DCs (developed countries) and LDCs (less-developed countries). In the DCs, agricultural price and income policies subsidize the farm sector. Tariffs, quotas, and price supports in the DCs protect their farmers from foreign competition and thus restrict the growth of international agricultural trade, including exports from the DCs. At one extreme, Japan heavily subsidizes its agriculture by imposing a combination of tariffs and nontariff barriers (NTBs) on food imports, by giving tax breaks to farmers, and by encouraging development and growth of agribusinesses that supply variable inputs to farmers at favorable prices. Estimates by the Organization for Economic Cooperation and Development (OECD) and the United States Department of Agriculture (USDA) indicate that Japanese farmers have received as much as 80 percent of their gross income from such policies and programs. In the 1980s, government programs and subsidies accounted for some 60 percent of gross farm income in the European Economic Community (EC) and some 30 to 40 percent in the United States and Canada. (Australia subsidizes its agriculture to a much smaller extent, and New Zealand has eliminated almost all agricultural subsidies.)

At the same time, economic policies in developing economies (including the United States in its early years as a nation) generally tax agriculture. An understanding of this phenomenon of a shift from taxation of to protection of agriculture as an economy develops can help to explain why international trade negotiations have excluded agriculture until recently. It also helps to predict what the chances are for agricultural trade liberalization.

To explain the phenomenon of changing agricultural protection, we start with the obvious. You can eat only so much. Many very poor individuals want to eat more than their budget constraints will buy. For poor countries, this means demand for agricultural products grows as income rises. But even for moderately affluent individuals, an increase in income has little effect on the amount of food consumed. In other words, the income elasticity of food demand eventually approaches zero as income increases. In countries where most consumers are affluent, aggregate demand for food does not increase much as incomes rise. Consequently, as a country develops, resources are shifted out of farming into production of industrial goods and services. For example, in Nepal, Bangladesh, and Zambia, the share of the labor force in farming is well over 50 percent, while in the United States and Germany it is below 5 percent. Similarly, the share of gross domestic product (GDP) arising in agriculture falls as a country develops.

If a poor country wants to collect taxes to finance development activities, the farm sector is an obvious target; it has most of the resources of the economy, and its members are widely scattered and have difficulty coordinating their efforts to bring about policy change. For a rich country, on the other hand, farming uses a small share of the economy's resources. Furthermore, farmers are a smaller proportion of society and therefore less costly to subsidize, and farmers and agribusiness leaders have developed effective political action organizations. Also, many nonfarm, rural residents, and agribusinesses share the farmers' political goals.

In many democracies, the allocation of elected representatives also gives farmers and other rural residents political influence greater than their numbers. In the United States, for example, every state has two U.S. senators, regardless of the population of the state. This means that states with low populations and high percentages of rural residents have just as much influence in the Senate as states with large and more urban populations. These factors help to account for higher levels of protection to agriculture as a country develops. They also help to explain why agricultural trade has been difficult to liberalize, even though agriculture in the DCs, and world agriculture in general, has much to gain from freer trade.

Early in 2001 the leaders of all democracies in North, Central, and South America signed an agreement to work toward the creation of the Free Trade Agreement of the Americas. The target date is initially set for 2005.

Agricultural Trade and the WTO

After World War II, representatives of the major developed countries met in Bretton Woods, New Hampshire, to create international institutions to ensure that the world economy would not see a repeat of the chaos of the Great Depression. Three bodies were initially proposed: the International Monetary Fund (IMF) to deal with ensuring easy exchange of national currencies, the World Bank to provide long-term loans to help countries recover from the war, and the International Trade Organization (ITO) to foster the development of more open international trade. The United States objected to the ITO, and it was not created. Instead twenty-three nations, including the United States, signed an agreement to set up a secretariat based in Geneva, collectively called the General Agreement on Tariffs and Trade (GATT), to supervise periodic multilateral trade negotiations. The GATT provided a forum for nations to work together to reduce or eliminate protectionist policies. Since World War II, major gains have been made in reducing the tariffs and NTBs to industrial trade among nations. These reductions were accomplished largely through international negotiations carried out by the GATT and its successor organization, the World Trade Organization (WTO). The general results of the

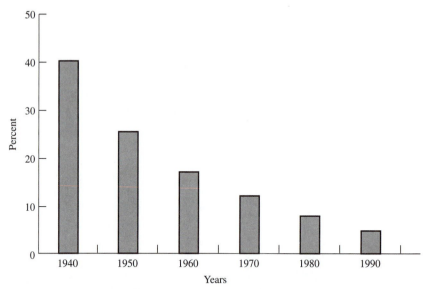

FIGURE 8-10   Average tariff rates in industrial nations. (*The Economist,* September 22, 1990.)

negotiations are impressive indeed. As Figure 8-10 indicates, since World War II, tariffs in the industrial nations have been reduced from 40 percent to less than 10 percent.

The WTO came into existence on January 1, 1995. Whereas GATT dealt mainly with trade in goods, the WTO and its agreements were expanded to cover trade in services and intellectual property. Some member countries would like to add international rules on investment as well.

The Uruguay Round   In 1986, the first meeting of the eighth round of negotiations under the GATT began in Punta del Este, Uruguay. These negotiations are known as the Uruguay Round. Before the Uruguay Round, agricultural trade had not been included. Originally this exclusion was requested by the United States and by many other countries to protect their agricultural sectors. In the mid 1980s, due to export competition from Europe, the United States took the position that agricultural trade should be included in the negotiations. The United States recognized its comparative advantage in the production of most temperate-climate crops. Also, several other major exporting nations exerted significant international political influence to include agriculture in the negotiations. The initial U.S. proposal called for the complete elimination of trade barriers and domestic subsidies to agriculture.

The Uruguay Round culminated, after long and difficult negotiations, with the new WTO agreement in 1994, which was subsequently ratified after divisive debate by the U.S. Congress and by other governments. The agreement puts in place a continuation of the reduction of average industrial tariffs to a new low of 3.5 percent. For the first time an agreement on agriculture was reached. Significant reductions in agricultural trade protection are being phased in over a period of years. Agreement was reached in four areas: *export subsidies, market access, internal supports,* and *sanitary and phytosanitary restrictions* (phyto is a term meaning plant, flora, and vegetation).

The *export subsidy* provisions require the reduction of these subsidies. The largest reductions are required in the European Union because it had the highest subsidies of agricultural exports. For example, in 1990 the EU spent about $25 in export subsidies for every $100 of exports while U.S. subsidies were $1 per $100.

The *market access* provisions require that all non-tariff barriers, such as quotas or bans, eventually be converted to tariffs. Initially quantitative barriers are to be converted to *tariff-rate quotas.* Imports under the quota face a low tariff, but additional imports have a high tariff. Over time, the quota is increased and the high tariff is reduced. One of the most difficult areas in this phase of the negotiations was the gradual lifting of the Japanese and Korean rice import bans. The United States will gain access to this market as well as to the markets for beef, cotton, sugar, peanuts, and dairy products.

Internal support provisions affect policies such as the U.S. deficiency payments and the EU's price support program. These policies increase farm incomes. Countries are required to lower their total internal support by 20 percent from a 1986-1988 base period. For the United States this provision will have relatively little effect, but in the EU, Japan, and Korea, substantial changes will be required.

The *sanitary and phytosanitary* provisions are intended to ensure that imported agricultural products are chemically safe, and free of pests. Some countries have used these measures to restrict trade. Under the agreement only restrictions based on scientifically valid concerns may be imposed.

DIFFERENTIAL IMPACTS OF TRADE LIBERALIZATION

A study by Hertel, et al. suggested that under the GATT the world trade volume will increase by $356 billion from 1992 to 2005. The impact on the rate of production growth will be dramatic in some countries. For example, agricultural production will fall significantly in Korea, Malaysia and the Philippines while textile production will increase dramatically in Korea and processed food production will expand greatly in the Philippines. While the negative effect on agricultural production was not great on a percentage basis for Japan and the European Union, it was large enough to stimulate strong opposition to the agreement by farmers in those countries. In contrast, while the percentage increase in agricultural production is expected to be small in the U.S. and Canada, the absolute magnitude was large enough to lead to support by many U.S. farmers while the drop in textile production lead to opposition from that industry.

The same study estimated the gains in welfare due to the Uruguay Round to be very large for most regions. The largest gainers are the European Union, Japan, and the United States and Canada. On a percentage basis, Malaysia is among the biggest gainers. Countries in Latin America and Sub-Saharan Africa show a small loss.

At the end of the Uruguay meetings a new round of trade liberalization discussions was agreed upon, to start before the end of 1999. The kick off meeting for these negotiations was held in Seattle. Thousands of protesters stormed the streets and some smashed windows. Negotiators failed to make any progress toward setting up a framework for future discussions. Since that time talks about agricultural negotiations have continued but the future of such negotiations remains uncertain.

SUMMARY

In this chapter, we have broadened our discussion of markets to the world level in response to the general question: Why do nations trade? We explained how international trade changes domestic consumption, production, and price of a commodity. We then showed why nations gain from international trade. We described the concept of comparative advantage, and showed that differences in resource endowments and production technologies mean that all countries can benefit from trade. When trade is unfettered by government intervention, domestic and world prices are the same. If a government is dissatisfied with the market outcomes that result, it may choose from a variety of trade policies to alter the domestic market equilibrium. The effects of three trade policy instruments—tariffs, export subsidies, and import quotas—were discussed.

Finally, we examined the history of U.S. agricultural trade, briefly considering the potential for liberalizing trade and some of the limits that have been placed on it. Expanding world trade may be essential for the continued prosperity of agriculture, especially among the large, developed trading nations. Movement to freer trade is still a major challenge to be faced in this century.

IMPORTANT TERMS AND CONCEPTS

absolute advantage   190
adjustment costs   196
ad valorem tariff   198
closed economy   187
comparative advantage   190
consumption possibilities frontier   192
exchange rate   186
Export Enhancement Program   201
export subsidy   199
exports   187
General Agreement on Tariffs and
  Trade (GATT)   205
import quota   201
imports   188
international trade   185
market access   207

marketing loan program   201
open economy   189
production possibilities frontier   191
quota rent   202
small-country assumption   186
social indifference curves   192
specific tariff   198
strategic goods   196
tariff   198
terms of trade   193
trade policies   198
unfair competition   196
Uruguay Round   206
world market   186
world price   186
World Trade Organization (WTO)   205

QUESTIONS AND EXERCISES

Name That Term

Read the following sentences carefully and fill in the missing term or terms.

1 _____ are domestically produced goods that are sold in the markets of a foreign country.
2 _____ are goods and services purchased by businesses, individuals, or governments from a foreign country.

3 A _____ is a tax imposed on an imported good.

4 An export _____ is a payment by a government to encourage the exporting of domestically produced goods and services.

5 An import _____ is a limit on the amount of a good or service that can be brought into a country.

6 _____ advantage is the ability of one country to produce a good more efficiently than another country.

7 _____ advantage in the production of a good or service means that a country can produce that good with relatively greater productive efficiency.

8 An _____ tariff is imposed as a percentage of the value of the imported product.

9 A _____ tariff is imposed as a fixed amount per unit imported.

10 The amount of one good or service that must be given up to obtain one unit of another good or service is called the _____.

11 The costs that are incurred when a firm or nation shifts from one equilibrium situation, or policy, to another are _____ costs.

12 An _____ economy allows international trade.

13 _____ is the sale and purchase of goods across national boundaries.

14 _____ goods, such as those used in national defense, play an important role in the survival or well-being of a nation.

True/False

Read the following sentences, then decide whether each statement is true (T) or false (F) and mark it accordingly.

T F 1 A country with an absolute advantage but not a comparative advantage in producing good X will export good X.

T F 2 The U.S. farm sector tends to be prosperous in periods of expanding agricultural exports.

T F 3 Freer trade benefits all consumers and producers.

T F 4 What a nation has (its resource endowments) and how it uses what it has (its production technology) determine its comparative advantage.

T F 5 Since the United States is a relatively small factor in the world for most agricultural commodities, changes in U.S. production and policies have little effect on the world prices in the long run.

T F 6 In an open market, the equilibrium domestic price for a good must be equal to the world price.

T F 7 A closed economy whose domestic price for a good is above the world price for that good would import the good if trade were permitted.

T F 8 An increase in the tariff on a good will cause the amount of the good imported to increase as exporters find they need to sell more units to keep profits high.

T F 9 An export subsidy for a good lowers the domestic price of that good by encouraging more domestic production.

T F 10 Imposing an import quota on a good will increase both domestic price and production of that good.

T F 11 A growing world population and continuing economic development mean increased demand for food and animal feed and the potential for expanding U.S. exports.

T F 12 While the early 1990s were a "boom" period for U.S. exports of agricultural commodities, the late 1990s were "bust" years.

Multiple-Choice Questions

Circle the letter of the response that best answers the question or completes the statement.

1 Imposing a tariff on an imported good causes the price of that good for domestic consumers to
   a increase.
   b remain the same.
   c decrease.
   d be uncertain.

2 An increase in domestic demand for an exported good causes the amount exported to
   a increase.
   b remain the same.
   c decrease.
   d be uncertain.

3 An export subsidy for a good causes the amount of that good exported to
   a increase.
   b remain the same.
   c decrease.
   d be uncertain

4 The terms of trade reflect the
   a exchange rate.
   b ratio at which nations will exchange two goods or services.
   c distribution of gains between producers and consumers.
   d distribution of gains from trade between trading nations.

5 Comparative advantage suggests that a good should be produced in the nation where
   a the money cost of production is the least.
   b the absolute cost in terms of resource use is least.
   c the production possibility frontier is furthest from the origin.
   d the cost is least in terms of alternative goods that might have been produced.

6 If two nations have straight-line production possibility frontiers
   a there will be a basis for trade only if their slopes differ.
   b there will be no basis for mutually advantageous trade.
   c there will be a basis for trade only if the frontiers cross.
   d there will be a basis for trade only if the nation's indifference curves are different.

7 Free trade based on comparative advantage is economically beneficial because
   a it promotes an efficient use of world resources.
   b it increases competition.
   c it provides consumers with a wider array of products.
   d all of the above.

8 A tariff is best described as
   a an excise tax on an exported good.
   b a government payment that allows producers to be more competitive in world markets.
   c an excise tax on an imported good.
   d a limit on the amount of a good that can be imported.

For the next four questions, refer to the following figure. P_3 is the world price.

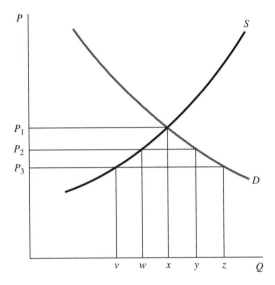

9 If the economy is closed to trade, the domestic price and quantity produced will be
  a P_1 and x.
  b P_2 and w.
  c P_3 and v.
  d P_2 and y.
  e P_3 and z.
10 If the economy is open to trade, the domestic price and quantity consumed will be
  a P_1 and x.
  b P_2 and w.
  c P_3 and v.
  d P_2 and y.
  e P_3 and z.
11 If a tariff of $P_2 - P_3$ is imposed, the domestic price and quantity produced will be
  a P_1 and x.
  b P_2 and w.
  c P_3 and v.
  d P_2 and y.
  e P_3 and z.
12 If a tariff of $P_2 - P_3$ is imposed, the amount of imports is
  a $z - v$.
  b $y - w$.
  c $z - w$.
  d $x - v$.
  e $z - x$.

Technical Training

The graph on the next page shows the aggregate domestic supply and demand for gadgets in the open economy of Oz. Use this graph to answer questions 1–6.

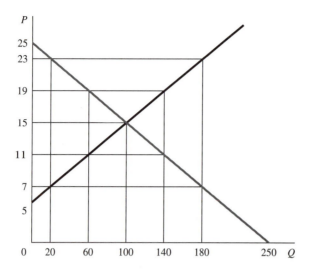

1 At a world price for gadgets of $7,
    a how many gadgets do the citizens of Oz consume?
    b how many gadgets are produced in Oz?
    c how many gadgets are imported into Oz?
2 Suppose the world price for gadgets is $7 and a new $4 specific tariff is imposed on gadgets. As a result of this tariff,
    a by how much does the price of gadgets increase?
    b how many fewer gadgets do the citizens of Oz consume?
    c how many more gadgets are produced domestically?
    d how many fewer gadgets are imported into Oz?
    e how much revenue does the tariff generate for the Oz government?
3 Suppose the world price for gadgets is $7 and the Oz government imposes a quota of 80 on gadget imports. As a result of this quota,
    a by how much does the price of gadgets increase?
    b how many fewer gadgets do the citizens of Oz consume?
    c how many more gadgets are produced domestically?
    d how many fewer gadgets are imported into Oz?
    e how much revenue does the quota generate for the Oz government?
4 At what world price does Oz neither import nor export gadgets?
5 At a world price for gadgets of $19,
    a how many gadgets do the citizens of Oz consume?
    b how many gadgets are produced in Oz?
    c how many gadgets are exported from Oz?
6 Suppose the world price for gadgets is $19 and a new $4 export subsidy for gadgets is offered. As a result of this export subsidy,
    a how much does the domestic price of gadgets change?
    b how much does the number of gadgets consumed by the citizens of Oz change?
    c how much does domestic production of gadgets change?
    d how many more gadgets are exported from Oz?
    e how much does the government of Oz spend on export subsidies?

Questions for Thought

1 People sometimes refer to the situation where a nation exports more than it imports as a "favorable" balance of trade. In what sense is it "favorable" to a nation to send away more things than it takes in?

2 Think about your hometown. What kinds of services does it import from other communities? Suppose that in order to protect local businesses from "unfair competition" from other communities, your hometown decided to destroy the roads leading to other communities. This policy, like a trade barrier, would increase the cost of importing goods from other places and protect local firms from outside competition. Would such a policy make economic sense? Why not?

3 Take stock of your personal endowments and technologies. In what activities do you have a comparative advantage? What do you "export" to the rest of the economy in exchange for your "imports"?

4 Suppose you were the manager of a firm. If your costs were below the world price for your product, would you favor freer world trade policies? How about if your costs were above the world price?

5 This chapter indicates that the export restraints on Japanese car manufacturers caused U.S. car buyers to pay about $500 (on average) more for a car in 1983 and 1984 or about $150,000 for each job created. Suppose U.S. car buyers could have gotten together and offered $100,000 for every autoworker willing to give up his or her job. Do you think anyone would have been willing to give up the job? Would the car buyers have been better off? Would the workers who gave up their jobs be better off?

6 How does an outward shift in the demand curve affect consumption in a closed economy? In an open economy?

7 What is absolute advantage? Are there circumstances in which a country will import a commodity in which it has an absolute advantage?

8 What is comparative advantage? Does it apply to all nations?

9 If the world can achieve its highest level of consumption through free trade, what prevents any or all nations from having free trade?

10 Country X imports coffee. X's coffee demand curve shifts out because of population growth. What happens to domestic price, consumption, production, imports and exports, and government revenues if the country has an import tax or an import quota? What happens to those five variables with an export tax and an import subsidy?

ANSWERS AND HINTS

Name That Term   **1.** Exports; **2.** Imports; **3.** tariff; **4.** subsidy; **5.** quota; **6.** Absolute; **7.** Comparative; **8.** ad valorem; **9.** specific; **10.** terms of trade; **11.** adjustment; **12.** open; **13.** International trade; **14.** Strategic

True/False   **1.** F; **2.** T; **3.** F; **4.** T; **5.** T; **6.** T; **7.** T; **8.** F; **9.** F; **10.** T; **11.** T; **12.** T

Multiple Choice   **1.** a; **2.** c; **3.** a; **4.** b; **5.** d; **6.** a; **7.** d; **8.** c; **9.** a; **10.** e; **11.** b; **12.** b

Technical Training   **1.** a. 180 b. 20 c. 160; **2.** a. $4 b. 40 c. 40 d. 80 e. $320; **3.** a. $4 b. 40 c. 40 d. 80 e. $0; **4.** $15; **5.** a. 60 b. 140 c. 80; **6.** a. increases $4 b. decreases 40 c. increases 40 d. 80 e. $640.

9

AGRIBUSINESS ORGANIZATION, MANAGEMENT, AND FINANCE

CHAPTER OUTLINE

OVERVIEW

This chapter expands the discussion of the forms of market competition, which were introduced in Chapter 7, by focusing on agribusiness organization, management, and finance. It also examines the significant role played by government in agribusiness as it provides information, sets grades and standards, and develops basic economic infrastructure. We show that firms create utility by changing the form of products, moving the products to a desired location, or holding the product until an appropriate time, that is, by creating form, space, and time utility. The accounting systems and measures of performance used by firms to assess the financial condition are briefly discussed. Many of these firms operate in markets that more closely approximate monopolistic competition or oligopoly than pure competition, and we give several important examples from the agribusiness sector.

LEARNING OBJECTIVES

This chapter will help you:

- Learn the basic principles of managing an agribusiness
- Improve your understanding of the production and marketing strategies of agribusiness as influenced by the economics of imperfect competition
- Broaden your knowledge of agribusiness, its organization and degree of concentration, and how it determines price and output
- Understand the measures of financial performance used by managers and investors

AGRIBUSINESS ORGANIZATION

As discussed in Chapter 1, the three major components of the agribusiness sector are as follows:

1 the input sector provides inputs to the farm sector
2 the marketing and processing sector moves farm products to the final consumer
3 the public sector interfaces with private firms.

In this chapter we focus on the management and operations of input and marketing and processing businesses. While these are much larger and more complex you will learn that the basic economic principles apply.

Input Sector

The farm-input component of the agribusiness sector consists of the firms that produce machinery, chemicals, fertilizer, capital, seed, and many other inputs and transport them to farmers. It includes many small firms such as local banks, soil testing services, short-line equipment manufacturers (firms that produce a small number of specialized equipment items), feed producers, and distribution firms. A smaller number of large firms also operate in this component of the sector, such as the major machinery manufacturers and the chemical companies that sell pesticides and fertilizers. It also includes significant portions of corporations that serve a broader market, such as automobile and truck manufacturers and the suppliers of energy. This component of the sector receives, indirectly, about 15 percent of the expenditures by consumers for food and other agricultural products.

Processing and Marketing Sector

The processing and marketing component of the agribusiness sector consists of the firms that buy farm products, process them using other factor services, and sell final products to consumers, both domestic and foreign. For example, a transportation firm moves wheat to a country elevator where it is dried and stored until it is shipped. A broker purchases the wheat and contracts with a shipping firm to move it to a central elevator. Then a firm that produces flour for sale to a bakery purchases it. The bakery ships the bread to a grocery wholesaler, which distributes it to numerous retail grocery stores. This example indicates that many firms are involved in the shipping, storage, and processing of the wheat produced by the farmer into the bread we consume. The marketing and processing sector, in total, receives approximately 75 percent of the consumer's food dollar. Figure 9-1 shows how the revenues are divided among the major expenditures. Labor accounts for approximately one-half of the total expenditures. In 1997 labor accounted for $216.6 billion of the total $441.4 billion expended by the processing and marketing sector.

FIGURE 9-1   Components of the Food Marketing Bill, 1997.

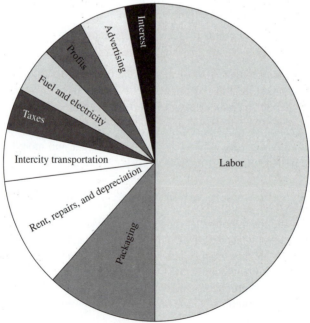

FARMERS' MARKETS

An interesting, and in some countries important, form of marketing is the **direct market,** sometimes called a farmers' market, free market, or bazaar. In these markets the *farmers, or their agents, bring their products to a central location where consumers congregate to make purchases.* In many of these markets prices are not posted; the buyer and seller negotiate the price for each transaction. Prices can change quickly in response to market supply and demand conditions. In these markets the costs of searching, the lack of standardization of products, and the cost in time of bargaining are higher than in more structured markets that regularly offer a wide range of products at a fixed price. Generally, as an economy develops, the relative importance of these less-structured markets declines. However, the weekend farmers' markets that serve a small segment of the population are becoming more common in the United States.

Important characteristics of direct markets are (1) a central location, (2) a fixed schedule, (3) open access to both sellers and buyers, (4) lack of standardization, (5) and flexible prices. Generally, all sales are final—there are no guarantees.

Given the number of firms and processes involved to produce a product as simple as bread, it is not surprising that the processing and marketing sector of the agribusiness industry is very large. The approximately half of the consumer's food dollar that goes to the marketing and processing sector is for labor, which accounts for about 35 cents of the consumer's total food dollar. Processing and marketing alone provided employment for more than 12 million individuals. Other important expenditures are for packaging, rent, repairs, depreciation, advertising, and intercity transportation. Another indicator of the size of this sector is the fact that the sales of just the twenty-five largest restaurant chains in the United States are over $45 billion annually.

Public Sector

The public sector—federal, state, and local governments—plays several important albeit less apparent roles in the agricultural and food economy. It establishes the rules and regulations under which the economic system functions, including enforcing contracts and property rights. It builds or encourages the development of the infrastructure used by the private firms, including roads, railroads, harbors, and communication networks. The public sector also provides market information, sets grades and standards, inspects agricultural commodities, determines agricultural policy, and supports conservation programs. Finally, the public sector, through the various levels of government, provides support for public research, education, and extension activities that develop knowledge and enhance the productivity of the human capital involved in agriculture.

ALTERNATIVE BUSINESS FORMS

In this chapter we focus on the private firms in the agribusiness sector. There are three basic forms of business organization in the private enterprise system: the individually owned business, the partnership, and the corporation. Most corporations are investor-owned. Some businesses are patron-owned and are called cooperatives.

TABLE 9-1   ALTERNATIVE BUSINESS STRUCTURES UNDER PRIVATE ENTERPRISE

	Types of business			
			Corporation	
Features	Individual	Partnership	Investor-owned	Cooperative
Owner	Individual	Partners	Stockholders	Member-patrons
Users	Nonowner/ customers	Nonowner/ customers	Nonowner/ customers	Chiefly owner/ patrons
Policy establishment	Individual	Partners	Common stock-holders and directors	Member-patrons and directors
Voting rules	Not necessary	Usually by partners' share of investment	By shares of common stock	Usually one member, one vote
Earnings	Individual	Partners by share of investment	Stockholders in shares held	Patrons in proportion to patronage

A typical growth pattern is for a firm owned by an individual is to form a partnership when a firm expands beyond the ability of one individual or family to provide the financial and management resources required. If growth continues, the firm may incorporate to have access to the major financial markets and to take advantage of the corporate tax codes. The characteristics of these forms of business organization are given in Table 9-1.

Individual and Partnership Ownership   Although there are notable exceptions (for example, the Cargill Company is one of the largest grain companies in the world, but it was until recently a family-owned business), most *individually owned businesses* are relatively small. Some farms and ranches are incorporated (often with the farm family retaining control), but most are controlled by individuals or **partnerships.** As noted in earlier chapters, purely competitive markets are made up of many small firms, usually individually owned and controlled. As indicated in Table 9-1, the ownership, control, and earnings, if any, are in the hands of the individual or individuals who own the business. The model of decision making implied in Chapters 3 and 4 approximates the experience in these small firms since the owner makes the purchasing, production, finance, and marketing decisions. We will find a quite different management model in those agribusinesses organized as corporations, even though profit maximization remains a fundamental goal.

Corporations and Cooperatives   Although there are many individual proprietorships in the input and marketing subsectors of the agribusiness sector, the predominant forms of organization are the corporation and the cooperative. Ownership is the basic distinction between corporations and cooperatives. **Corporations** are owned by stockowners, individual and institutional investors, most of whom provide no input into operating decisions. Shareholders may or may not purchase products produced by the corporation. The patrons of the **cooperative,** on the other hand, are its owners. Thus the primary customers of the cooperative also set its policies, and receive the income it earns. However, the management structures of corporations and cooperatives, discussed in the next section, generally are quite similar.

The cooperative is an important form of business organization in agriculture. A major objective of the federal act that created the rules under which cooperatives may form (the 1922 Capper-Volstead Act) was to provide market competition to the investor-owned corporations selling to or buying from farmers. Farmers in some markets felt that they were being exploited, and this legislation was one means of overcoming such exploitation. The establishment of a cooperative increased the number of competitors in a market. Also, if a cooperative achieved a position of market power, any earnings were paid to the owner patrons. Often several local cooperatives are organized into a larger cooperative to compete more aggressively against the large investor-owned agribusinesses operating in the market.

Farmers' cooperatives have approximately a 30 percent share of the market both in processing-marketing services and in farm service supply. Over the last thirty years cooperatives operating in the input sector have captured an increasing share of the market. For example, the cooperatives selling fertilizers, petroleum, and farm chemicals have increased more than 10 percent over the period and are now in control of 30 to 40 percent of the market. The cooperatives' share of petroleum sales to farmers has reached an all-time high of 50 percent in 1998.

The share of market held by processing and marketing cooperatives has varied over time and commodity. They market approximately 85 percent of milk, 40 percent of grain, oilseeds, cotton, and cottonseed, 20 percent of fruits and vegetables and 15 percent of livestock.

For grain, marketing cooperatives market share fell from a high of nearly 40 percent to about 30 percent, but has returned to about 40 percent. For livestock, there was a decline from about 15 percent to less than 10 percent but a subsequent return to the 15 percent level. The cooperative share of the dairy market has risen from about 60 percent to about 85 percent, and for cotton the increase has been from 20 to 40 percent over the same thirty-year plus period. Thus, the cooperative is an important, and in some markets growing, form of business organization.

Cooperative Bargaining Associations   Most of the farmers' cooperatives in the United States were organized to operate as a business. However, some cooperatives operate as collective bargaining associations seeking to improve farmer's prices and terms of trade. Federal law allows farmers to organize bargaining associations under state statutes. Also, in some instances, several cooperatives work together on their farmer-owners' behalf—which is allowed under the 1922 Capper-Volstead Act providing there is no violation of antitrust laws.

The primary purpose of these farmers' cooperatives is to serve as **bargaining associations,** and nearly all bargaining associations are organized as cooperatives. These have been most active in fruits, vegetables, dairy products, sugar beets, and some specialty crop markets. A few have been active among poultry producers. Generally, cooperative farmers' associations only attain their goal of substantially higher prices for their products if they control a large enough proportion of the available supply to force buyers to bargain with them. If successful, the farmers' association no longer faces a perfectly elastic demand curve, and the cooperative may be able to increase revenues by offering a smaller quantity and receiving a higher price. Fruit farmers in California and other states have used such bargaining associations to get higher prices for their commodities than they would have realized in competitive markets. Prices of some vegetables and nuts have also been raised and stabilized under such marketing orders.

Efforts to develop large regional or national bargaining associations in grains and livestock, such as the attempts of the American Farm Bureau Federation (AFBF) in the early 1920s and the National Farmers Organization (NFO) in the late 1950s, have had limited success. In these instances, the farmers and their organizations could not gain control of enough supply to force buyers to bargain with them collectively. Generally, efforts to organize bargaining associations nationally have been discontinued after some activity. However, the cooperatives of the AFBF have gained prominence both regionally and nationally, and the Farmers Union grain cooperatives are important in some grain-producing areas of the United States.

Marketing associations commonly use **market segmentation,** *control of the quantity of product being placed on the several markets,* such as the fresh, processed, and foreign markets to increase returns. This is illustrated in Figure 9-2. The fresh market segment is characterized by quite inelastic demand, the processed market is rather inelastic and the export market is perfectly elastic. If the industry is made up of competitive firms all of the product will be sold at the export price as suggested in Panel *d* of Figure 9-2. In this case the inelastic portions of the demand curve are not exploited by the industry.

By segmenting the market and allocating one-third of the product to the fresh sector as indicated in Panel *a*, one-third of the product to the processed sector in Panel *b*, and the remaining one-third sold at export, total revenue is higher. Producer cooperatives that market milk and many specialty crops such as fruit, vegetables, and nuts use market segmentation. It also occurs in other industries when firms charge different prices for essentially the same good. For example, utility firms charge residential customers higher rates than they charge industrial customers. Also, airline prices for first class, business class, tourist, and frequent flyers differ significantly, as do their fares from route to route.

FIGURE 9-2   Market segmentation.

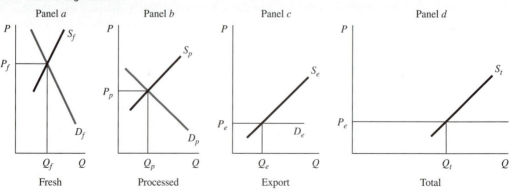

AGRIBUSINESS MANAGEMENT

So far we have presented the decision making in the firm as a relatively simplistic process of an individual deciding what and how much to produce and sell. This model approximates the decision process in family farms and in many small agribusinesses, but it does not capture the essence of the management structure of the large agribusiness firm. However, the same concepts of marginality, efficiency, and specialization versus diversification that are important to the small firm are also basic in the decisions of large, complex firms. The primary distinction is in the size and complexity of the large agribusiness firm.

Because the management of agribusiness is an important economic function, we introduce the principles of agribusiness management, the steps in the management process, and the roles of managers. The management activities of an agribusiness firm—sometimes with thousands of employees, hundreds of products, plants in several countries and markets in many others—are dramatically different from those of the classical family farm. Even in a business with a dozen or so employees, the management activity differs significantly from the process that occurs in a business with a single decision maker. An individual decision maker must secure information from many sources and become an expert in the physical aspects of the production of one or a few commodities. While the owner-operator is generally a price taker in both input and output markets, marketing decisions, such as timing of transactions, also must be made. Large agribusiness firms, in contrast, operate with numerous employees who specialize in various activities, thus requiring sophisticated communication and coordination functions. Although these firms seldom operate in purely competitive markets as defined earlier, they do compete aggressively with other firms.

The sole proprietor makes numerous decisions including what to produce, how much, what combination of inputs to use, when to sell, whether to borrow funds, and so on. In some cases the manager decides to "outsource" some of these decisions by employing others to perform some of these tasks. In a large agribusiness, no one individual can make all these important decisions. Dozens, or perhaps thousands, of people make them. The challenge is to develop and maintain an effective team of individuals, each with a specific task, but all working toward common objectives.

Goals of Agribusiness

Agreement on a set of common goals for the agribusiness is central to the management process. These goals—including profit, growth, market share, and social responsiveness—shape the firm's basic competitive strategy. Some short term goals, such as growth and increasing market share, may be pursued to increase profits in the long run. **Market share,** the percentage of total market sales controlled by the firm, is often an important short-term goal. The goal of social responsiveness may play several roles in the agribusiness. For example, it can buttress the profit goal by enhancing the image of the business (and thus increasing sales) and by improving the morale of the firm's employees. Perhaps more important than the identification of precisely "correct" goals is the articulation of a set of goals shared throughout the agribusiness. The lack of a shared set of goals can lead to individuals in the firm working at cross purposes.

The Global Environment

Agriculture has been a global enterprise since the first international transactions when spices, dyes, fabrics, other commodities were traded. However, the extent of globalization has dramatically increased, especially with the advent of the Internet and other forms of electronic commerce. A number of major changes have occurred.

1 There has been continued volatility and growth in the quantity of international trade, as discussed in the previous chapter.

2 Many of the major agribusiness corporations now operate globally. They operate production facilities in many countries and produce products tailored to culture-specific markets. This implies that these businesses have market opportunities around the globe—but they also face international competitors. Numerous mergers and acquisitions now occur across national borders.

3 The financial markets operate with almost no regard for national borders. Investments in new and existing businesses within most countries are dominated by investors from all other countries. Movements of funds can dramatically affect the fortunes of businesses in a country and even in an entire region, as evidenced by the Asian financial crisis of the late 1990s.

4 New products, be they inputs or consumer products, are being developed in many nations and in multinational research facilities rather than flowing almost exclusively from U.S. firms and institutions such as the universities.

Operating Functions of Agribusiness

A firm must develop an organizational structure to make decisions and carry out the activities of the business. The firm may establish divisions of purchasing, finance, research and development, production, personnel, marketing, and controlling. Each of these divisions has responsibilities suggested by its title and reports to the central management of the firm as it develops the operating policies to accomplish the firm's objectives. Other firms may organize around the products they produce. Each division of the agribusiness may have responsibility for one or several products. Still other firms may organize on a geographic basis, with divisions responsible for certain areas of the nation or of the world. Of course, some agribusinesses use a combination of several of these structures. Each firm must determine the structure that best fits its situation and be open to the possibility of changing the structure as the firm and its environment grow and change.

Four Functions of Management

There are numerous ways of developing an effective management team to accomplish the objectives of an agribusiness. The many views concerning this topic are evidenced by the constant stream of books and articles suggesting more effective management approaches. Movement has been from top-down directives, whereby individuals are assigned simple and specific tasks, toward more open, two-way communication and the assignment of teams with significant functions. A basic approach includes just four major functions: planning, organizing, controlling, and directing.

Planning is charting a course for the future, be it the next day, week, year, or decade. The planning process should include all the separate components of a business, and for the firm to be successful these components must come together into a unified plan.

Organizing is the process of transforming plans into action. Its function is to ensure that the work gets done efficiently and effectively, and in a way that will maintain good employee morale. To reduce the organizational activities to a manageable level, firms may divide the tasks to be accomplished by function (for example, sales, manufacturing, distribution, purchasing), by products (chemicals, feeds, consulting services), or by region (Midwest, Southeast, Europe, South America).

Controlling refers to management's assessment of the extent to which the firm is meeting the goals that have been established. Generally, each unit in a firm should assess the performance of the individuals or subunits reporting to it and communicate the results of these assessments to the next higher level of management. Clearly, the larger or more complex the firm is, the more complex the controlling function will be.

Directing is the activity of bringing the planning, organizing, and controlling together to accomplish the objectives of the firm. Management must coordinate the plan, the physical and human resources, and the controls into an effective team working to accomplish the objectives. Usually, the most critical resource is the employees. One of the strategies used by firms to develop capable managers is to change an individual's responsibilities over time to provide experience in several areas of management and in several of the firm's facilities, including locations in more than one country if the firm is a large international company.

Business Strategies

Given a set of goals and a structure, the firm must determine its *business strategy*. It may decide that the economic environment and its current market position make significant expansion a reasonable goal. In this case, the firm might move new products from its research and development division into production, requiring the opening of new manufacturing facilities and the purchase of additional inputs. Simultaneously, the sales and distribution activities will be expanded into target markets, perhaps by marketing internationally. The marketing department must develop a marketing strategy, including pricing and advertising strategies to meet its goal. For example, the introduction and rapid increase in popularity of the microwave oven resulted in the opportunity for food-processing firms to develop new food products for microwave use.

Another firm currently in a strong position may determine that the preferred strategy is to continue operating in the same general manner by marketing a well-accepted product. The firm might try to reduce the costs of production, improve product performance, or streamline its marketing activities.

In some cases, a firm will determine that purchasing or merging with another company that has complementary capabilities is the best way to develop a given activity. Clearly, the management of the firm faces a major challenge in coordinating all these planning activities to develop a coordinated, effective, and efficient plan.

Decision Making in the Firm

The firm managers must implement goals and strategies. Thus, a key to the success of the firm is employing individuals with the needed managerial skills and providing the training necessary for them to improve those skills. An effective member of the management team must be:

- Knowledgeable about the job responsibilities, the firm, its products, the market, and the broader economic environment in which it operates
- Effective and efficient, with the ability to reason and communicate
- Technically proficient in several areas of the firm's organization, be they functional areas—such as purchasing, marketing, and production—or product lines or geographic areas
- Able to make decisions that have a high probability of achieving the firm's short- and long-term goals
- Able to work in large and small teams, as well as individually

Individuals who are successful in the environment of an agribusiness, or in any other organization, have good **decision-making skills.** Generally, this involves attacking a problem in a systematic fashion. One approach is to follow these steps:

1 Specify the problem clearly in terms of the organization's goals
2 Specify all the alternatives, including the option of not responding to the problem
3 Evaluate each alternative in terms of the effects on the firm's goals
4 Choose the best alternative
5 Implement the alternative expeditiously
6 Evaluate the effect of the decision as early as possible and at reasonable intervals thereafter
7 Revise the decision if the alternative chosen does not achieve the established objectives, and review the decision-making process

TIME, SPACE, AND FORM UTILITY

Since consumers make purchasing decisions based on the utility provided by goods and services; the firm strives to offer a product that provides utility. There are three aspects of a product or service that can provide utility: time, space, and form. A product has value to the consumer only if it is available when, where, and in the form wanted. The consumer wants a loaf of fresh bread, a frozen and prepared meal, a pair of blue jeans, or a sumptuous dinner in a full-service restaurant. In the agribusiness sector, transformations from raw materials to finished products are made by the firms that provide inputs to farms and ranches, by the farms and ranches themselves, and by the numerous firms that move the products to the final consumers. The function of the agribusiness sector, and all economic activity, is creating utility in time, space, and form. The agribusiness input supply sector produces and delivers the nonland inputs needed for production to farmers. The marketing firms in the agribusiness sector process the agricultural commodities into food products and transport and sell them to the consumer. All this must happen in a timely fashion through the markets in time, space, and form as discussed in Chapter 5. Firms in the agribusiness sector, operating to maximize their profits, provide these activities and many others. We begin with a discussion of time utility.

Time Utility Agricultural products are consumed throughout the year but in temperate zones most crops are harvested just once during the year, usually six months or more after making the production decisions. Retail firms often purchase goods several months, or seasons, before they are offered to consumers. Forests are harvested after several years or decades of growth. Time is an important factor that must be considered explicitly in the development of every business plan.

All firms make purchasing and production decisions in anticipation of consumer decisions that will be made some time later. The farmer makes decisions before planting and must wait a full season before selling the harvested crop. In the intervening time, market conditions can change dramatically. At the beginning of the season, some farmers contract with buyers and set a price for their planned output. This avoids the risk of selling at a lower price when the crop is harvested but also forgoes the possibility of selling at a higher price. Futures markets were developed to enable grain merchandising firms, farmers, and other market participants to deal more effectively with the risk and uncertainty associated with future price changes. Grain-handling firms are an example of an agribusiness that creates utility by storing grain for sale at an appropriate time. These grain-handling firms also create utility by shipping grain from producing regions to domestic and foreign markets.

Timeliness is also an important consideration in the marketing plans of many firms. Some firms strive to develop a transportation-delivery system that enables the firm to deliver the product to the consumer rapidly. Of course, one of the major elements explaining the substantial growth of the "fast-food" industry is timeliness.

Space Utility Resources must be transported to the desired location to provide utility to the consumer. Agricultural commodities produced in rural areas must be transported to urban areas for sale to consumers. Ores can be mined only where they are found and must be shipped to where they are used. Coffee and bananas are produced in countries with warm climates and are transported around the world. Cotton may be produced in one country, shipped to a second for processing into cloth, shipped to a third to make clothing and shipped to other countries for sale. Thus, transporting raw materials and finished products is an important function provided by firms as they provide utility to consumers.

In many countries, this seemingly simple function of the economic system is a major constraint on the efficient use of basic resources in providing goods and services to consumers. Many underdeveloped countries do not have adequate systems of roads, bridges, trucks, railroads, harbors, barges, and airport facilities to move the products of agriculture from the producing areas to the cities where they are consumed or to the ports for exporting. In many countries, a high percentage of farm products is lost resulting from spoilage in inadequate transportation and storage systems. Likewise, it is difficult to move the raw materials needed to support production to the farms. As a result, fertilizers may be unavailable, or may be prohibitively expensive due to poor transportation and storage systems.

The development of a transportation system generally involves both the public and the private sectors. The public sector usually develops roads, bridges, railroads, and similar major facilities. The private sector usually develops the firms that own and operate the trucks, railroad equipment, ships, and so on. In most countries important parts of the transportation industry are regulated by government to prevent exploitation of users through monopoly or oligopoly power.

Form Utility   Most resources cannot be consumed in their natural state. A lump of coal, a barrel of crude oil, an acre of land, or a bag of fertilizer has little use to a consumer. Thus these goods have little value until they are transformed into goods and services valued by consumers or, in some cases, by industries. The economic system converts such raw materials into energy, food, and a multitude of other goods. **Form utility** is of particular relevance to the agribusiness-marketing firm because of the need to plan for the life cycles of the products. Sometimes the industries themselves experience life cycles, in which case the firms must produce a different product or go out of business.

SOLID WASTES

The conversion process does not end with the production of the goods and services. The generation of large quantities of scrap and human wastes is associated with consumption. When the many products—cars, computers, books, paper clips, clothing, and bottles—have served their function, they do not cease to exist. Some of this material can be reused, some can be recycled, but much of it is discarded. The choices on how to manage the waste stream involve both the market economy and the political sector. We return to this topic in Part Five.

Product Life Cycles

Some agribusiness firms deal with basic agricultural commodities—such as wheat, rice, and flour—which are relatively homogenous. In these cases the firm need not be concerned about product life cycles, even though there may be changes in the relative sales of commodities. For example, the consumption of chicken in the United States has increased substantially relative to red meats over the last several decades.

Most brand-labeled products, however, have a life cycle. The **product life cycle** has a major effect on the business plan. Over time, most industrial products, and even industries, move through four stages: initiation, growth, maturity, and decline.

Initiation   After a new research concept is transformed into a marketable product, the firm can move into production, develop and initiate a marketing plan, and commence operation. If a successful marketing effort is developed, the sales of the product grow. Initially, sales may be low, while the marketing channels are developing and as consumers become aware of the product and begin to buy it. If the product is successful, this stage is followed by the growth stage, in which the sales grow rapidly.

PRICING STRATEGIES

One crucial element of the firm's strategy is the price charged for the product. In markets with little or no competition, perhaps because of patent protection, the price will be determined by the estimated elasticity of the demand curve. If the demand is inelastic, a high price will maximize revenues; if the demand curve is elastic, a lower price will maximize revenue. Even with an elastic demand curve the firm might introduce the product at a high price and lower the price to increase sales as its production capacity increases.

In a monopolistically competitive market, a firm may charge a high price as part of a strategy to convince consumers that the product is high in quality, and the firm will expect a lower sales rate. Another firm may price a product near its production cost in an attempt to capture a larger share of the market. Sometimes a firm might sell a product at a price below the cost of production to capture a share of the market and then later increase the price in an attempt to maximize profits over the long run.

As the firm formulates its pricing strategy, it must consider the reactions of present and potential competitors. A high price that generates significant profits in the short run might entice numerous competitors and lead to smaller profits in the long run. The firm must assess the market elasticity and consider the other elements of the marketing strategy (advertising, for example) when developing the pricing plan.

Growth   During the initiation and growth stages, the firm's marketing effort attempts to identify the unique and desirable characteristics of the product, or to create the impression of uniqueness, through its marketing and advertising strategy. If this strategy is successful, the firm's product is differentiated from the products offered by competitors. The firm can expect to realize a larger share of the market sales during the remaining life of the product.

Maturity   The sales of the product begin to stabilize or plateau when the product reaches its full sales potential. Sales may remain at this level for some time in the mature phase of the product's life cycle. Eventually, however, most products go into a declining phase and are replaced in the marketplace by newer products. Most firms that have successfully marketed a product for a number of years must plan to update the product or phase it out in favor of another product.

Decline   Because of the short life cycles of some products and the constant change in economic conditions, a product may enter a stage of decline. In this stage, the firm will continue to monitor indicators of its performance, such as sales and market share. If the decline continues, the product will be discontinued.

Because they recognize the importance of product life cycles, food companies are quite aggressive in the introduction of new products. Approximately 10,000 new food products are introduced in American markets each year. Some of these introductions are the result of research and development efforts that produce new products. For example, the development of low-calorie sweeteners and fat substitutes generated many new products that are reformulations of existing products incorporating the new ingredient.

Sometimes firms must respond to a change in tastes and preferences. For example, research indicating that a food ingredient has adverse health effects results in firms developing new consumer products containing less of the ingredient. In such cases, the market is responding to make desirable products available to consumers.

Sometimes the only change from the old to the "new" product is the package. Often the firm changes a product in an attempt to differentiate it from another product. This may be done in conjunction with an advertising campaign, to gain an advantage over competitors. In these cases, while the change may be profitable for the firm, it does not necessarily represent an improvement in consumer welfare.

Changes in tastes and preferences, advances in technology, and advertising pressures produce consumer product life cycles. The firm must plan effectively to modify and change its product offerings over time. And, many new products fail to meet targeted sales goals and are taken off the market soon after introduction because of a lack of consumer acceptance.

Market Development

The marketing function of an agribusiness is often one of the most crucial to the development and growth of the firm. One goal of the firm is to sell more of its product in the existing markets, in other words, increase the firm's share of a market. In addition the firm will consider other strategies to increase sales. These include:

- Expanding markets. The firm might sell existing products in new geographical areas, be they a neighboring city or county or an international market. Or a firm may attempt to develop a market over the Internet. Another strategy is to expand into a new market segment; for example a producer of poultry feed might expand into the livestock feed market.
- Modifying or developing new products. A food manufacturer might add a low-fat line of products. Chemical firms are constantly seeking to develop new pesticides.

In some cases the firm modifies its product mix through a research and development activity and in other cases it acquires the product by purchasing or merging with another firm. The firm attempts to develop a mix of products that are complementary in both production and in marketing.

Each product in each market will face a product life cycle. Thus the firm may have products in the initiation phase in one market and in a declining phase in another. Clearly the firm must coordinate the marketing and production functions so that it is able to produce the products as they can be sold in the various markets. And the firm must have the financial resources available necessary to build the production facilities to produce existing, as well as modified and new, products and product lines.

FINANCIAL ANALYSIS OF AGRIBUSINESSES

The firm must develop a record system that allows it to determine profits and to assess the general performance of the company. We turn to a brief presentation of the basic record systems and the types of measures of financial performance used by firms.

Financial analysis depends on the maintenance of a system of accounts that record the firm's performance. The accounting system of the firm includes several financial statements that serve the following purposes:

- Indicate the firm's profits or losses
- Indicate the firm's current financial position
- Allow projection of the firm's future financial position
- Provide financial information for management, owners, and potential investors
- Provide records of past financial activity for use by owners, tax agents, and other reporting requirements

The two basic financial statements discussed here are the balance sheet and the income statement. As firms become larger, there is a need for more information, such as cash flow budgets, statements of change in financial position, and numerous records specific to the needs of the firm. Our purpose is not to explain in any detail the development of accounting systems or the financial management of the firm, tasks well beyond the space available here, but rather to make you aware of these important topics in management.

Balance Sheet

The **balance sheet** provides a "snapshot" of the firm's financial condition at a given point in time. In the balance sheet, the firm's **assets** (things of value owned by the firm such as cash, buildings, inventories, and goodwill) are in "balance" with the firm's **liabilities** (obligations owed to others) and **equity** (the value to the owners if all assets were sold and all debtors paid). In the hypothetical example of Lovely Green Acres, Inc. (Table 9-2), if the owners were to sell all the firm's assets, they would net $1,058,000 after paying all the outstanding obligations. If the firm's obligations were greater than its assets, so that the net worth of the firm was negative, the firm could be forced by creditors to file for bankruptcy.

The balance sheet generally includes more detailed information than is presented here, and is generally categorized to reflect the firm's management structure. For example, a firm organized into several product divisions would have a balance sheet for each division and a consolidated balance sheet for the entire firm.

Income Statement

The **income statement** (the profit and loss statement) summarizes the revenues and expenses of the firm over a given period. If revenues exceed expenses, the firm realizes a profit in that period. Again, the income statement for the hypothetical Lovely Green Acres, Inc. (Table 9-3) is not as detailed as would be generated for an actual corporation. The corporation would maintain detailed income statements for the various operating divisions and a consolidated statement for the corporation. Because most firms purchase and sell many dollars' worth of different commodities during the year, the purchase, sale, and inventory changes are shown in a separate section of the income statement.

TABLE 9-2   BALANCE SHEET FOR A HYPOTHETICAL FIRM

Balance Sheet
Lovely Green Acres, Inc.
September 30, 200___

Assets

Current assets:		
Cash	$ 25,000	
Accounts receivable	50,000	
Inventories	200,000	
Subtotal		$275,000
Intermediate assets:		
Vehicles	$ 75,000	
Machinery	175,000	
Notes receivable	60,000	
Subtotal		$310,000
Long-term assets:		
Land	$ 750,000	
Buildings	150,000	
Investment in subsidiary..................	100,000	
Goodwill	75,000	
Subtotal		$1,075,000
Total assets		$1,660,000

Liabilities and Equity

Current liabilities:		
Accounts payable	$ 25,000	
Taxes payable	75,000	
Installment payments due	12,000	
Subtotal		$112,000
Intermediate liabilities:		
Intermediate note payment	$ 140,000	
Deferred accounts payable	50,000	
Subtotal		$190,000
Long-term liabilities:		
Long-term notes payable	$ 300,000	
Subtotal		$300,000
Owner's equity or net worth	$1,058,000	
Subtotal		$1,058,000
Total liabilities and owner's equity		$1,660,000

Note that the ending inventory of $200,000 appears on both the income statement and the balance sheet. This example presents the following result: the firm realizes $220,000 of net income on nearly $4 million in sales. However, it would be earning approximately 20 percent return on the owner's equity in the firm of $1,058,000. Not surprisingly, there are numerous measures of the financial health of a firm. Some of these are presented in the following section.

TABLE 9-3   HYPOTHETICAL INCOME STATEMENT

Income Statement
Lovely Green Acres, Inc.
October 1, 200__, to September 30, 200__

Revenue from sales..............................		$3,950,000
Cost of goods sold:		
Beginning inventory	$ 236,000	
Goods purchased..............................	2,600,000	
Cost of goods available for sale........	$2,836,000	
Ending inventory	200,000	
Total cost of goods sold		$2,636,000
Gross margin		$1,314,000
Operating expenses:		
Salaries and wages	$ 646,000	
Interest...	55,000	
Insurance	82,000	
Rent	77,000	
Advertising	105,000	
Utilities and energy	21,000	
Depreciation......................................	31,000	
Miscellaneous	21,000	
Total operating expenses......................		$1,038,000
Income before taxes	276,000	
Allowance for taxes	56,000	
Net income		$220,000

Indicators of Firm Performance

In this discussion, we go beyond our earlier analysis of the firm's decision-making processes, which concentrated on the basic economic question of determining the appropriate combinations of inputs and output to maximize profits. Implicitly we assumed that the firm purchased and paid for inputs and then sold and received payment for products produced. In this chapter, we present a somewhat more complex and realistic picture of the financial records maintained by the firm and, in doing so, suggest that the measure of firm performance is more complex. At any point, the firm is making payments on borrowed funds, has inventories of inputs to be used in future production, has products to be sold later, owes suppliers for previous purchases, has accounts receivable, and owns assets of substantial value.

The magnitude of the firm's varying commitments in relation to its ability to meet those commitments determines its health. In some cases, the firm may be strong in several measures and vulnerable in others. In this situation, it is more difficult to determine how well the firm is performing. Conventional accounting practices provide a means of recording and assessing performance. Financial analysts use this information in managing the firm. Banks, other lenders, and investors rely on this information to assess the

TABLE 9-4   SELECTED INDICATORS OF FIRM FINANCIAL PERFORMANCE

Indicator	Definition	Indication
Rate of return on assets	After-tax income + interest paid ÷ average total assets	Higher => more profitable
Rate of return on equity	After-tax net income ÷ net worth	Higher => more profitable
Gross ratio	Total expenses ÷ gross revenue	Lower => more efficient
Turnover ratio	Gross revenue ÷ average total assets	Higher => more efficient
Current ratio	Current assets ÷ current liabilities	Higher => more liquid
Cash flow coverage ratio	Excess available cash ÷ cash required for interest and principal payments	Higher => more liquid
Debt-servicing ratio	Cash required for interest and principal payments ÷ total cash available	Higher => less liquid
Leverage ratio	Total debt ÷ total equity	Higher => less solvent
Net capital ratio	Total assets ÷ total liabilities	Higher => more solvent
Debt-to-asset ratio	Total liabilities ÷ total assets	Lower => more solvent
Interest coverage ratio	After-tax net income + interest paid and accrued ÷ interest	Higher => more solvent

Adapted from P. J. Barry, J. A. Hopkin, and C. B. Baker, *Financial Management in Agriculture,* The Interstate Printers and Publishers Inc., Danville, IL, 1983, p. 81.

viability and expected profitability of the firm. Table 9-4 gives some financial measures used in these analyses.

The rates of return on equity and assets indicate the firm's profitability in the immediate accounting period. The measures of efficiency assess the firm's cost of operations in relation to its size and should be compared to similar firms to assess performance. The measures of liquidity assess the ability of the firm to meet its financial obligations. It is possible for a firm to have assets much greater than its obligations and still be in financial difficulty because the assets are in land, equipment, and other materials that cannot be used to pay short-term obligations such as payroll as they become due. The solvency measures indicate the basic financial health of the firm. The interest coverage ratio indicates whether the firm's interest payment obligations are increasing or decreasing.

The role of the financial analyst is to assess the firm's performance on all of these measures and advise the management, owners, and potential investors on the expected performance of the firm relative to other firms in the industry and relative to other firms in the economy.

SUMMARY

The agribusiness sector consists of the firms that provide inputs to the farm sector, the firms that market the commodities produced on the farms, the public sector that provides infrastructure information, and other services that facilitate the operation of the private sector firm. In very general terms, the agricultural industry is concentrated and oligopolistic in the inputs sector, is competitive in the farm production sector, and has monopolistically competitive markets in the processing and distribution sector. Several market strategies are used to try to offset some of the short-run costs to the farm sector of imperfect competition within agricultural industries. These include farmers' cooperatives, farmers' cooperative bargaining associations, production and marketing contracts, and vertical integration. The great bulk of total farm output is not handled through these systems, however, but rather through more perfectly competitive markets.

The management of the corporation or cooperative business is a complex process. It involves basic economic considerations with significant importance placed on the development of decision-making teams and procedures that allow a firm to pursue a clear set of goals. The systems of financial accounts used by managers and investors to assess the performance of the firm were introduced.

IMPORTANT TERMS AND CONCEPTS

assets   229
balance sheet   229
bargaining associations   219
business strategies   223
controlling   223
cooperative   219
corporation   218
decision-making skills   224
directing   223
equity   229
farmers' market   217
financial analysis   229

form utility   226
income statement   229
individually owned businesses   218
liabilities   229
market segmentation   220
market share   221
organizing   223
partnerships   218
planning   223
product life cycle   226
space utility   225
time utility   225

QUESTIONS AND EXERCISES

Name That Term

Read the following sentences carefully and fill in the missing term or terms.

1 A _____ is a monetary debt owed by a firm or an individual.
2 An _____ is anything of monetary value owned by a firm or an individual.
3 A _____ is a form of business owned by the customers.
4 _____ is the net worth of a firm determined by deducting all debts from the value of property.
5 An unincorporated business firm owned and operated by two or more people is a _____.
6 _____ is a firm's percentage of the total sales in a market.
7 _____ utility is the utility created by firms that store products for future use.
8 _____ utility is the utility created by firms that transport raw materials, intermediate products, and final goods to locations where buyers desire them.
9 _____ utility is the utility created by a manufacturing process that converts raw materials and intermediate products into a form with more value to buyers.
10 A _____ is a legal entity (person, firm, or specified agency) chartered by a state or the federal government.
11 A _____ association is a group of firms (e.g., farmers) that works for higher prices or other favorable economic conditions for member firms.

True/False

Read the following sentences, then decide whether each statement is true (T) or false (F) and mark it accordingly.

T   F   **1** The three major components of the agribusiness sector are the input sector, the marketing and processing sector, and the public sector.

T   F   **2** Direct markets, farmers' markets or bazaars are characterized by a central location, a fixed schedule, open access to other sellers and buyers, lack of standardization, and flexible prices.

T   F   **3** Agriculture is a sector that functions well with no involvement from the public sector.

T   F   **4** A cooperative's earnings are usually divided among patrons in proportion to their purchases from the coop.

T   F   **5** Groups like the American Farm Bureau Federation (AFBF) and National Farmer's Organization (NFO) have tried to develop large regional or national bargaining associations in grains and livestock.

T   F   **6** Four major functions of management are planning, organizing, controlling, and directing.

T   F   **7** Marketing involves intermediaries taking money from both farmers and consumers while not providing anything of value.

T   F   **8** The higher a firm's turnover ratio the more efficient the firm.

T   F   **9** The income statement summarizes the firm's revenues and expenses over a given period of time.

T   F   **10** The balance sheet indicates the firm's financial condition and expenses over a given period of time.

Multiple Choice

Circle the letter of the response that best answers the question or completes the statement.

1 Generally speaking, the supply curve of an imperfectly competitive agribusiness firm
   a is identical to the marginal cost curve above the average total cost curve.
   b tend to be more elastic when prices are high than when they are low.
   c cannot be shifted by a change in the price of the firm's products.
   d is generally more elastic when the prices of the products are rising than when they are falling.
   e none of the above.

2 Which of the following is not a function of management?
   a controlling
   b directing
   c producing
   d planning
   e organizing

3 Which form of utility does the marketing and processing sector of agribusiness add to agricultural products?
   a time
   b space
   c form
   d all of the above

4 Which of the following is not a stage of a product life cycle?
   a introduction

    b research and development

    c growth

    d decline

    e maturity

5 Imperfectly competitive agribusiness firms

    a tend to operate in a range of inelastic down-sloping demand curves.

    b face inelastic demand curves.

    c generally experience elastic up-sloping demand curves.

    d tend to operate in a range of downward-sloping demand curves.

    e none of the above.

6 The agribusiness sector in the United States

    a contains many examples of vertical integration but few examples of horizontal integration.

    b tends to exhibit many examples of economies of large scale.

    c contains few examples of oligopsony.

    d is dominated by monopolies.

    e all of the above.

7 Farmer cooperatives in the United States

    a generally have been organized to carry out a business operating under farmer-owner control.

    b have been most successful as large regional or national bargaining organizations.

    c tend to conform to the model of pure competition rather than oligopoly or oligopsony.

    d generally have more than one-half of the total market share in both farm service supply and marketing/processing industries.

    e none of the above.

8 The largest sector within agribusiness is

    a processing and marketing.

    b public.

    c futures markets.

    d input.

    e cooperative.

9 Which of the following measures is based on a comparison of a firm's income to its assets?

    a liquidity

    b rate of return

    c solvency

    d profitability

    e none of the above

10 Which of the following measures is based on a comparison of a firm's assets to its liabilities?

    a liquidity

    b rate of return

    c solvency

    d profitability

    e none of the above

11 A major difference between a small firm, such as a farm, and a large cooperative or corporation is/are that

    a development of an efficient management team is more important and difficult in the large firm.

    b economic concepts are more important in the small firm.

    c the large firm is more likely to make pricing decisions.
    d both a and c.

Technical Training

Below are abbreviated versions of the condensed consolidated balance sheet and statement of income for a real firm. Use these data to answer the technical questions in this section.

BALANCE SHEET (in $1000)

Assets
Current assets:

Cash and marketable securities	70,842	
Other marketable securities	15,074	
Accounts receivable, net	94,381	
Inventories	10,800	
Prepaid expenses and other current assets	21,444	
Total current assets		212,541
Other assets	12,203	
Property, plant and equipment, net	335,058	
Intangible assets, net	81,997	
Total assets		641,799

LIABILITIES AND STOCKHOLDERS' EQUITY

Current liabilities:

Accounts payable and other current liabilities	39,508	
Current portion of long-term debt and leases	2,907	
Total current liabilities		42,415
Long-term debt and leases	299,508	
Other long term liabilities	9,744	
Total long-term liabilities		309,252
Total stockholders' equity	290,132	
Total liabilities and stockholder's equity		641,799

INCOME STATEMENT (in $1000)

Total revenues		406,968
Operating expenses	313,849	
Interest expense	10,836	
Provision for income taxes	15,333	—
Other Expenses	37,182	
Total expenses		377,200
Net income		29,768

1 What is the firm's accounting profit?
2 What is the firm's rate of return on assets?
3 What is the firm's rate of return on equity?
4 What is the firm's gross ratio?
5 What is the firm's turnover ratio?
6 What is the firm's current ratio?
7 What is the firm's net capital ratio?
8 What is the firm's debt-to-asset ratio?

Questions for Thought

1 What are the major components of the agribusiness sector, and how are they linked? Which is the largest subsector?
2 Give examples of firms that primarily provide time, space, and form utility.
3 How do farmers' cooperatives differ from other corporations in terms of ownership and control, return on investment, and price policies?
4 Generally, what accounts for the success or failure of cooperative farmers' bargaining associations? In what commodity situations have they been the most successful? The least successful?
5 In what areas are production and marketing contracts most widely used? In what areas are they used least?
6 Describe the differences and similarities in the management of a family farm and a large agribusiness firm.
7 What is included in a business plan? What procedures are followed in plan development?
8 What are the functions of a system of financial accounts? Why are several measures of financial performance used? What do these measures of performance indicate about the goal of profit maximization?
9 Visit your local grocery store. Buy a food item and list how that item has changed in time, form, and space since it left the farmer's gate.
10 Make a list of ten management decisions you must make in your life this week and how you use the skills discussed in class to make good choices.
11 Think about the management of a firm where you have worked or are working. How do they determine their pricing strategy?
12 People who work for large businesses often seem more concerned with market share than economic profit. Why?

ANSWERS AND HINTS

Name That Term   1. liability; **2.** asset; **3.** cooperative; **4.** Equity; **5.** partnership; **6.** market share; **7.** Time; **8.** Space; **9.** Form; **10.** corporation; **11.** bargaining

True/False   1. T; **2.** T; **3.** F; **4.** T; **5.** T; **6.** T; **7.** F; **8.** T; **9.** T; **10.** F

Multiple Choice   1. c; **2.** c; **3.** d; **4.** b; **5.** d; **6.** b; **7.** a; **8.** a; **9.** d; **10.** c; **11.** d

Technical Training   1. $29,768,000; **2.** 6.33%; **3.** 10.26%; **4.** 0.9269; **5.** 0.6341; **6.** 5.0110; **7.** 2.075; **8.** 0.4818

10

ECONOMIC PERFORMANCE OF AGRIBUSINESSES

CHAPTER OUTLINE

OVERVIEW

In this chapter we discuss several important agribusiness industries, ranging from farm inputs to food retailing. This should help you develop an appreciation of these industries, the structure and competition within them, and the factors affecting their performance. Most of them operate under imperfectly competitive market conditions. In most industries, there are both large and small firms. In many cases, there are only a few buyers or sellers in a market. Firms differentiate their products, and there are other barriers to entry. However, in many industries, firms become more efficient as they become larger, and much of the industry growth can be attributed to this.

LEARNING OBJECTIVES

This chapter will help you:

- Learn about several major subsectors of the U.S. agribusiness sector
- Broaden your knowledge of agribusiness, its organization and degree of concentration, and its determination of price and output
- Improve your understanding of the production and marketing strategies of agribusiness as influenced by the economics of imperfect competition

INDUSTRY PERFORMANCE

In this chapter, we consider the performance of agribusiness industries from a national welfare perspective, not whether the firm meets its own objectives. Indicators of performance include price responsiveness, lack of excess profits, innovation, and productive efficiency. The benchmark for comparison is the performance of a competitive industry. A purely competitive industry generally achieves the preferred social outcomes of efficient production and allocation of resources among individuals in society. In some cases, usually due to economies of large scale, other forms of market organization have the advantage of operating with less expense. Thus, they may deliver products to producers and consumers at lower costs than if the industry consisted of many small firms operating in a perfectly competitive market.

Structure, Conduct, and Performance

The structure of the industry and the conduct of the firms in the industry influence performance. The **structure** of the industry refers to the number of firms, **concentration ratios** or market shares (the proportion of the market sales held by the largest firms), **barriers to entry** (patents and high start-up costs are important in agriculture), and the extent of vertical integration. There is a strong relationship between the number of firms, measured in terms of concentration ratios, and the price and profit levels in an industry. The **conduct** of firms includes the actions that firms take to increase the demand for their products or to change the elasticity of demand in ways favorable to them. Conduct may also include collusion with other firms in areas such as pricing.

As explained in Chapter 9, in oligopolistic markets a firm's decisions on pricing are influenced by consideration of how competitors will react to its marketing strategies. If a firm has (or a few firms have) a large share of the market—as in soft drinks, soups, and breakfast foods, for example—firms may follow a rigid price policy supported by an aggressive advertising effort. When hard times occur, a firm might follow a strategy of not reducing its price if it expects that other firms may match the reduction, thus eliminating any advantage expected from the price reduction. Instead, the firm may lay off workers and produce less. Another strategy it may follow is to lower prices by offering coupons, sales, and special promotions, which also may be matched by competitors.

Individual oligopolistic firms pricing decisions depend on their judgments concerning the elasticity of their demand curves and how the demand for their product can be increased or otherwise changed given their expectations of how and when competitors are likely to react. What they do also depends on their cost structure and their estimate of the cost structures of other firms in the industry. An aggressive expansion policy by a large firm might push up the prices of inputs used by all firms in an industry. For this reason alone, a large firm may have a less aggressive growth policy than a small firm. However, this may be good for society because it leaves open the possibility of other firms getting started and growing.

All of the above factors affect how imperfectly competitive firms compete, especially in oligopolistic industries. Thus, the structure of the industry and the conduct of the firms determine industry **performance,** that is, whether the industry efficiently delivers quality products at prices that reflect the costs of the products. Vertical coordination and integration are important aspects of the market structure in agriculture. We will discuss this topic before considering several major agricultural industries.

Vertical Integration and Coordination in Agribusiness

Vertical integration occurs when two or more firms operating at different stages of production, processing, and marketing combine under a single ownership and management, or enter into a contract specifying that one firm will deliver to another a product at a specified price. Examples include a sugarcane plantation having its own sugar processing plant, a meat processing plant operating a cattle or hog feeding operation, a food retailer operating a food processing plant, and a firm that both grows and markets fresh vegetables. Vertical coordination is accomplished through ownership of production facilities or marketing contracts. A **marketing contract** involves one firm contracting to supply a certain quantity and quality of output to a processing firm at a stated price. Under a marketing contract, production is generally undertaken according to specifications set by a processing firm, which may also provide financial and other management services. The contracts reduce uncertainty for both the producer and the processor by ensuring a market for the product at a mutually satisfactory price.

Generally American farms and ranches produce and sell most of the aggregate output without production and marketing contracts. However, such contracts may be important in setting standards for the industry. Since the early 1970s, approximately 20 percent of total output was produced under production contracts, generally between the farmer and the first buyer. A portion of these contracts are set through bargaining associations. The others are private, mutual contracts between the farmer and the processing-marketing firm.

Even though the percentage of farm produce raised or sold under production and marketing contracts is generally low, there are important exceptions, such as sugar beets, vegetables for processing, and citrus fruits. Contracts are used to varying degrees for feed cattle and lambs and extensively for broilers, turkeys, and fluid-grade milk. Recently the amount of contract production has increased in the hog industry, especially for feeder pigs. The poultry and hog industries are discussed in more detail below.

There are numerous reasons for vertical integration, including a more reliable supply, economies of large scale, reduced price uncertainty and transaction costs, assurance of desired product characteristics, and diversification or reduction of risk. For example, a processing firm that owns or contracts with farms may provide the farmer with the young animals, feed, veterinary services, and management advice. The farmer is also assured of a market for the product produced at a set price. Thus, the farmer receives financial and management support, and a reduction in the risk of losses resulting from low prices (in some cases the farmer does not take ownership of the animals). However, the farmer forgoes the possibility of larger profits that higher market prices would generate. The major advantages to the processing firm are the assurance of a supply of the inputs when needed to meet its production and marketing schedules and the control over quality of the inputs.

The broiler industry is the most completely integrated in agriculture. About 90 percent of all broilers are produced under contract or on farms owned by the integrated firm. Figure 10-1 depicts the extent of vertical coordination in a fully integrated broiler

FIGURE 10-1   Functions of a typical integrated broiler firm. (From E. A. Lasley, H. B. Jones, E. H. Esterly, and L. A. Christensen, "The U.S. Broiler Industry," USDA, ERS, Agricultural Economics Report No. 591, November 1988.)

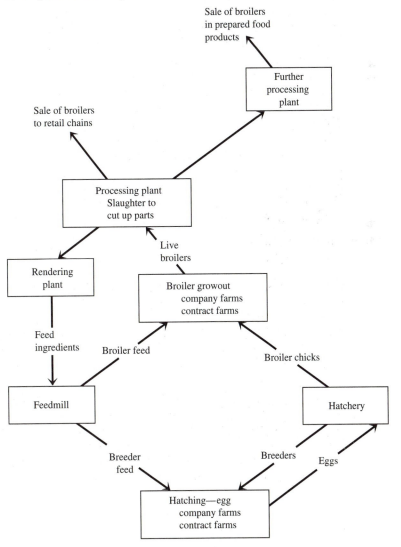

TABLE 10-1   PERFORMANCE IMPLICATIONS OF MARKET COORDINATION IN THE BROILER AND PORK INDUSTRIES

Firm goal	Firm outcome	Consumer effect
Improved access to capital	Technological advance	Increased quantity and quality; lower prices
Improved risk management	Lower costs	Increased quantity; lower prices
Lower measuring and sorting costs	Improved animal quality and uniformity	Improved quality and convenience
Assured input supply and market outlets	Stable input flow; reduced risk	Increased quantity; lower prices
Market power	Increased profits	Ambiguous

producing and processing firm. Such integrated firms have grown rather rapidly in some additional areas, such as sugarcane, fresh vegetables, potatoes, some other fruits and nuts, eggs, and turkeys.

Recently, there has been an increase in the proportion of hogs produced on farms owned by feed producers or under contract to other firms in the industry. In 1980 about 3 percent of hogs were produced under contract. By 1993 this had increased to 11 percent and by 1998 it had increased to 40 percent. The shift to a more vertically integrated structure is controversial. The major conclusions of a 1999 USDA study by Martinez[1] are summarized in Table 10-1. It indicates that firms have a number of incentives to increase the level of coordination. Pursuing these incentives has affected both the industry and the consumer market for the meat products. For example, the firms may integrate in order to get access to capital. Access to capital allows the firms to develop or purchase new and technologically more advanced equipment. By adopting new technology the industry can then produce more and higher quality product, and do so at lower prices. Similarly, improved risk management, such as lowering the incidence of disease through improved sanitary conditions, can lower costs to the firm and generate higher quality products at lower costs to the consumer. According to Martinez's study, consumers have benefited by receiving larger quantities of better products at lower prices.

AGRIBUSINESS INDUSTRIES

Agribusiness industries differ in terms of their structure, forms of competition, and performance. Therefore, it is useful to consider the characteristics of some of these industries to increase our understanding of their performance. The industries selected suggest the diversity among agricultural industries.

Fertilizers—A Stable Industry

United States use of the major plant nutrients—nitrogen, phosphate, and potash—increased from 7.5 in 1960 to about 20 million tons in the late 1970s. Use remained

[1]Martinez, Steve W., "Vertical Coordination in the Pork and Broiler Industries: Implications for Pork and Chicken Products", Food and Rural Economics Division, ERS, ESDA, Report #777.

at that level to the end of the century. World use has reached about 70 million tons (150 million metric tons) per year and is projected to continue to grow, especially in the developing nations.

The fertilizer industry is composed of firms marketing nitrogen, phosphate, potash and other trace minerals. While there are numerous linkages, they can be considered as three separate industries. In each case they sell an undifferentiated product. The nitrogen industry is the most competitive of the three with the largest four firms controlling less than half of the market. At the other extreme, the potash market best fits the oligopoly model with nearly 80 percent of the market controlled by the largest four firms, according to the USDA.

The United States is the world's largest producer and exporter of phosphate, with most of the production occurring in Louisiana and Florida. In 1996, IMC-Agrico accounted for 30 percent of U.S. production (13 percent of world production) and the next three largest firms accounted for another 40 percent. Because the phosphate deposits are concentrated in a few locations, entry is restricted. Thus, the phosphate industry also fits the oligopoly model.

While the national nitrogen market is not as highly concentrated as the phosphate or potash markets, in most local markets three or four firms account for the major portion of the total supply, and in some local markets one or two firms sell all of the fertilizer.

Chemical Pesticides—An Industry Based on Innovation

The pesticide industry also approximates an oligopoly. It has followed a pattern of mergers and acquisitions similar to that of the fertilizer industry. Presently there are a small number of American and foreign firms developing new pesticides. The fundamental basis of competition in the industry is *innovation*. Firms support large research and development efforts to find new pesticides that are less expensive, more effective, safer to handle, or that have less potential to adversely affect the environment. Where several pesticides can be used to control a given pest, there is substantial advertising by firms in an attempt to increase, or at least maintain, their market shares. Often, however, these products are marketed under patent protection, which may give a firm the opportunity to charge monopoly-level prices until the patent expires. Whether the firm can or will charge a monopoly price depends on such factors as the available alternatives, chemical or other, and the extent and type of government approval and control. The high costs of developing, testing, and registering new pesticides are a significant barrier to entry, which prevents new or small firms from entering the industry. Because there are significant economies of large scale, many products are unique or do not have close substitutes.

The development of plant varieties that are not adversely affected by broad-spectrum herbicides has led to new weed-control strategies and to new business arrangements. Farmers have the alternative of purchasing the new and higher priced varieties of seed and using the broad-spectrum herbicides to achieve yield increases sufficient to offset the higher seed costs. In the input sector, this has led to some chemical firms purchasing seed producing firms.

Farm Machinery and Equipment—An Industry Serving a Changing Market

The farm machinery and equipment industry has undergone significant changes over the last several decades. Mergers and acquisitions have reshaped the industry during a period that saw sales decrease significantly due to lower farm incomes and changes in technology. Technological advances improved the quality of the equipment leading to a longer life span and thus less frequent replacement.

A few very large firms offer a full line of machinery and equipment, and supply the major share of the market. These firms are both price and service leaders. They differentiate their products through research, new inventions, and patents. They market aggressively on a global, national, and regional level with advertising and company service. The core of their agricultural equipment market is the large family farm and the larger corporations where operators buy most of the new tractors, implements, and self-propelled machines. The smaller farms, including part-time farms, constitute a market for lower-powered tractors and equipment. In addition to the full-line manufacturers, there are numerous small firms that specialize in the production of a small number of special purpose equipment items.

The farm machinery and equipment industry, like the pesticide industry, is highly concentrated and closely approximates an oligopolistic market in terms of competition among the major producers of full lines of machinery, including large tractors and combines.

Inventions of new machines and significant improvements of earlier models occur frequently. This changes the market position of individual firms and gives some segments of the industry a structure approximating monopolistic competition. Also, many firms that are small, in comparison with the large manufacturers, offer maintenance and repair services. This enables farmers to keep existing machines running in years of poor crops or low prices. Hence, the farmers' demand for new machinery is sensitive to market conditions, and this often leads manufacturing firms, or their local dealers, to offer discounts that modify the oligopolistic price behavior of the large manufacturers. That is, the farmers' demand for new machinery is much more variable from one year to the next than the demand for inputs, such as fertilizer and pesticides. The demand for fertilizer is closely correlated with the total acreage seeded to crops, and the demand for pesticides is often dependent on expected or realized growing conditions.

In nominal terms, the sales of farm machinery grew from about $2 billion in 1950 to a peak of nearly $12 billion in 1979. Sales then fell with the depressed agricultural economy to about $4.5 billion in 1986 and have since rebounded and leveled off at about $8 billion annually. In real terms, the sales have fallen by 50 percent from 1975 to 1995. This sales experience was reflected in plant capacity utilization rates that went from over 80 percent in the 1970s to about 30 percent in the early 1980s. As a result of plant closings and the rebound in sales, the capacity utilization reached 60 percent in the mid 1990s.

The small-farm market in the United States has continued to shrink, as the economies of large-scale farm production favor large-scale farm machinery and equipment. The 50 to 100 horsepower, two-wheel drive tractor has been replaced with the more powerful four-wheel drive tractors on many larger U.S. farms. The decrease in demand for the smaller tractors, and barriers to trade have resulted in the shift of production of all of the smaller two-wheel drive tractors to plants outside the United States.

Recently there have been numerous mergers and acquisitions. Ford-New Holland purchased Case International and Deutz purchased the Allis-Chalmers line, in both cases merging major competitors. Some years ago the heavy-equipment manufacturer Caterpillar entered the industry with a line of track-driven tractors sized to compete with the existing firms. Also, Kubota, a Japanese manufacturer, has made some inroads in the tractor market. As a result of these and other mergers, the four-firm market share has grown from 60 to 80 percent.

Other industry developments include the leasing of the highly technical and expensive equipment used for precision agriculture. Also, some financial institutions purchase farm equipment for lease to farm operators. If leasing by financial institutions becomes more prevalent, banks may be able to wield some countervailing market power to the oligopolistic farm equipment manufactures.

Globally there are many farms in the developing countries that are too small to use large equipment. Markets serving these farmers have very different demand characteristics than those faced by American manufacturers. Several Japanese multinational firms have developed and are marketing equipment sized for small farms. Also, in a number of the less-developed countries there are firms that produce farm equipment for domestic use.

Even so, the individual farm machinery manufacturer in the United States faces a demand that is characterized by the kinked demand curve discussed in Chapter 7 and reproduced in Figure 10-2. The kinked demand curve influences the individual manufacturer's conduct and performance. For example, suppose "hard times" hit the farm

FIGURE 10-2   Stable prices with decreasing demand in an oligopoly with a kinked demand curve.

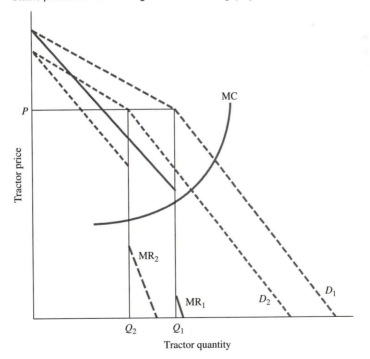

sector. The demand for a firm's output drops from D_1 to D_2 and the firm's marginal revenue curve drops from MR_1 to MR_2. As indicated in Figure 10-2 the firm's MC curve does not change and it cuts MR_2 through a gap caused by the kink in the demand curve at the existing price level. The typical manufacturer will lay off workers and reduce production to point Q_2 in Figure 10-2. This example shows why the typical oligopolistic firm reduces the quantity supplied but maintains a relatively fixed price.

If the reduced sales continue for some time, and if some of the firms in the industry are operating with ATC higher than AR, financial difficulties may force strategic adjustments. Firms may exit the business, reorganize, merge with another firm, terminate product lines, select new advertising agencies, and so on.

Transportation—An Industry That Experienced Deregulation

The U.S. transportation industry serving agricultural input and the output markets has undergone several important stages of development. In the earliest years the absence of an efficient transportation system slowed the development of agriculture because moving goods from producing to consuming areas was prohibitively expensive. Both the federal and the state governments responded by supporting the development of the transportation system. The costs of construction of roads, canals, and railroads were publicly subsidized. Because the railroads were monopolies in many areas, they could charge high transportation rates and provide poor service. In the late 1880s complaints by farmers and others about such rates and service led to the establishment of the Interstate Commerce Commission (ICC), with the power to regulate rates and services of the railroads involved in interstate commerce.

With the development of publicly financed roads and highways it became economical to ship some agricultural products by truck, using farmer-owned vehicles or commercial shipping firms. This provided competition for the railroads, especially on short hauls, and led to limited deregulation of the transportation industry in terms of routes served and rates charged (the Staggers Act in 1980). After deregulation (from 1980 to 1991) the number of Class I railroad companies, the major long-distance haulers, fell from forty-eight to thirteen. The largest four firms carried 35 percent of the freight in 1980 and 64 percent of the freight in 1991. By 2000 the largest five Class I railroads handled 90 percent of all agricultural products and there were just two Class I railroad firms west of the Mississippi River.

Similar results of deregulation occurred in the airline industry, with the share of the market for the three largest firms increasing from 40 to 52 percent over this period. With modern developments, the major segments of the nation's transportation system have taken on the economic structure and many of the competitive practices of an oligopoly. The ICC retains the authority to maintain regulations concerning weight, dimensions, and safety standards for transportation firms. The result is a market that has elements of oligopoly for long-distance hauling and monopolistic competition for shorter-distance hauling. Within the trucking industry, only a few firms compete for long-distance hauling contracts. But for short hauls, and in areas where both rail and water transportation are available, the market approximates monopolistic competition.

The differences in cost structure among firms in the transportation industry significantly affect the competitive nature of the industry. Railroads and air transportation firms have high fixed costs (railroad track, stations, and rolling stock; aircraft and terminal space), while barges and trucking have high variable costs (the public sector maintains waterways and highways). Because of the low proportion of variable costs, railroads and airlines often compete with short-term rate reductions in an attempt to drive out competitors so that they can charge higher monopolistic prices in the longer term.

Truck and barge firms can enter the market with a small investment but incur proportionately higher variable costs per mile for labor and fuel. (Trucks now haul about 45 percent of agricultural commodities.) The railroads can force these firms out of a market in the short term, but they may reenter if rail rates increase. In the airline industry it is much more difficult and expensive for new competitors to enter, implying that as the number of airlines decreases, the prices may revert to higher levels and remain there.

In comparison, in many developing nations, most agricultural commodities are transported relatively short distances and mainly by truck. The trucking industry is generally unregulated with many individual truck drivers operating in competitive markets, although in some of these countries poorly regulated, oligopolistic or monopolistic trucking industries operate. In contrast, in Europe, an extensive railroad system—which is regulated, subsidized, and efficient—is augmented by truck and water transport somewhat analogous to the situation in the United States.

Grain Marketing—Globally Competitive, Locally Concentrated

The grain industry in the United States and other exporting countries consists of a large number of cooperatives, independent entrepreneurs and a few large multinational companies. Economies of scale have encouraged consolidation of firms through merger, purchase, and closure. Although four large firms dominate the international markets, monopsony power is limited by the presence of small independents and cooperatives. Even in regions with only one or two firms, market distortions are limited because most farmers have access to several buyers within a reasonable distance of their farm. However, a recent rapid increase in the number of mergers among small firms as well as multinationals (e.g., Cargill purchased most of the grain facilities of Continental Grain in 1999) continues to raise questions as to how much concentration can exist before market power creates opportunities for excess rents and profits. Grain cooperatives were created in the 1930s to counter monopsony power by the large grain firms at a time when transportation by wagon from farm to country elevator limited the farmer to one or two choices. These cooperatives continue to offer alternatives in terms of services and prices and, through patronage refunds, they return profits to the farmer. These firms operate with the same strategies and the same environment as the large multinationals and are generally indistinguishable in their day-to-day operation from independently owned firms. However, the presence of the cooperatives and ready access to market information still serves as some insurance against monopsony power.

Prices paid by country elevators generally follow prices on the Chicago Board of Trade, which in turn respond to supply and demand forces in the international market. Destination prices (domestic or foreign) minus transportation costs and merchandising margins determine the prices received by farmers. Marketing margins differ widely among firms and among countries in response to differences in efficiency and local supply and demand.

One of the important marketing functions is the standardization of quality through the use of grades and standards. In the United States, nationally uniform grades and standards were established by law in the 1916 U.S. Grain Standards Act. Although the regulations and the standards have been changed many times, they still provide the basis for uniform, third-party, objective evaluation of quality, and provide the information by which buyers and sellers determine value and negotiate price. Importers in other countries can rely on USDA grade certificates because federal employees must inspect all exported grain. In contrast, many other exporting countries give inspection authority to private firms or government agencies that also own and sell grain, reducing credibility of their results.

In some nations (Australia, Canada, China, and Indonesia, for example) the government plays a major role in buying, selling, storing, and grading grain. Government-operated grain marketing boards may operate as state monopolies without the discipline of profit incentives, creating price distortions and inefficiencies, and raising transactions costs. In several countries (Mexico, Columbia, South Africa, and Argentina, for example) the state agencies are being replaced by private firms. The privatization process differs among countries, but all are moving toward a substitution of private ownership for government operation of merchandising activities. It remains to be seen if the strategies generate a competitive industry, or simply substitute private monopoly for government monopoly.

Grain firms are continuing to adjust to accommodate the increasingly specialized requirements of domestic and international buyers of grains. Processors (especially those who are converting grains directly into food products) are increasingly concerned about controlling quality and composition and are contracting for special varieties, pesticide free, and organically grown grains. Genetically engineered crops present a problem because of the difficulty of maintaining the identity and purity of these grains in the complex bulk-market channel. The use of identity preservation, in which the identity of the crop is maintained from seed to final user, can meet the needs of these specialized processors, but only at a higher cost than the high volume, generic, bulk-commodity market channel. Firms and government agencies in other countries, concerned about the future safety of genetically modified crops, have placed restrictions on the importation and the production of some of these crops creating an effective barrier to trade. The possibility of identity preservation as grain moves through the market channel is appealing to some. However, this will require new contractual arrangements and will mean higher costs of transportation, segregation, storage, and processing. The higher costs mean there is a trade-off with identity preservation. This question is likely to generate conflict in future trade negotiations.

Seed Industry—Global Competition and Controversy

The seed industry is interesting in that it is truly global. The eleven largest seed companies are headquartered in seven different nations, with the four largest in the United States, Switzerland, France, and Japan. The largest thirty-five companies control about 50 percent of the market. Thus, when viewed in global context, the industry is competitive. However, since many of the companies specialize in the production of certain seeds, the market for the seed for a particular crop may be less competitive than is implied by industry-level concentration ratios. Also, as firms attempted to improve their competitive position there have been mergers and acquisitions leading to some increase in concentration. And, if viewed from a country perspective, there may be less competition because not all seed companies operate in all countries.

Another interesting feature of the seed industry is that the companies compete in large measure through their research activities. The firms allocate substantial resources to their research and development efforts to identify seed with improved performance characteristics. Firms then seek patent protection for the new variety of seed, which can be sold at a premium price.

The type of research conducted by the seed industry has drawn considerable attention. For many years, new seed varieties have been developed through extensive cross-breeding trials and through the search for desirable genetic mutations. When a seed with desirable genetic characteristics is identified, it is placed on the market. Essentially all seed planted for food production has been modified through such research. The relatively recent development of techniques to transfer genetic material from one organism to another (discussed in more detail in the development chapter) allows firms to more efficiently develop and introduce the desired characteristics. Seed produced in this way is termed a genetically modified organism (GMO) and has become the subject of considerable public controversy. Those in opposition often pressure the industry to stop using these techniques or to label all foods produced using (or not using) such techniques. Not using the techniques may slow the development of improved varieties and make feeding a growing population more difficult, especially in the long run.

Meatpacking Industry—Oligopoly and Oligopsony

The meatpacking industry is relatively competitive in its advertising and sales, but it approximates an **oligopsony** (few buyers). There are often questions about the type and level of competition among the firms in their purchases of livestock from producers.

Recently the meatpacking industry has seen numerous mergers and acquisitions and the movement toward larger and more specialized facilities. In the 1990s the largest four cattle slaughtering firms controlled over 70 percent of the beef market. For hogs, chickens, and turkeys, the largest four firms controlled only 40 to 45 percent of the market. Additional firms, especially those currently operating in related industries, could enter the market if excessive profits were being realized.

There is little product differentiation in beef and hogs. Several firms have developed differentiated products in the chicken (Tyson, the leading firm worldwide, and Perdue) and turkey (Butterball) industry. Thus, these latter two industries can be considered monopolistically competitive in their sales while the cattle and hog industries more nearly approximate the competitive model even though the cattle industry is more concentrated.

Concerns about the competitive nature of the industry are more pronounced on the buying side. In a competitive market the farm and retail prices of meat are set at equilibrium levels by the interactions of supply and demand at the several stages in the marketing chain. Supply is the amount of livestock delivered to market by farmers. Demand is the amount purchased by consumers. Meatpackers and distributors in the marketing chain play a passive role in determining market price, with their long-run profits set at normal levels by the entry and exit of firms. However, because of the large, specialized, and widely separated plants, as well as the costs of transportation, livestock farmers cannot move their livestock long distances to a slaughter facility. As a result, producers often have a limited number of buyers for their products. Farmers have often contended that firms follow an oligopsonistic buying policy that keeps prices low. That is, the meatpacking firms accept a rather constant share of the market instead of trying to increase their share by more aggressive bidding and buying. This, some farmers argue, keeps cattle prices lower than if the competition were more aggressive. If the industry continues to consolidate to a smaller number of firms operating larger plants the possibility of oligopsonistic behavior will also increase.

HOW FIRMS CHANGE OVER TIME

The Purina Company provides an interesting example of how firms change over time. The Robinson-Danforth company was founded in 1894 with the marketing slogan of "Where Purity is Paramount." The slogan led to the name Purina. During WWI the Chow Division was created to handle the animal feed operations. In 1984 this division became Purina Mills. In 1986 Purina Mills was sold to BP Amoco, which in turn sold it to an investment firm, the Sterling Group in 1993. It was again sold in 1998, this time to Koch Industries. In 1999, due to the downturn in the agricultural economy, Koch filed for bankruptcy protection.

Food Processing and Marketing—Multidimensional Change

Both the processing and the retailing sectors of the food industry have experienced significant changes. There are constant changes in the products produced and marketed. There have also been dramatic changes in the structure of the market. We give attention to concentration ratios in food processing and retail marketing because of the strong link between concentration and the price and profit levels in this sector of the food industry. For example, studies have found that as the share of the local market controlled by the four largest retail food marketing firms increases, so do their prices and profits.

Food Processing

The market share data for selected food processing industries is complex and indicates some tendency over time toward higher degrees of concentration. If viewed as an industry total, the largest fifteen firms control about one-third of the U.S. market. The same is true in Europe.

When viewed in terms of product categories, nationally, the degree of concentration of the market share held by the four largest and the eight largest firms is generally not high compared with other American industries. Although the four largest food processing firms usually have less than 50 percent of the total market, the eight largest firms frequently have more than 50 percent. Between 1970 and 1990 there was a significant (nearly 40 percent) drop in the number of food processing firms and this consolidation has continued. Even though the number of small firms operating in the food processing industry fell significantly, there was only a small increase in the market share held by the largest firms. These trends may indicate that the smallest firms are leaving the food processing industry, or are combining with other small or medium-sized companies. If so, this would move a major segment of the food processing industry toward a monopolistic competition structure enabling newly created firms to compete more effectively with the larger and older oligopolistic firms.

When individual products are considered, in many cases the largest firms control most or all of the market. Small numbers of firms sell significant portions of the coffee, soft drink, breakfast cereals, and soups. In these cases competitive pressures are from the potential entry by other large firms in the food industry. A common example of this competition is the development by retail firms of private label products. Often these products are produced on contract by the firms producing the brand name products. But there is a possibility that the retail firm would open its own production facility, giving the retail firm considerable market power.

Retail Food Industry

Dramatic changes have occurred in the retail food industry with more away-from-home consumption, development of the fast-food industry, and the expansion of niche marketing. There has been merger, acquisition, and divestiture activity, particularly in the last several decades. For example, in one five-year period there were approximately 450 acquisitions of food processing firms by wholesale and retail firms. In the same period there were 200 acquisitions and nearly 300 divestitures of food firms by companies in other industries. The **acquisitions** occur for a variety of reasons. Firms may seek to expand a product line, expand their market area, acquire a profitable firm, absorb a competitor, or more effectively utilize existing facilities. The activity has included a significant number of international **mergers** as this industry also becomes more global in character. **Divestitures** often occur when firms in financial difficulty attempt to generate additional revenue by eliminating a segment of the company that is not as profitable as desired or that does not "fit" the firm's long-range plans. Some divestitures occur because of pressure from the Federal Trade Commission to avoid excessive concentration or monopoly.

In the United States there has been a consistent rise in the concentration of market share held by the four largest grocery firms. By 1997 the four largest food retailers controlled over 75 percent of U.S. grocery sales. In addition, because consumers shop in the metropolitan area in which they live, concentration ratios in these areas are important. Generally, there is a clear increase in prices and profits as the share of the market controlled by the dominant retail grocery chains operating in a given metropolitan area increases from 40 to over 70 percent. There is a consistent increase in prices and a dramatic increase in profits as the percentage of sales held by the largest of the four major firms in a metropolitan area market increases from 10 to 55 percent. At the extremes, profits are almost ten times higher when the major firm is dominant (i.e., one firm controls more than 55 percent of a four-firm 70 percent or greater market share) than when the largest firm controls only 10 percent of a four-firm 40 percent market share.

The market share of the four largest retailers operating in metropolitan marketing areas or **standard metropolitan statistical areas (SMSA)** has increased in each reporting period in all sizes of SMSAs. Also, as might be expected, the concentration ratios are higher, on average, in the smaller market areas. The larger markets provide an opportunity for several large firms to operate successfully, whereas in a smaller market four firms with one or two stores each may saturate the market. Generally, increasing concentration in the retail food industry means higher prices. However, it is also clear that the major marketing chain firms provide a wider variety of products and at lower prices than the small independent retail stores that were common decades ago. The case for government regulation of competition in the food industry is ambiguous.

Food scientists say the U.S. food supply is the best and safest ever produced. However there is a clear case for more attention to nutritional factors, research and education, and continuing government enforcement of safety standards in domestically produced and imported foods. The development of appropriate standards for residues from agricultural chemicals is a particularly difficult issue. There is also a case for directing more rigorous attention to food service in depressed areas.

FOOD INDUSTRY IN DEPRESSED BIG-CITY MARKETS

From the 1970s to 1990, many large supermarkets closed stores located in economically depressed neighborhoods of major metropolitan areas. For example, the number of supermarkets located in Los Angeles, Chicago, and Manhattan fell from 2,373 to 1,340. Reasons for the closings included an inability to acquire the space to expand older and smaller operations in order to offer the full range of items, significant bureaucratic hurdles to opening new stores, and the perception of a less hospitable socioeconomic environment. The substantial reduction in stores in some areas has resulted in some new firms entering this market; however, in depressed areas food prices have remained high and services have been relatively poor. Questions were being raised about whether new ventures would provide competition for the older chains, and whether this would occur without government assistance. Census data from 1992 and 1997 show an increase of nearly 100 stores in New York and Los Angles, suggesting a leveling off of the numbers of stores, and the possibility of a rebound.

Several more general inferences can be drawn from the above observations about the food industry. First, the American food industry is dynamic, with numerous changes in ownership and control as well as an almost constant proliferation of new products. Second, there is increasing concentration in segments of the industry, which may signal a reduction of effective competition and a movement toward oligopoly. However, the consumer may gain from large-scale economies. Third, the substantial degree of involvement by foreign firms suggests that the "globalization" of the food industry is occurring. Wal-Mart is the largest worldwide food retailer and has sales three times larger than the German retailing firm that is in second place. This trend can speed the adoption of new technology and business practices, such as Internet sales, which would cross international borders and thus increase competitiveness.

Textiles and Textile Products—Where Farms and Chemical Firms Compete

The textile industry is global in nature and has several different forms of market competition. Two types of fibers are used in the production of textiles, natural and manufactured. The natural fibers (wool, cotton, linen, and silk) are produced in the agricultural sector and under market conditions that approximate perfect competition. The manufactured fibers (cellulose-based fibers such as rayon and acetate, and synthetics such as nylon and polyester that are derived from petrochemicals) are produced by large corporations such as DuPont and Monsanto in markets that are oligopolistic. Because the fibers are highly substitutable the competition tends to be intense.

The millers purchase these raw fibers and transform them into fabric by spinning, weaving, knitting, or bonding. They are also "finished" through the application of dyes and other treatments that tailor the fabric to specific uses. This segment of the industry has, over the decades of the 1980s and 1990s seen considerable consolidation through mergers and acquisitions and the failure of some weaker firms. Associated with this consolidation was a 60 percent increase in pounds of fiber produced per worker during one ten-year period. From the mid 1970s to the mid 1990s, U.S. output from this industry grew from $32 billion to $74.2 billion. This segment of the industry has moved from a structure that approximates monopolistic competition toward an oligopoly. Some of the major firms are Springs Industries, Burlington Industries, West Point Stevens, Cone Mills, and Concord Fabrics.

The products of end-use firms are apparel, home furnishings (carpeting, draperies, upholstery), and textile for industrial applications. We focus on the apparel segment that accounts for almost half of the sales. This segment, in contrast to the production and milling sectors, is labor intensive because the sewing machine and a skilled operator is best suited to transforming a two-dimensional fabric into a three-dimensional article of clothing. This segment of the industry has been shifting from the capital-intensive developed economies to the less-developed economies where labor costs are lower. This has occurred in spite of relatively high tariffs and some import quotas by the United States. As the world economy moves toward freer trade due to the reduction in trade barriers, one must expect a continuing shift to the countries with low labor cost. And one must also expect the U.S. textile workers to continue to be vocal in their opposition to this trend and thus toward freer textile trade.

Other dramatic changes in the structure of the industry are occurring. Many of the large and well-capitalized firms such as Levi Straus and Liz Claiborne are vertically integrating into the upstream and downstream sectors of the industry. They have also used computer and communications technologies, along with geographical proximity, to develop quick response networks that allow domestic producers to compete with foreign suppliers. At the same time, many of these firms are opening their own fabrication facilities in the developing countries.

The significant changes in the retail sector have been most noticeable to us as consumers. Most firms are small businesses that fall in the monopolistic competition model. However, the mergers, acquisitions, and buyouts that began in the 1980s have resulted in huge vertically integrated, international companies. Kmart, Target and Wal-Mart are the best examples. The success of these firms also is indicative of a change from traditional department and specialty stores to discount stores and outlet stores. There has also been a recent growth in sales by catalog, television, and the Internet.

SUMMARY

Our study of agribusiness competition in production and marketing reveals a broad range and variety of competitive situations, with imperfect competition being the dominant characteristic. Even though both monopoly and monopsony are rare, markets approximating these forms are important. Oligopoly and oligopsony, as well as monopolistic competition, are typical in both farm service supply and processing-marketing; this contrasts sharply with the general situation of purely competitive firms in the farm sector. Under given costs, a firm in imperfect competition will produce less and charge more than a firm in pure competition. This alone results in higher social costs; however, the economies of scale associated with large firms are a factor that can offset the effects of imperfect competition in most agribusiness industries.

The geographical aspects of agricultural markets are important. First, both farmers and final consumers generally conduct their market transactions in a small area. Firm concentration ratios are often significantly higher in these local areas than at a national level. Thus markets may be more concentrated than it appears. Second, global competition is increasing in most of these industries. Thus, the national concentration ratios may underestimate the level of competition in the market.

IMPORTANT TERMS AND CONCEPTS

acquisitions   251
barriers to entry   239
concentration ratios   239
conduct   239
divestitures   251
farm machinery and equipment industry   244
fertilizer industry   242
grain industry   247
marketing contract   240
meatpacking industry   249
mergers   251

oligopsony   249
performance   240
pesticide industry   243
production contract   240
seed industry   248
standard metropolitan statistical
  areas   251
structure   239
textile industry   252
transportation industry   246
vertical integration   240

QUESTIONS AND EXERCISES

Name That Term

Read the following sentences carefully and fill in the missing term or terms.

1 A _____ involves the combining of two or more firms into one operating unit.
2 An _____ refers to the purchase of one firm by another firm.
3 _____ refers to a group of plants engaged in different stages of the production of a final product and owned by a single firm.
4 A _____ contract is a contract between a producer and a processor (or other seller and buyer) that assures a market for a commodity at harvest at a negotiated price.
5 The _____ of an industry refers to the number, size, and thus market share of the firms in an industry.
6 The _____ refers to the percentage of the total sales of an industry made by the four (or some other number) largest sellers (firms) in the industry.
7 Firm _____ refers to the joint activities of firms in an industry with regard to the establishment of prices, quantities, advertising, and other economic decisions.
8 _____ refers to the sale of a division of a corporation to another firm.
9 _____ includes the introduction of a new product, the use of a new method of production, or the employment of a new form of business organization.
10 Production _____ refers to producing a good in the least costly way.

True/False

Read the following sentences, then decide whether each statement is true (T) or false (F) and mark it accordingly.

T F   **1** Firms in many industries become more efficient as they become larger.

T F   **2** When an individual firm has a large share of a market it is likely to follow a policy of rapidly adjusting prices to maximize profit and to eliminate advertising to cut costs and increase profits.

T F   **3** Over 80 percent of the aggregate output of U.S. farms and ranches is sold through production and marketing contracts.

T F   **4** Production and marketing contracts are used extensively for sugar beets, vegetables for processing, citrus fruits, broilers, turkeys, and fluid-grade milk.

T F   **5** The broiler industry is among the most integrated in the agribusiness sector.

T F   **6** The U.S. fertilizer industry is characterized by monopolistic competition.

T F   **7** The U.S. pesticide industry is a monopoly.

T F   **8** Because of the kinked nature of the demand curve for their products, firms in the U.S. farm machinery and equipment sector have been forced to respond to decreased demand for their products by dramatic reductions in their prices.

T F   **9** The "homogeneity" of grain products is enhanced by a system of nationally uniform grades and standards established in the United States in 1916.

T F   **10** There is considerable evidence that prices and profits fall as a few firms begin to dominate the retail food market in a regional market.

Multiple-Choice Questions

Circle the letter of the response that best answers the question or completes the statement.

1 Which of the following are indicators of desirable industry performance from a national welfare perspective?

a innovation

b no excess profits

c productive efficiency

d price responsiveness

e all of the above

2 Among the reasons for vertical integration are

a assuring a more reliable supply of inputs.

b economies of large scale.

c reduced price uncertainty and transactions costs.

d assurance of desired product characteristics.

e reduction of risk.

f all of the above.

3 The American food industry is characterized by

a a stable pattern of ownership and control.

b a lack of new products.

c decreasing concentration in most segments of the industry as firms split into competing units.

d all of the above.

e none of the above.

4 The most commonly used measure of market structure is

a the market concentration ratio.

b location of production facilities.

    c elasticity of demand.

    d domestic and foreign sales.

5 The agribusiness industry that is the most vertically integrated is the

    a grain industry.

    b food processing industry.

    c broiler industry.

    d fertilizer and pesticide industry.

6 The retail food industry is characterized by

    a an increasing amount of food consumed away from home.

    b large numbers of new products.

    c increased concentration at the national level.

    d increased concentration at the local level.

    e all of the above.

7 The market for nitrogen fertilizer is characterized by

    a highly differentiated products.

    b not as highly concentrated as the markets for phosphate or potash.

    c high levels of product innovation.

    d highly variable prices.

    e all of the above.

8 Patent protection is the most important aspect of which of these industries?

    a grain

    b livestock

    c pesticides

    d fertilizers

    e textiles

9 An important factor in the transportation industry is

    a the lack of effective rules and regulations.

    b the importance of foreign firms entering the market.

    c the high variable costs and the low fixed costs of the truck and barge firms versus the low variable costs and the high fixed costs in rail firms.

    d the high levels of concentration in the trucking industry.

    e none of the above.

10 Which of the following industries is the most global in nature?

    a livestock

    b farm machinery

    c seed

    d textiles

Technical Training

Below are data on sales (in $10,000) for three different industries.

	Industry A	Industry B	Industry C
Largest	10	120	1,500
2nd largest	9	90	1,400
3rd largest	7	80	1,200
4th largest	6	70	1,100
5th largest	2	60	1,000
All other firms	4	100	23,000
Total	38	520	29,200

1 What is the CR4 (concentration ratio for top four firms) in industry A?
2 What is the CR4 ratio for industry B?
3 What is the CR4 ratio for industry C?
4 By the CR4 measure, which industry is the most concentrated?
5 By the CR4 measure, which industry is the most competitive?

Questions for Thought

1 Does "competition" mean the same thing to economists and businesspeople? Why or why not?
2 Is cooperative behavior more likely to occur among firms in a perfectly competitive industry or firms in an oligopoly? Consider two extreme cases. If a neighboring farmer has trouble getting the crop to market, is a farmer likely to volunteer help? If Ford has trouble getting cars to market is GM likely to volunteer help?
3 Why do industries with very high fixed costs, such as railroads, electric companies, and water companies, tend to end up as monopolies? Hint: Think about marginal costs and firm behavior in the face of competition.
4 What is vertical integration? Where is it most prevalent? Why hasn't vertical integration become more widespread?
5 There are varying forms of competition in U.S. agricultural industries. List the reasons for these differences.
6 Even though competition within the farm service supply sector is imperfect, progress in this sector appears to be a major source of development and growth in agriculture. If this is true, what are the major reasons?
7 Describe the evolution of the transportation industry in terms of the nature of competition in the industry and the regulation by the government. What do you expect will happen in this industry? Why?

ANSWERS AND HINTS

Name That Term **1.** merger; **2.** acquisition; **3.** Vertical integration; **4.** marketing; **5.** structure; **6.** concentration ratio; **7.** conduct; **8.** Divestiture; **9.** Innovation; **10.** efficiency

True/False **1.** T; **2.** F; **3.** F; **4.** T; **5.** T; **6.** T; **7.** F; **8.** F; **9.** T; **10.** F

Multiple Choice **1.** e; **2.** e; **3.** e; **4.** a; **5.** c; **6.** e; **7.** b; **8.** c; **9.** c; **10.** c

Technical Training **1.** 84.2%; **2.** 69.2%; **3.** 17.9%; **4.** A; **5.** C

PART FOUR

MACROECONOMIC RELATIONSHIPS

INTRODUCTION TO MACROECONOMICS

CHAPTER OUTLINE

OVERVIEW

The previous chapters have dealt largely with microeconomics—the study of individual decision making or single-commodity markets. The center of attention was a single entity—a consumer, a firm, or the market for a single good. On the demand side, we looked at how an individual decides what to consume in order to maximize his or her utility. On the supply side, we analyzed what and how much a firm produces to maximize profit. We summed individual supply and demand curves to generate market demand and supply curves to explain price formation.

This chapter and the next three provide an introduction to macroeconomics, the study of how a nation's economy performs. This performance is of utmost importance to you. For example, it determines whether jobs are available when you and other students graduate, your salaries, and how much those salaries will buy. It determines what you can enjoy now and your children's economic prospects. As you will see in the following chapters, what happens to the economy as a whole has a major impact on the agricultural sector of the economy. Successful farmers and agribusinesses watch the news about interest rates, government deficits, and our balance of international payments with as much interest as individuals and firms in the rest of the economy.

LEARNING OBJECTIVES

This chapter will help you learn:

- How national income accounts are used to measure the performance of an economy
- Four ways of measuring the total goods and services produced in an economy
- How to construct price indexes and use them in measuring the rate of inflation

MACROECONOMIC CONCEPTS

Macroeconomics developed partly in response to the terrible hardships of the Great Depression of the 1930s. For many years, the most fundamental question for macroeconomics was, why does unemployment of potentially productive resources exist? When the word *unemployment* is used, we typically think of workers. But other resources are unused as well. Equipment stands idle or underutilized. Buildings are left vacant. Farmland is left unplanted. All nations have unemployed resources, and sometimes the proportion unemployed is large and persistent. A central challenge for macroeconomists is to explain why unemployment happens and to suggest policies to reduce unemployment of labor and other resources. In studying macroeconomic performance, economists try to answer questions such as the following:

- Why did the U.S. economy (and the whole world) sink into the Great Depression of the 1930s, and how did it recover?

• What factors caused the U.S. rate of inflation to reach 14 percent in the late 1970s? Did different factors cause the Indonesian rate of inflation to reach 600 percent in the mid 1960s and again in 1998?

• Why did the United States experience a simultaneous rapid growth in both its federal budget deficit and trade deficit in the early 1980s and then experience a budget surplus and a trade deficit in the 1990s?

• What factors account for the long period of prosperity and growth in the U.S. economy in the late 1990s and why did the farm sector not participate in that prosperity?

• Has the agricultural sector (or any other) become more sensitive to changes in macroeconomic conditions? If so, why?

We start the chapter with a discussion of the economic goals societies strive for, how an economy works and then turn to how to measure the economy's progress in achieving those goals. We give examples of the United States and a developing nation to give you a feel for what these measures mean to you. We also describe how we adjust economic data to account for changes in price levels and show that nations vary dramatically in the levels of inflation.

Macroeconomic Goals

The members of an economy determine its goals in an often-complex process that varies from country to country. Four basic economic goals are common to most countries.

1 The first is full and productive use of the economy's resources so that the economic "pie" is as large as possible. Another way of stating this first goal is to reduce unemployment of an economy's resources.

2 The second goal is to provide for the future. In macroeconomics, "saving" means not consuming part of current output in order to increase the future productive capacity of the economy. That is, some current production is devoted to creating production facilities, such as buildings and machinery.

3 A third goal, of paramount importance in many countries, is to control inflation. High rates of inflation reduce the productivity of a nation's resources because rapidly rising prices are less effective in allocating resources to their most productive use. This is because individuals and businesses make decisions to avoid the consequences of rising prices. In addition, high inflation rates can alter income distribution. It hurts individuals who live on fixed incomes and helps those who have income or assets that increase in value with price inflation.

4 A fourth goal, long-term balance in international transactions, has become especially important as the world economy becomes more interdependent.

As consumers and producers, we make decisions knowing that our actions alone are unlikely to have a measurable effect on the performance of the economy. The *ceteris paribus* assumption cannot be made in macroeconomic analysis where decisions of governments and industries do make a difference. We need to visualize all the interrelationships that occur among the different economic actors in an economy. To help clarify those relationships, we start with an extended metaphor.

The Economy as a Factory: An Extended Metaphor

Think of a nation's economy as a giant factory. This factory buys natural resources—iron ore, petroleum, agricultural land—and processes them using its plant and equipment. The factory hires workers to operate the machines and provide management and other services. The resources can be processed in the factory to produce either goods for current consumption or new machines to increase the factory's productive capacity in the future. Four groups buy the output of the factory: consumers, investors, the government, and foreigners.

1 The nation's consumers buy for current consumption.

2 Investors buy output to build new machines and repair the existing machinery because they expect to produce and sell more output in the future.

3 The government buys output to provide public goods and services for citizens.

4 Foreigners buy for consumption, for investment in their own factory, and for their government's use.

We can use this factory metaphor to think about some of the macroeconomic goals described above. A first goal is to increase the likelihood that the resources of the economy—including the natural resources, the labor, and the machinery in the factory—are used productively. Clearly, if a resource is not used (unemployment exists), the output of the factory is not as high as it could be, and the owner of the unutilized resource will not receive income to purchase output. Also, if the factory is mismanaged and operates inefficiently or does not use the latest technology, the factory will produce less output than possible.

The issue of who gets the output of the factory has two parts. The first part is about distribution among members of society today. The second part is about distribution between consumption today and investment to meet the consumption needs of tomorrow.

We can extend the metaphor and think of the world as many such factories: a German factory, a Japanese factory, a Malaysian factory, and so on. Each of these factories use domestic resources to produce output to meet domestic demands. In addition, the factories trade with each other, buying and selling output (such as cars, pineapples, and canned ham), intermediate products (such as spark plugs and soybean meal), or raw materials (oil and ores).

The fourth goal, long-run balance in international payments, means that one factory cannot be forever in debt to another factory. If a factory buys goods from another factory now, in the future it must send goods to the other factory.

Now let's take a look at what happens inside the factory. As we peer into the window, we see millions of assembly lines. Some are lines producing automobiles, some are fields growing rice, some are producing new machines, and some are classrooms producing an educated work force. For each line there is a set of machines and workers. Some of these lines use natural resources and intermediate inputs from other assembly lines. For example, the farming assembly line uses labor, land and water, and fertilizer and tractors from other assembly lines. Some of these lines run in parallel. Others are sequential; that is, the output of one line becomes an input for the next line. For example, most of the output of the farming line is used as input for the agribusiness lines. The output of the agribusiness lines flows into the wholesale and then retail food assembly lines. Also, some inputs come from the other countries' factories and some output is sold to the other factories.

Prices and money are critical to the efficient operation of the factory. With millions of assembly lines the number of transactions among them and to resource owners and consumers is enormous. Money facilitates exchange and serves as a common denominator for all prices. Money allows workers to sell their labor and allows resource owners to make their resources available, knowing that money received can be used to buy goods. Prices of the resources, intermediate goods, and final output of the factory are all denominated in money: $30.00 per hour of work by a skilled carpenter, $2.50 per bushel of corn, $22,950 for a new car. Each of these prices is determined in a market for the services used and goods produced in the factory. (Without money, managers of assembly lines would need to set up extremely complicated barter arrangements with resource suppliers, managers of other assembly lines, and output buyers. A manager of an automobile assembly line might trade one car for 200 tires. A farmer might offer 350 hogs for a new car.)

Money also serves as a measure of the value of the economic activities in the factory. The macroeconomic goal of stable prices is important in that it allows assembly line managers to make decisions without having to gather price information constantly. Also, people who receive money in payment for goods and services do not have to worry about its losing value. We will return to the important topic of money and its creation in Chapter 13.

Another important element of an economy is personal savings. People save, say in a bank, part of their income for future use. (Without money, people could only save by not consuming some goods and services.) The financial sector (including banks) lends the savings to investors who use the money to buy some of the goods and services. The financial sector specializes in identifying the most profitable investment opportunities.

With this brief overview of how the pieces of an economy fit together, let's turn to how macroeconomic performance is measured.

MEASURING MACROECONOMIC PERFORMANCE

All modern economies are extremely complex. Firms are the real-world equivalents of the assembly lines in the factory metaphor, and there are many of them. For example, the United States alone has millions of enterprises! Some employ one or two people, use equipment worth less than $10,000, and have sales of a few thousand dollars each year. Others, such as General Motors, employ thousands of people, have assets worth billions of dollars, and generate billions of dollars of sales annually. Most firms produce many goods and services so that the number of different products is difficult to imagine. Every transaction—whether it is the purchases made by college students, the wages paid to steelworkers, the profit earned by farmers, or the payment of a health insurance bill—is part of the nation's macroeconomic activity.

With all this complexity, how is it possible to measure macroeconomic performance? We need some way of reducing this complexity in order to measure what happens in an economy. For a firm, preparation of the income statement provides a convenient summary of all its transactions and allows the creation of other measures of performance such as liquidity, profit, and solvency. For a national economy, a similar accounting system is used, called the *national income accounts* system. In principle, the monetary value of all economic transactions in a period is recorded and then totaled to produce a few key measures of macroeconomic performance. The most commonly used measure of macroeconomic performance, **gross domestic product (GDP),** *is the value of all output (or income) produced within a country.* A related statistic, **GDP per capita,** *is the value of GDP per person.* GDP is analogous to the amount of the single output produced by the factory described above. It is a single measure of all the goods and services produced by an economy that can be either consumed today or invested to produce more goods and services in the future. An economy's GDP can be measured for any period of time, a month, a quarter, or a year. Unless specified otherwise, GDP is a measure of the value of all goods and services produced by the economy in a year.

Why is it called gross domestic product? The words in the definition of gross domestic product also can help to explain its meaning.

- *Gross*—because it includes all production, even that which replaces old equipment (depreciation). **Net domestic product (NDP)** *is GDP minus depreciation.*
- *Domestic*—because it is produced in that country. GNP, another measure of the economy, measures income and output produced in facilities owned by U.S. citizens, be they located in the United States or another country.
- *Product*—because it measures what is produced, not what is consumed.

NATIONAL INCOME ACCOUNTS

The basic idea behind national income accounting is the following: the economy has a **stock** of resources (also called **factors of production**) that includes skilled and unskilled workers, buildings, machinery, land, and many other things. The economy uses a **flow** of services from those resources to produce a flow of goods and services: the gross domestic product (GDP). This output is either used for current consumption or saved and invested to increase productive capacity. **National income accounts** are *measures of the sources and destinations of flows of current income, expenditure, and production.*

FLOWS AND STOCKS

Flow—an amount measured over a specific period of time; a rate (e.g., income is measured per week, per month or annually).
Stock—a fixed amount that doesn't depend on time (e.g., the stock of textbooks you own doesn't depend on the time period in which you measure it). To understand the difference between a stock and a flow, think of the money in your savings account as a stock that generates a flow to you of money in the form of interest earned. Or, you can think of your student loan as a stock of money the bank has provided to you for which you must make a flow of payments to pay the interest.

We can return to the factory metaphor to make these ideas clearer. The factory and its machinery (capital), the managers and workers (labor), and the natural resources (land) are the economy's stock of resources. The number of hours worked in a given period by the workers is the flow of labor services. The time the assembly lines operate is the flow of machinery services. The iron ore, petroleum, and other natural resources brought to the factory are the flows of natural resource services from the stock of resources in the ground. These inputs are combined to produce a flow of output from the factory. The workers, the owners of the assembly lines, and the owners of the natural resources are all paid for their inputs. The sum of these payments is national income. Part of the income is used to buy output from the factory for consumption; part of it is saved and invested in the factory. With these basic ideas in mind, we now turn to a more detailed discussion of the national income accounts.

The Circular Flow of Goods and Payments

We start with a simple economy that consists of only **households** and **businesses** to illustrate the basic concepts used in national income accounting. We can describe the economy as **circular flows** of goods and services, and of payments for those goods and services (see Figure 11-1). Households own all the stocks of factors of production. They sell a flow of services from their factors of production to businesses. They receive payments for these services in the form of wages (payments to workers), interest and

FIGURE 11-1   The circular flow of economic activity.

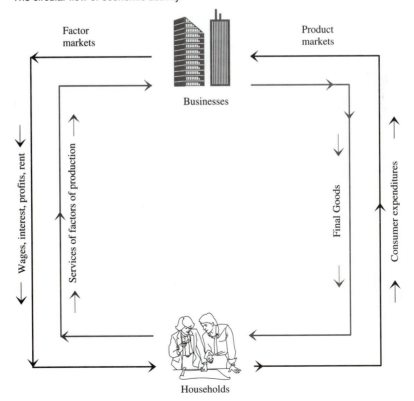

profits (to owners of capital), and rents (to natural resource owners). Businesses use these factor services to produce the goods and services sold to households.

In Figure 11-1, the left side of the diagram represents the flow of factor services to businesses and the compensating flow of payments to households. The right side represents the flow of consumer goods to households and the compensating flow of payments back to businesses. The inner loop represents physical flows—factor services to businesses and outputs to households. The outer loop consists of payments for the physical flows—payments for factor services (wages, interest, profits, and rent) and payments for outputs.

It is important to understand that in this economy, without government, saving, investment, or trade, the values of the four flows are equal. In the product markets, the right side of Figure 11-1, the goods produced in the businesses are valued at the amount the households were willing to pay for them. Thus the value of final goods exactly equals consumer expenditures. On the factor markets, the left side of Figure 11-1, payments for wages and other factors services are valued at the amount businesses were willing to pay households for those services. Thus the value of factor services exactly equals household incomes. In the factory metaphor, income received by resource owners (the assembly lines, the natural resources, and the workers) is equal to the value of the output of the factory. All of the income earned by producing goods and services is used to buy these goods and services.

To illustrate, if the economy in Figure 11-1 has a GDP of $1 million then the four arrows would represent:

- households spending $1M to purchase $1M worth of goods from businesses
- businesses spending $1M to purchase $1M of labor and management services from households
- businesses receiving $1M from households for $1M worth of goods sold
- households receiving $1M from businesses for $1M of labor and management services

Because all of these flows are equal we can measure GDP using any one of these flows.

How would agricultural production fit in the circular flow diagram? Farms, ranches, and agribusinesses can be thought of as assembly lines in the macroeconomic factory. They buy services of natural resources and inputs from other assembly lines (e.g., fertilizer makers, implement manufacturers). These businesses also sell products both to other assembly lines (wheat to flour mills and flour to bakeries) and to final consumers (bread, fruits, and vegetables). Farming is unusual because it is the primary user of some natural resources, land and water, but it also uses machinery and labor like other parts of the economy. In the circular flow diagram, farming and agribusiness production are in the "businesses" box. Of course the farm family's consumption would be included in the households.

OWNERSHIP

While businesses hold title to many resources, one or more individuals (stockholders, partners, etc.) own all businesses. The owners of the businesses are, then, the actual owners of the resources controlled by businesses. In the terminology of national income accounting, households are made up of all individuals and own all resources.

Circular Flows with Financial Intermediaries, Government, and International Transactions

A real economy is, of course, more complex than suggested by the simple circular flow diagram. First, households don't use all their income for current consumption; they save part of it. These savings flow through the financial sector to individuals and firms who make investments in production facilities and infrastructure (see Figure 11-2). That is, some of the economy's production is not sold to consumers, instead it is devoted to increasing the capacity of the business sector. Financial intermediaries—banks, insurance companies, mutual funds, and other financial institutions—provide a market in which household savings are loaned to investors for this purpose. The profits from these investments flow back to the savers through the financial institutions.

The second addition to the circular flow diagram shown in Figure 11-2 is the government. Government purchases outputs from businesses and hires services of factors from households. It provides services, for example, to build and operate schools, construct roads and harbors, and provide national defense. The resources for these activities come from tax revenues and other charges (collected from households and businesses) and by borrowing from the public and from foreign investors. When government

FIGURE 11-2   Circular flows with financial intermediaries and government.

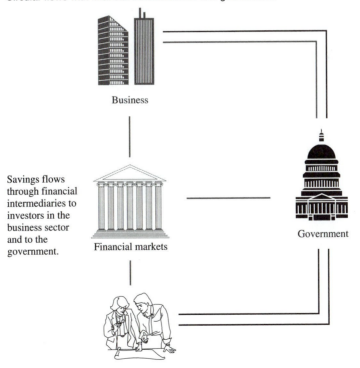

Business

Savings flows through financial intermediaries to investors in the business sector and to the government.

Financial markets

The government taxes households and businesses; purchases factor services from households and final goods and services from businesses; and produces services for businesses, financial markets, and households.

Government

Households

operates in this way it is analogous to a business, except it is "owned" by the public. We will show later that the government also functions in ways significantly different from a business.

Government revenues are collected from three sources: (1) *Direct taxes,* based on wages, profits, and property rents, are collected from income recipients. (2) *Indirect taxes* are collected on sales transactions, for example, sales taxes, value-added taxes, and excise taxes. (3) Governments receive payments in the form of fees and user charges from individuals and businesses that use resources owned by the government. The services of these resources are sometimes sold to individuals or business firms, generating fee and user-charge revenues. For example, the governments of many countries own some or all of the forests and sell logging rights to the private sector. In the United States, the federal government and some state and local governments own grazing land, forestland, and other property and charge user fees or rents. Also, when you buy a stamp you are paying for a service provided by the government.

Our final addition to the circular flow diagram is the foreign sector (see Figure 11-3). An economy's output is sold domestically to households and businesses but also through international markets to foreign households and businesses. *Sales to foreigners are called* **exports.** Similarly, a portion of domestic purchases are **imports,** *purchases*

FIGURE 11-3   International transactions in the circular flow diagram.

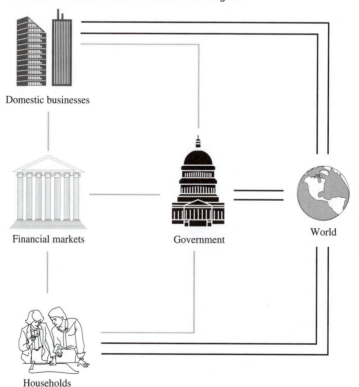

Domestic businesses

Financial markets

Government

World

The rest of the world purchases goods from businesses (exports) and sells goods to households, businesses, and the government (imports).

Households

by consumers, investors, and the government from foreign markets. Instead of treating exports and imports as two separate items, we often combine them into a single value. **Net exports** *are exports minus imports.* **Net imports** *are imports minus exports.*

MEASURES OF GROSS DOMESTIC PRODUCT

The fact that an economy can be represented by a circular flow means there are several ways to measure GDP. In this section, we present four equivalent ways of measuring it. Why do we want several measures? First, because each of the four measures gives the same answer we can check the accuracy of our estimates. Second, it helps to understand how the economy and the system of measures work. Third, and most importantly, the different approaches use different ways of aggregating transactions and provide us with information useful in economic analysis. Four measures of GDP are the following:

1 **Final-expenditure approach**—*the values of all final goods and services.* Intermediate inputs—goods produced and then used to produce other goods—are not counted directly in this measure.

2 **Value-added approach**—*the value added by factor services used by all producers, including those that make intermediate inputs.* In this approach, we don't include the value of final goods directly.

3 **Sources-of-national income approach** (or the national-income approach)—*all income received by owners of factors of production.*

4 **Uses-of-national-income approach**—*the uses of household income for consumption, savings, and tax payments.*

Next we look at the first three methods of measuring GDP in more detail and see how they are related. We use a hypothetical production chain, from wheat to bread, shown in Figure 11-4 to illustrate each of these approaches. We will then discuss the fourth measure, the uses of national income.

The Final-Expenditure Approach

The final-expenditure approach sums the total value of final goods and services produced in the economy. A **final good** *is any good purchased for final use.* **Intermediate inputs** *are goods to be resold or used as inputs into other goods.* The values of intermediate goods are not included in the final-expenditure approach calculations because their value is included in the final good. For example, the price you pay for an automobile includes the cost of the tires and all of the other inputs used to produce it. Adding the cost of the tires to the cost of your new car would double count the value of the tires.

Consider a hypothetical bakery that sells bread directly to consumers. In this example, bread is the final good. To make this final good, a farmer purchases inputs and uses them to produce wheat that is sold to a flour mill, which produces and sells flour to a bakery, which produces the bread for sale to the consumer, as shown in Figure 11-4. All of these transactions are necessary to provide the consumer with bread. Intermediate inputs in this example include fertilizer on the farm, wheat to the flour mill, and flour and yeast to the bakery.

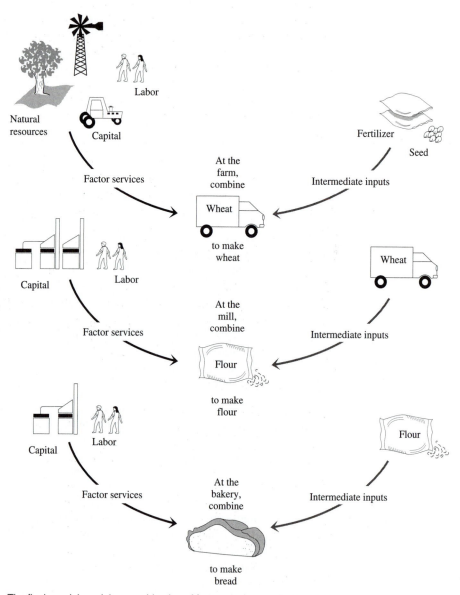

FIGURE 11-4   The final good, bread, is a combination of factor services and intermediate inputs.

INTERMEDIATE OR FINAL GOOD

If the bakery sold the bread to a grocery store, the bread would be an intermediate good in the transaction between the bakery and the grocery store. The bread would not become a final good until the grocery store sold it to a final consumer.

TABLE 11-1 GROSS DOMESTIC PRODUCT OF THE UNITED STATES, MEXICO, AND INDONESIA

	United States*	Mexico**	Indonesia**
GDP (billions of dollars)	9191.8	483.7	151.2
GDP per capita (dollars)	33,544	4,900	580
Percent consumption	67.6	68	72.5
Percent investment	19.3	23	13
Percent government	17	10	6.5
Percent exports	11.8	31	35
Percent imports	−15.9	−32	−27

*U.S. Department of Commerce, Bureau of Economic Analysis, first quarter 2000 data annualized.
**World Bank Group, Country data tables for 1999.

But who, then, are the final goods consumers in an economy? All final outputs are bought by one of four groups:

1 Households that use final goods for consumption (C),
2 Businesses that use final goods for investment (I),
3 The government (G) which buys final goods to provide services, and
4 Net exports that occur when foreigners buy final goods for use abroad ($X - M$).

Thus using the final-expenditure approach, GDP can be expressed algebraically as:

$$GDP = C + I + G + (X - M)$$

In words, gross domestic product equals consumption + investment + government + exports − imports. This equation is crucial to understanding macroeconomics, you will use it many times in the following chapters.

Table 11-1 gives the final expenditure and GDP data for the United States, Mexico, and Indonesia. The GDP and the GDP per capita of the United States is far larger than Mexico and Indonesia. However, the percent of the GDP that is devoted to personal consumption is approximately 70 percent in all three countries, as is the case in most countries. Government expenditures are higher in the United States than in Mexico and Indonesia. Trade, both imports and exports, represent a larger share of the economy in Mexico and Indonesia than in the United States.

The Value-Added Approach

At each stage in a production chain like that for bread, a firm combines intermediate inputs it purchases from other firms with services from its resources: equipment, hired labor, managerial skills, land, and so on. The payments for these services are called value added (because value is added to the intermediate inputs).

Value added *is equal to the value of the firm's output minus the value of intermediate inputs purchased. Value added is also equal to the firm's payments for factor services.* In the bread example, the farmer combines intermediate inputs like fertilizer and seed with services from hired workers, capital equipment, and land to produce wheat. At the flour mill, the wheat is an intermediate input, and payments to the workers and payments for buildings and equipment are elements of value added. At the bakery, the flour is an intermediate input, and payments to the workers and payments for buildings and equipment are elements of value added.

If we were to add all the transactions, including sales of intermediate inputs, we would count value added by the wheat farmer, the flour mill, and producers of other intermediate inputs at least twice. Hence, the final-expenditure approach includes only the sales of the final goods (like bread) not used in other production activities. In the value-added approach, the value added to intermediate goods purchased in all production activities is aggregated.

Another important difference between the final-expenditure approach and the value-added approach to measuring GDP is in how the data are collected and aggregated. In the final-expenditure approach, we collect data from final users of goods. In our example, using the value-added approach, we divide an economy into several sectors. For example, there might be an agricultural sector, a manufacturing sector, and a retail sector. The value added by each firm in a sector is summed to arrive at the total value added by that sector. GDP is then equal to the value added in all sectors of the economy.

We now return to our bread example. Assume the economy has been divided into agriculture, manufacturing, and retail sectors. The value added by the wheat farmer is included in the agricultural sector, the value added by the flour mill is included in the manufacturing sector, and the value added in the making and selling of bread is included in the retail sector. To compute GDP, we sum the value added in the three sectors. Equivalently, the value-added approach is the sum of all the factor services represented on the left side of Figure 11-4.

Whether we use the value-added approach or the final-expenditure approach to measure GDP, the result is the same. The different approaches to data collection give us a way to check our calculations. Value-added data are collected from each firm and then summed. Final-expenditure data are collected by examining purchases of consumers, the government, investment expenditures, and export and import data. They also give us different perspectives on the economy.

Table 11-2 gives the sources of consumer income in the United States. It shows that the bulk of income is from salaries and wages. Proprietor's income, corporate profits, and interest payments are all significant contributors. Rent receipts is the smallest contributor.

TABLE 11-2   SOURCES OF NATIONAL INCOME, UNITED STATES

Percent salaries and wages	70.4
Percent proprietors income	8.9
Percent rent	1.2
Percent corporate profits	12
Percent interest	7

*U.S. Department of Commerce, Bureau of Economic.
  Anaysis, first quarter 2000 data annualized.

The Sources-of-National-Income Approach

In this approach to the measure of national income the *payments made to the resource owners are summed.* These include the wages received by workers, the rents received by resource owners for the use of land, buildings, and equipment, and the profits and depreciation reserves received by business owners. Thus while the final-expenditure and value-added approaches measure the amounts of money businesses receive, this approach measures the amount businesses pay for the resources they use. Referring again to Figure 11-4, it is the sum of the payments for factor services received by the owners of the resources represented on the right side of the diagram (wages and profits + rent + interest + depreciation + taxes or government income).

$$GDP = NI = W + R + i + D + T$$

A Numerical Example

We now extend our hypothetical bread example to illustrate three ways to measure GDP (Table 11-3). We have four firms: an input producer, a farm, a flour mill, and a bakery. The input firm produces the intermediate inputs such as fertilizer used by the farmer; the farmer produces wheat, an intermediate input to the flour mill. The flour mill sells flour to the bakery. The link between the firms is that the gross revenue of one firm is a cost of intermediate goods to the next firm in the production chain. The three GDP measures are as follows:

1 The final-expenditure approach, we count only the value of the bread sold which is $175,000.

2 The value-added approach, we sum the value added in each of the four firms (in thousand dollars, 41 + 59 + 34 + 41) which is $175,000.

TABLE 11-3   EXAMPLE OF CALCULATION OF GDP USING THE FINAL-EXPENDITURES, VALUE-ADDED, AND NATIONAL-INCOME APPROACHES

Cost item	Input firm	Farm	Flour mill	Bakery	Total sources*
Intermediate inputs		41	100	134	
Factor services					
Rents	10	30	5	4	49
Wages	12	8	5	25	50
Interest	6	6	7	4	23
Depreciation	8	7	7	2	24
Profit	5	8	10	6	29
Value added†	41	59	34	41	175
Revenues	41	100	134	175	

*The total sources is an example of another measure of national income used by economists. In this case national income = wages + rents + profits + interest + depreciation = $175,000.
†Value added equals revenues minus cost of intermediate inputs.
Note: Hypothetical data in thousands of dollars.

TABLE 11-4   GROSS DOMESTIC PRODUCT AND PERCENT OF VALUE ADDED BY SECTOR, UNITED STATES, MEXICO, AND INDONESIA

	United States*	Mexico**	Indonesia**
Percent agriculture	1.3	5	17.3
Percent industry	21.8	28.2	43.1
Percent services	77.7	66.8	39.6

*U.S. Department of Commerce, Bureau of Economic Analysis, first quarter 2000 data annualized.
**World Bank Group, Country data tables, 1999.

3 The sources-of-national-income approach, we sum all income received by the factors of production (rents, wages, interest and depreciation, and profits). The sum of income received at all levels is also $175,000 (in thousand dollars $49 + 50 + 23 + 24 + 29$).

Thus in all three approaches, the contribution to GDP of these four firms is $175,000.

Table 11-4 shows dramatic differences in the makeup of the economy. The U.S. economy is a predominately service economy with a significant component in industry, but less than 2 percent in production agriculture. (The agribusiness activities are included in the industry and service sectors.) Indonesia, a low-income, developing country, has over 17 percent of its activity in production agriculture and only 40 percent in services. Mexico, with a per capita income well above Indonesia but far below the United States, is also midway between the United States and Indonesia in both agriculture and services. This pattern is consistent with Engle's law, which states that as per capita income increases, production agriculture's percent of GDP decreases. This is illustrated in Figure 11-5, which shows these data for various countries as of 1998.

FIGURE 11-5   Per capita gross domestic product and agriculture's share of gross domestic product, 1998.

For selected countries, first number is per capita gross domestic product and the second is value added by agriculture as a share of gross domestic product.
From World Bank, World Development Report, 1999/2000.

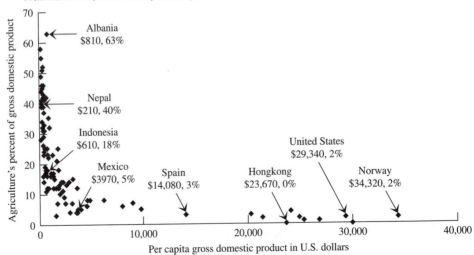

TABLE 11-5   USES OF NATIONAL INCOME, UNITED STATES, MEXICO, AND INDONESIA

	United States*	Mexico**	Indonesia**
Percent consumption	81.6***	68	72.5
Percent savings	0.1	11.9	14.5
Percent taxes	15.6	20.7	16.3

*Percent of personal income, U.S. Department of Commerce, Bureau of Economic Analysis, 2000.
**World Bank Group, Country data tables, 1999.
***Equal to 67.8 percent of GDP.

The Uses-of-National-Income Approach

Another important method of measuring GDP is to sum the uses of the income households receive for the use of the factors of production. The income received by households is used in one of three ways: consumption, savings, or taxes. **Consumption** expenditures—to buy food, clothing, transportation, leisure activities, and so on—takes most of our income. Various levels of government take part of our income as **taxes.** The remainder is **savings,** whether kept as cash, deposited in a bank, used to buy corporate stocks or bonds, or used to buy government bonds. The formula for this approach is as follows:

$$GDP = NI = C + S + T$$

So, gross domestic product = national income = consumption + savings + taxes.

Table 11-5 gives the percent of GDP consumers allocate to consumption, savings, and taxes. Note that largest share of GDP is allocated to consumption by consumers in all three countries. Taxes take 15 to 20 percent of consumer's resources in all three countries. The most notable datum in the table is the approximately 0% United States savings rate. This indicates that as of early 2000, the net addition to savings by consumers was negligible, which is unusual.

PRICE INDEXES AND INFLATION: NOMINAL AND REAL INCOME

In measuring the performance of the national economy, we use monetary or money values—quantity times price—to avoid the problem of trying to sum goods such as apples and oranges. However, some price changes alter money values without changing the underlying real value of the good or the performance of the economy. For example, if all prices doubled, the (nominal) value of the output of the economy would also double, but the performance of the economy would not change, *ceteris paribus.* To analyze the true performance of an economy, we adjust for the effects of general price increases or decreases through the use of price indexes. **Price indexes** are constructed *to provide a measure of how much prices have changed over a period.* There are a variety of ways to calculate a price index, but all have the following characteristics:

1 They include prices from a representative collection of goods and services, usually called a market basket of goods when applied to consumer goods. The **market basket** *reflects a typical expenditure pattern of some group in the economy.* The consumer price index is a basket of goods that a typical consumer would buy. The "prices paid by farmers" index is an index of prices of goods purchased by a typical farm operator.

2 The price of each good is weighted by the proportion of that good in the market basket.

3 The index is calculated with respect to a reference period, a "base" year (or set of years).

The value of the index in the base year is set equal to 100. The general formula for the price index in year n is:

$$P_n = \frac{C_n}{C_b} \times 100$$

where C_n is the cost of the basket of goods and services in the nth year. C_b is the cost in the base year.

To illustrate how a price index is calculated, we have constructed a hypothetical student price index (or SPI) (Table 11-6). The index includes a representative set of the goods and services that a typical student might purchase in a month. In our example, the first year is the base year, and so the price index for that year is set equal to 100. In the second year, prices of most goods increased. The value of the index is equal to the cost of the bundle in the second year divided by its cost in the first year times 100 [e.g., $(931.7 \div 857.3) \times 100 = 108.68$].

AGRICULTURAL PRICE INDEXES

Two price indexes for agriculture, prices received by farmers and prices paid by farmers, are sometimes used when making judgments about the economic well-being of the farm sector. Suppose that the prices received by farmers increased by 20 percent over several years, while prices paid for inputs increased by 30 percent in the same period. If nothing else changed during that period of time, income would fall. However, incomes could have risen if farms were larger and if they used less input per unit of output due to new technology.

TABLE 11-6   A HYPOTHETICAL STUDENT PRICE INDEX

Item	Quantity purchased per year	Year-end prices ($)		
		Year 1	Year 2	Year 3
Milk (quarts)	40	0.65	0.69	0.72
Telephone charges	50	1.65	1.79	1.84
Shirt	10	19.95	21.49	21.99
Cereal (boxes)	30	2.45	2.45	2.69
Sandwich	200	1.65	1.79	1.84
Notebook	20	3.59	3.79	3.79
Bus rides	200	0.50	0.60	0.75
Cost of student basket		$857.30	$931.70	$986.40
Price index (Year 1 = 100)		100.00	108.68	115.06

Nominal and Real GDP

A **nominal** economic statistic refers to *the cost, price, or value that is observed in the economy.* A **real** economic value refers to *an observed price, cost, or value adjusted to compensate for the effects of price changes.* Over time, real GDP provides a better measure of the performance of the economy than nominal GDP because adjustments for the effects of changes in the general price level have been made.

A price index can be used to adjust nominal GDP data (i.e., based on current prices). The adjusted number, called real (or constant) GDP, is calculated with the following formula.

$$GDP_{real} = \frac{GDP_{nominal}}{\text{price index}} \times 100$$

Inflation

The **inflation** rate *measures the extent to which prices in an economy have increased.* The **deflation** rate *measures the extent to which prices in an economy have decreased.* A price index is used to calculate the rate of inflation. The formula below shows how to calculate an inflation rate expressed as a percentage, based on the price index P:

$$\text{Inflation rate} = \frac{(P_n - P_{n-1})}{P_{n-1}} \times 100$$

For example, we can use the SPI from Table 11-6 to calculate the effect of inflation on student living costs. From year 1 to year 2, the annual increase was 8.67 percent [100 × (108.68 − 100) ÷ 100]. From year 2 to year 3, the annual increase was 5.87 percent [100 × (115.06 − 108.68) ÷ 108.68].

Experiences with Inflation: The United States and Brazil

Around the world, experiences with inflation vary widely. Some countries perennially have had high inflation rates, while others have had little inflation for long periods. Still other countries have experienced highly variable inflation rates. While deflation seldom occurs, it also disrupts the functioning of the economy. For example, in the late 1990s Japan experienced deflation and instances of zero interest rates.

To show how different inflation rates can be, we present data for two countries: the United States and Brazil. The United States generally has had low inflation except for a period in the late 1970s. Brazil has had extended periods of very high inflation.

Figure 11-6 presents the U.S. consumer price index (CPI) from 1965 to 1999. With 1982 to 1984 as the base period, the CPI rose from 30 to over 160. Average consumer prices in 1965 were only 30 percent of what they were in the base period. A common way of describing this change in prices is that a dollar's worth of goods in the base period (1982 to 1984) cost $0.30 in 1960 and $1.60 in 1999. Notice that the CPI increased most rapidly from the mid 1970s to the early 1980s. This effect can be seen more clearly in Figure 11-7. Inflation reached a high of over 13 percent in 1980 and fell dramatically

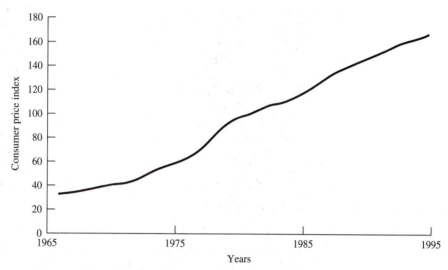

FIGURE 11-6   The U.S. consumer price index, 1965–1999 (1982–1984 = 100).

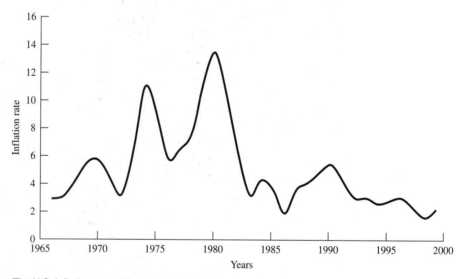

FIGURE 11-7   The U.S. inflation rate, 1965–1999 (percent).

from 1981 to 1985. Because of high inflation in the late 1970s, nominal GDP grew much more rapidly than real GDP, as shown in Figure 11-8.

Although the U.S. inflation rate has fallen back from the record high levels of 1980, it remained high by historic standards (about 3 percent) until the late 1980s. This high inflation hurt people on fixed incomes, or with fixed interest-bearing assets, and created

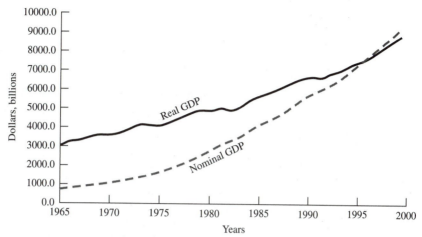

FIGURE 11-8   U.S. GDP 1965–1999 (billion dollars).

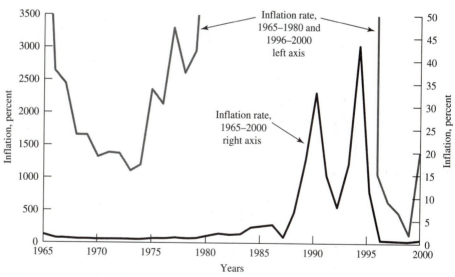

FIGURE 11-9   Inflation rate, Brazil, 1965–2000.

many other problems for businesses and consumers. We discuss the costs of high inflation again in Chapter 13.

Brazil's experience with inflation has been dramatically different from that of the
United States. We present here a graph of the Brazilian inflation rate from 1965 to 2000
(Figure 11-9). The left side vertical scale for Brazil ranges up to over 3000 percent.

That scale shows the hyperinflation that occurred in Brazil from the mid 1980s to the mid 1990s. If one were to calculate the Brazilian consumer price index from 1980 to 1989 with1985 as the base year, it would have grown from one to almost 100,000. In contrast, the U.S. CPI increased only three and one-half times from 1960 to 1989 (see Figures 11-6 and 11-7), a period almost three times as long. This rate of inflation means, for example, that a pair of Brazilian shoes that cost 20 cruzeiros in 1980 cost 2,000,000 cruzeiros in 1989.

The right hand scale of Figure 11-9 shows the rate of Brazilian inflation during years when it was below 50 percent. This scale indicates that the Brazilian economy has suffered levels of inflation much higher than those experienced in the United States during almost all of recent history.

These enormous rates of inflation make economic planning decisions very difficult. People try to convert money into goods as quickly as possible. It makes little sense to try to save or to hold cash or checking account balances because money quickly loses so much value. In some countries with high inflation rates, the currency of another country becomes widely used. But this only partly reduces the economic havoc caused by very high inflation rates.

GDP AS A MEASURE OF NATIONAL ECONOMIC PERFORMANCE

If GDP data are to be used to measure the performance of an economy, they should provide a picture of how well the economy is meeting the current and future needs of its citizens. GDP per capita is generally considered the single, best measure of economic welfare. However, it has several shortcomings.

First, it doesn't measure the benefits from leisure. One way to increase your income (and therefore raise GDP per capita) is to work more hours, perhaps by taking a second job. But everyone gets utility from leisure activities, whether they are sports, reading, family activities, or something else. Even workaholics need some sleep! At some point, the value of an additional dollar earned by working is not worth the loss in utility from less leisure time.

Second, several kinds of economic activity do not get counted. For example, goods and services that do not pass through a market are generally excluded from measures of GDP. A volunteer group that paints a house for charity contributes to someone's welfare, but that work involves no exchange of money and therefore is not counted. For example, the value of farm products consumed at home does not pass through a market and is not counted. In those developing countries where a large share of farm production is consumed on the farm, this omission is important, and efforts are made to correct for it in GDP statistics. Another example is child care. In developing countries, child care is often done by family members and thus is not counted in GDP. In developed economies, as more women have joined the labor force, a growing share of day care happens in day-care centers (and is paid for). In this case child care is counted.

Another example of uncounted economic activity is the "black market," or underground economy. If transactions are not reported to government agencies because they are illegal or to avoid tax payments, they are not counted in GDP. For example, marijuana is thought by some to be the most valuable crop in California. By some estimates, well over 25 percent of economic activity in Italy is not reported to avoid tax payments.

A third problem with using GDP per capita as a measure of economic performance is that it does not deal appropriately with externalities (externalities refer to pollution and other effects not captured by the market, see Chapter 15 for a discussion of this topic). The costs of dealing with a negative externality add to GDP, instead of subtracting from it. For example, the costs of cleaning up an oil-tanker spill increase GDP, but the damage resulting from the spill does not reduce GDP. Another example is the health cost associated with secondhand cigarette smoke and other forms of air pollution. The costs of treating the illnesses count as a contribution to GDP, but the costs of health care and lost work are not subtracted. Positive externalities are also not counted. For example, if Central Park in New York City were planted in corn, the corn harvested would add to the GDP, but national welfare would certainly go down.

Fourth, there is no measure for depletion of natural resource stocks, equivalent to depreciation for capital, to account for the use of natural resource stocks (see the box on page 284 for an example). National income accounts include a depreciation component to account for the fact that capital equipment wears out, but do not measure the fact that a nation that mines minerals, extracts oil, cuts forests, or allows soil to erode might be less able to provide for its citizens in the future.

Fifth, GDP doesn't consider income distribution. An added dollar earned by the richest person is counted the same as a dollar earned by the poorest person, although the contribution to the poor person's welfare might be much higher. If helping the poor were an important national goal, a nation would choose to implement policies that make many poor people better off at the expense of a few rich people, even if somewhat slower growth results.

INCOME DISTRIBUTION

If every person in the United States received the 1998 per capita GDP of $31,500, the average American family of 2.5 would have had an annual income of $78,750. If this seems high to Americans from average income families, it is! The average includes income received by the very rich and profits retained by firms rather than distributed to households. Income is not distributed evenly in the United States. The richest 20 percent of the population receives over 46 percent of the income (the richest 10 percent receive over 30 percent). The poorest 20 percent receive only about 5 percent (the poorest 10 percent only receive 1.8 percent).

Despite these limitations, the use of GDP and GDP per capita as measures of macroeconomic performance is widespread. Although they do not measure all aspects of economic performance, they are the best measures widely available. There have been, and are likely to continue to be, efforts to adjust these measures to reflect various aspects of social, resource, and environmental aspects of the performance of the economy.

WASTING ASSETS: NATURAL RESOURCES IN THE NATIONAL INCOME ACCOUNTS

Despite its importance in guiding and evaluating economic policies, the system of national income accounts currently used by many countries does not measure a nation's income correctly. This failure happens because natural resources are treated as gifts of nature instead of productive assets whose value must be depreciated if they are used up. National income accounts treat such assets as buildings and equipment as productive capital whose value depreciates over time as they perform valuable work. Natural resource assets are not so valued. A country could exhaust its mineral resources, cut down its forests, erode its soils, pollute its aquifers, and hunt its wildlife and fisheries to extinction, but measured income would not be affected as these assets disappeared. Such balance sheet asymmetries are particularly dangerous for developing countries, which usually depend heavily on their natural resource base for employment and exports.

If natural resources are not treated as productive assets, economic planners receive false signals that reinforce the unsound dichotomy between the economy and the environment. Confusing the depletion of valuable assets with the generation of income promotes the idea that rapid economic growth can be generated by exploiting the resource base, resulting in illusory gains in income and permanent losses in wealth.

Source: Adapted from *Wasting Assets: Natural Resources in the National Income Accounts* by Robert Repetto et al. World Resources Institute Publication Brief, June 1989.

SUMMARY

The goal of this chapter has been to describe how macroeconomic performance is measured. Performance is always relative to a set of goals. The chapter began with a discussion of four common macroeconomic goals: improving the current and future well-being of the citizens of the economy, and maintaining price stability and long-run balance in international transactions. Most of the chapter was devoted to measuring gross domestic product (GDP) and its various components.

We presented the price index and inflation rate, and showed how changing prices resulted in a difference between nominal and real GDP. Some of the problems with using GDP as a measure of an economy's performance were discussed. Finally, we compared national income accounts of the United States and Indonesia to understand better some similarities and differences between developing and developed economies.

IMPORTANT TERMS AND CONCEPTS

QUESTIONS AND EXERCISES

Name That Term

Read the following sentences carefully and fill in the missing term or terms.

1 _____ are measures of the sources and destinations of current income, expenditure, and production.
2 Natural resources, capital, and human capital are _____.
3 A _____ is a selection of items that is representative of purchases made by consumers that is used to determine whether average consumer prices are increasing or decreasing in the economy.
4 _____ are goods sold in the market of another country.
5 _____ are goods that are purchased for resale and for further processing or manufacturing.
6 _____ is a rise in the general (average) level of prices in the economy.
7 A _____ shows how the average price of a market basket of goods changes through time.
8 The U.S. _____ is the total market value of all final goods and services produced annually within the boundaries of the country, whether by U.S. or foreign-supplied resources.
9 _____ refers to extremely high rates of inflation.
10 The _____ is the most commonly used index of price changes.

True/False

Read the following sentences, then decide whether each statement is true (T) or false (F) and mark it accordingly.

T   F   1 Except for a period in the late 1970s, the U.S. economy has been spared extended periods of very high inflation.
T   F   2 The 2000 per capita GDP of $33,500 means that a typical U.S. family enjoys an income of over $70,000.
T   F   3 The real interest rate is the interest rate you must pay when you borrow money from a bank.
T   F   4 Under the uses-of-income approach, national income is the sum of consumption, savings, and taxes.
T   F   5 GDP is a good measure of national economic performance because it tries to account for leisure, nonmarket activities, and externalities.
T   F   6 U.S. citizens net savings was nearly zero in 2000.

T F   7 In the United States the percent of national income payed in taxes is less than in Mexico or Indonesia.

T F   8 The stock of a resource is the amount of that resource used per unit of time.

T F   9 Using the final-expenditure approach, consumption accounts for about 67 percent of GDP, investment for about 20 percent, government for about 17 percent, and net imports for about 5 percent for the United States in 2000.

T F   10 At the world level, just as imports must balance exports, inflation must balance deflation.

Multiple-Choice Questions

Circle the letter of the response that best answers the question or completes the statement.

1 The goal of macroeconomic policy is
  a the full and productive use of the economy's resources.
  b providing for the future by investing some of today's output.
  c keeping prices stable and reducing inflation.
  d long-term balance in international transactions.
  e all of the above.

2 Which of the following flows is not equivalent to the others?
  a factor services from households to businesses
  b income to households
  c total imports plus total exports
  d goods from businesses to households
  e payments for the goods in goods from businesses to households

3 Which of the following are not final uses of output under the final-expenditure approach?
  a consumption by households
  b investment by the business sector
  c consumption by the government
  d net imports
  e net exports

4 About what share of U.S. GDP is exported?
  a 5 percent
  b 10 percent
  c 20 percent
  d 30 percent
  e 40 percent

5 GDP per capita is not an accurate measure of what a person can spend on consumption items primarily because
  a it does not take into account inflation.
  b it does not reflect imports and exports.
  c it does not reflect volunteer activities.
  d it does not take into account taxes and saving.
  e none of the above.

6 Macroeconomics deals with
  a imperfectively competitive markets.
  b the long run adjustments to equilibrium in the economics.
  c a nation's economy as a whole, in aggregate.
  d individual units within the economy.
  e all of the above.

7 One of the implications of Engel's law is
  a as per capita income increases the share of GDP from farming decreases.
  b nominal incomes are generally higher than real incomes.
  c consumption is positively related to imports.
  d inflation is positively related to unemployment.
  e imports are positively related to economic growth.
8 National income accounts are measures of
  a personal income levels.
  b income on a real basis.
  c exports and imports, adjusted for inflation.
  d income adjusted by the appropriate price index.
  e sources and destinations of current income, expenditures, and production.
9 The gross domestic product is equal to
  a $C + I + G + (X - M)$.
  b NI.
  c $C + S + T$.
  d $W + R + i + D + T$.
  e all of the above.
10 Gross domestic product is not a perfect measure because it does not account for
  a the value of leisure activities.
  b volunteer and black market activities.
  c income distribution inequities.
  d resource depletion and environmental externalities.
  e all of the above.

Technical Training

A "market basket" is a combination of goods and services that is selected to be representative of what consumers actually purchase. Suppose you were given the cost data below for the items in that market basket on the last day of each of ten years. Suppose the base year used to calculate a price index was 1990.

1 By what percent did the price of the market basket increase between 1990 and 1999?
2 What was the price index in 1999?

Year	Cost of market basket of goods (current $)
1990	722.00
1991	750.88
1992	795.93
1993	859.61
1994	919.78
1995	928.98
1996	956.85
1997	1052.53
1998	1105.16
1999	1204.62

3 What was the price index in 1996?
4 By what percent did the price of the market basket increase in 1996?

5 By what percent did the price index increase in 1996?

6 In what year was the inflation rate highest?

7 In what year was the inflation rate lowest?

8 If gasoline cost 80 cents per gallon in 1990 and $1.25 per gallon in 1999, in which year was its real price lower?

9 If nominal GDP in 1990 was $4.9 trillion, what was real GDP?

10 If you accepted a new position in 1990 at $35,000, what would your salary have to be to give you the same purchasing power in 1999?

Questions for Thought

1 Explain why gross domestic product is equal to national income and is equal to the value added in an economy.

2 How are the following economic transactions included in GDP: (a) final sales to consumers (b) pollution and environmental degradation costs (c) value-added business transactions (d) after-tax wages and salaries (e) rental payments to property owners?

3 Is there a general relationship between the level of income and the share of consumption in total income? If not, why not? If there is, what is it?

4 If U.S. income were somehow redistributed so that every American received the same amount, would the mix of goods and service produced be affected?

5 Make a list of ways you have been engaged as a buyer or seller of economic goods and services that do not appear in macroeconomic accounts of economic activities.

6 The "underground economy" reduces the validity of aggregate economic accounts. What other harm do "underground" activities do? In what ways do "underground" economic activities benefit the economy?

7 Consider a job you have or once held. What were the intermediate inputs used in your firm? Does/did your firm produce final or intermediate outputs? How does/did your firm add value to the economic flow?

8 Everyone who has lived for several decades has been known to shake their head and marvel about how low prices once were: 25 cents per gallon for gas, $10,000 for a house, $1,200 for a new car, 5 cents for a candy bar, and so on. Is it accurate to conclude from these observations that "everything is more expensive these days?" Why or why not?

ANSWERS AND HINTS

Name That Term   **1.** National income accounts; **2.** factors of production; **3.** market basket; **4.** Exports; **5.** Intermediate inputs; **6.** Inflation; **7.** price index; **8.** gross domestic product; **9.** Hyperinflation; **10.** consumer price index (CPI)

True/False   **1.** T; **2.** F; **3.** F; **4.** T; **5.** F; **6.** T; **7.** T; **8.** F; **9.** T; **10.** F

Multiple Choice   **1.** e; **2.** c; **3.** d; **4.** b; **5.** e; **6.** c; **7.** a; **8.** e; **9.** e; **10.** e

Technical Training   **1.** 66.84%; **2.** 166.84; **3.** 132.53; **4.** 3%; **5.** 3%; **6.** 1997; **7.** 1995; **8.** 1999; **9.** $4.9 trillion; **10.** $58,394.00

12

GROSS DOMESTIC PRODUCT AND FISCAL POLICY

CHAPTER OUTLINE

OVERVIEW

In this chapter we develop the basic tools that allow us to show why unemployment might exist and what causes inflation. The explanations for both problems lie in the interaction of aggregate demand and supply. We identify the determinants of aggregate consumption. We also examine how households allocate income among uses. Finally, we begin to examine the effect of macroeconomic policies on employment and inflation, starting in this chapter with fiscal policy.

LEARNING OBJECTIVES

This chapter will help you learn:

- How aggregate demand and supply interact to determine equilibrium GDP
- Why an economy might reach macroeconomic equilibrium at a level below full employment
- How fiscal policy can affect employment and inflation

EMPLOYMENT AND INFLATION

The main goal of this chapter is to develop the tools of aggregate demand and aggregate supply to examine the determinants of unemployment and inflation. We begin with a definition. The economy is producing **full-employment output,** or full-employment GDP (GDP_{fe}), when all available resources are productively employed. At GDP_{fe} the owners of resources used as productive inputs receive incomes of Y_{fe}. Recall that in Chapter 11 we showed that we can measure the size of the economy in terms of the value of final goods or the total of all household income sources. If actual GDP is less than GDP_{fe}, the economy has some unemployed resources and therefore Y will be less than Y_{fe}. In practice, it is impossible for every resource to be fully employed. There is always some **frictional unemployment,** associated with people moving between jobs who are technically unemployed but for whom employment in the transition would not make economic sense. There is also some **structural unemployment,** referring to workers out of work because their skills are not demanded by employers or because they lack sufficient skills to obtain employment. For the United States, an unemployment rate of around 5 percent was considered to be "full employment" until the late 1990s. For several years at the end of the century, the country's unemployment was below this level. The ability of the economy to sustain this level of employment is unclear.

All economies experience periods when some resources are not employed and some potential GDP_{fe} is not being produced. The period of highest unemployment rate in the 20th century was the Great Depression of 1930 to 1939. During that period, one in five American workers did not have a job. Similar conditions existed in many other countries.

Figure 12-1 shows the general relationship between unemployment and **capacity utilization**. When unemployment spiked in the 1970s and 1980s, the percent of productivity capacity used also fell to relatively low levels. The decade of the 1990s was rather unusual in that unemployment fell to historically low levels without a simultaneous rise in percent of capacity utilized. This is likely an important factor explaining why price levels did not increase with such low levels of unemployment. Other factors that contributed to the low unemployment and inflation were the technological advances as-

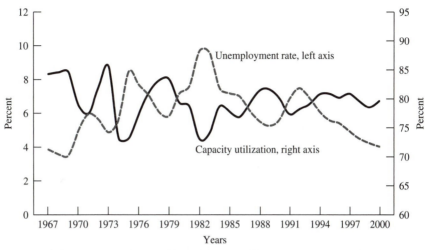

FIGURE 12-1 U.S. percent unemployment and percent capacity utilization, 1966–1999.
Source: U.S. Department of Commerce, Bureau of Economic Analysis.

sociated with the information age, the Clinton administration economic policy, and the increased competitive pressures from freer international trade.

The principal concern of macroeconomists in the 1930s was how unemployment could continue for extended periods and whether there was a role for macroeconomic policy to reduce unemployment. The conclusion of these Keynesian economists was that government policies could be devised to stimulate aggregate demand and reduce unemployment, and they advocated an active role for the government in macroeconomic management.

JOHN MAYNARD KEYNES AND "KEYNESIAN" ECONOMICS

The theory underlying much of macroeconomics draws from ideas developed by John Maynard Keynes, a British economist whose professional career spanned the great depression. Keynes wrote his most famous work, *The General Theory of Employment, Interest, and Money,* to provide a theory of why the Great Depression happened.

Few macroeconomists today accept all of Keynes' views, but most would agree that their work has roots in his insights on how an economy functions.

In the 1970s, the emergence of inflation as a serious problem in the U.S. economy spurred a new generation of macroeconomists to examine more carefully the assumptions made by Keynes and his followers and to rethink the role of government macroeconomic policy (see Figure 11-9). These neoclassical or new classical economists are skeptical about the effectiveness of any government actions to reduce unemployment. They are more concerned about the harmful effects of high inflation and argue that use of macroeconomic policies to stimulate aggregate demand is a principal cause of inflation.

Both issues—the causes of unemployment and inflation, and the role of macroeconomic policy—raise complex questions, and macroeconomists disagree about the
answers. In this chapter and the next, we present the basic concepts that underlie the
attempts to answer these questions. We start with an examination of aggregate supply
and then turn to aggregate demand.

We start the analysis assuming that the aggregate price level is constant. Firms are assumed to use as many resources as necessary to meet the aggregate demand for goods
and services, without affecting the price level. This assumption allows us to focus on the
mechanism that moves an economy to full-employment GDP, without worrying about
the complications of changing prices. Later, we relax this assumption and examine the
links among aggregate demand, aggregate supply, and the price level as the economy
functions at different levels of output and unemployment.

AGGREGATE SUPPLY

The **aggregate supply** (AS) of goods and services in an economy, GDP, is equal to the
value of final goods and services produced by the economy. Firms determine output as
they make the microeconomic decisions discussed in Chapters 3 through 5. If firms did
not produce anything, aggregate supply would be zero. As firms use more resources to
produce more output, both GDP and national income (Y) increase along the AS line as
shown in Figure 12-2. GDP increases as firms produce more goods. National income increases because the sources of household income grow as firms make higher levels of
payments to the owners of resource inputs for their services. As explained in Chapter 11,
we can measure the size of an economy using measures of final output or national in-

FIGURE 12-2   Aggregate supply.

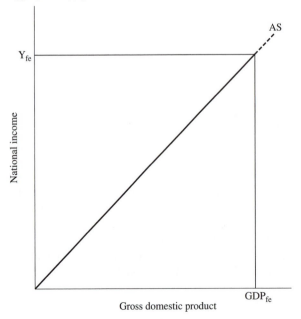

come. This means that GDP = Y and there is a one-to-one relationship along the AS curve.* This one-to-one relationship results in the AS curve being a 45-degree line from the origin if the scales on the axes are the same.

The full employment line (GDP$_{fe}$) in Figure 12-2 is the maximum amount that this economy can produce given its resources and its production capacity. It cannot produce more because all of the economy's resources are employed. The message of Figure 12-2 is that the economy can produce at any level of goods and services on the AS line up to GDP$_{fe}$. Later we will show that it is possible, over time, for an economy to shift the full employment output to a higher level by saving and investing and by improving productivity.

AGGREGATE DEMAND

Aggregate demand AD for an economy's output includes four components: consumption (*C*), investment (*I*), government (*G*) and foreign demand or net exports (*X − M*). The final expenditure measure of GDP is:

$$GDP = AD = C + I + G + (X - M)$$

To understand why aggregate demand and GDP are at a certain level, and why it might not be at full-employment GDP, we need to know what determines each of these components of aggregate demand. Remember from the previous chapter that we showed that GDP is the total output produced and AD is the total output consumed by the economy and that these are equal. In this chapter we will not consider any of the aspects of the international environment because it is not necessary to understand fiscal policy. Thus both imports and exports are zero and the (*X − M*) term falls out of our equation. We return to the international environment in Chapter 14. We start by considering consumption. We will then add savings and investment and conclude by adding government. This will give us a model that we can use to understand how fiscal policy, government taxes, and expenditures affect the economy.

Consumption

As indicated in Table 11-5, for the United States, Mexico, and Indonesia, the **consumption** component of demand makes up 60 to 85 percent of GDP. Therefore, factors that explain consumption behavior also explain the use of the largest percentage of GDP. A large body of evidence suggests that income is the main determinant of consumption. Your own experience probably reflects this fact. It is likely that if your income increased substantially, you would consume more. This is true for national consumption as well.

The information presented in Figure 12-3, on U.S. real per capita income and consumption expenditure for 1960 to 1999 shows the relationship between national income and aggregate demand for the United States. The graph would look similar for other countries. The relationship of consumption to income is called the **consumption function.** As you can see in Figure 12-3, a straight line is a good representation of the

*To simplify our analysis we do not distinguish between national income and aggregate personal income.

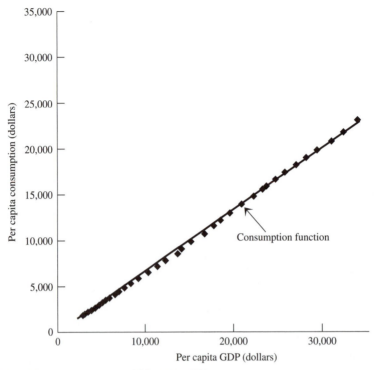

FIGURE 12-3   U.S. consumption and GDP, 1960–1999.

consumption function. This suggests that consumption is primarily a function of income; other economic variables have little effect.

Note in Figure 12-3 that the slope of the consumption function is less than 1 (it is
about 0.65). The slope of this line is called the **marginal propensity to consume (mpc).**
It shows the change in consumption that results from a change in income. The formula
is as follows:

$$mpc = \Delta C \div \Delta Y$$

In most countries, the mpc is between 0.6 and 0.7. This means that if national income
increases by $1000, consumption increases by between $600 and $700.

The line that runs through the plots of consumption and per capita income observations cuts the consumption axis (the y axis) above zero. The value at which the consumption function cuts the vertical or consumption axis is called **autonomous consumption.**
Autonomous consumption is the expenditure an individual would make if income were
zero, by reducing savings or borrowing. One way to think of autonomous consumption is
as a subsistence level. More accurately, autonomous consumption measures all of the factors other than income that determine the level of household consumption. Some of these
other factors include consumer expectations, inflation rates, and the wealth of the individuals. In equation form, the consumption function is written as follows:

$$C = a + mpc \times Y$$

This says that households will consume (C) an amount equal to their autonomous consumption (a) plus the marginal propensity to consume (mpc) times their income (Y). Thus, the consumption function tells us how much of the economy's output, GDP, households demand.

AVERAGE AND AUTONOMOUS CONSUMPTION

We can use the consumption function to see why the average consumption share in GDP ($C \div Y$) for countries with per capita incomes below $2000 is above 75 percent, while for rich countries it is about 60 percent. When income is small, autonomous consumption is a large component of consumption. This means that average consumption is roughly equal to autonomous consumption. For rich countries autonomous consumption is a relatively small share of total consumption, and so average consumption is nearly equal to the mpc ($\Delta C \div \Delta Y$).

We add a consumption function to Figure 12-2 to give us Figure 12-4. GDP is on the horizontal and uses of national income is on the vertical axis. So far in our discussion consumption is the only component of aggregate demand considered. That is, all national

FIGURE 12-4   Aggregate production, aggregate consumption, and savings.

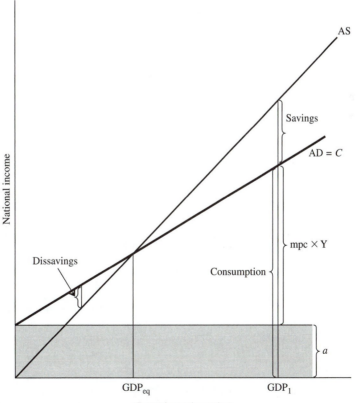

Gross domestic product

income is being used by households for consumption. That is, households use all national income for consumption. We will turn to the remaining components in the following pages. All measures are expressed in real (constant dollar) terms, which is the same as assuming that all prices are constant. This very simple economy would be in equilibrium (GDP_{eq}) where the C and AS lines cross. At this point households purchase all of the output being produced by the economy. This also implies that all income received by households is being spent to purchase all of the goods produced by all firms in the economy or, consumption is the only source of demand for the output of the economy. In other words, the economy supplies just enough output to meet consumption demand. The final-expenditure approach becomes:

$$GDP = C$$

Next we add savings and investment to our economy.

Aggregate Savings

Generally, households do not use all of their income to buy goods for consumption. They save some of their income. In effect the savings are loaned to others who, as we will explain below, invest those funds. In an economy with no government or trade, income will be either consumed (spent on goods and services) or saved. Thus:

$$GDP = Y = C + S$$

The amount of consumption and savings in an economy can be seen in Figure 12-4. In this figure we add the consumption function to the AS line from Figure 12-2. Autonomous consumption is the distance from the x-axis to the horizontal line, as indicated by a. The variable portion of the consumption function (mpc × Y) is added to autonomous consumption as indicated. In the remainder of the chapter we add investment and government to consumption to complete the AD function.

At levels of GDP above GDP_{eq} savings occur. At GDP_1 the amount of savings is the difference between the aggregate supply function and the aggregate demand function. If aggregate supply is below GDP_{eq}, negative savings, or dissaving, result. This is something that occurs only in the most desperate of economic conditions. It implies that some of the economy's assets are being converted to consumption goods. At a level of aggregate supply such as the GDP_{eq} level shown in Figure 12-4, no net savings or investment would occur in the economy. Savings are positive above this level and negative below it.

INTEREST RATES, SAVINGS, AND CONSUMPTION

Since consumption and savings are determined simultaneously, it seems that the interest rate would play an important role in determining savings and, therefore, consumption. In fact, the role of the interest rate in determining consumption and savings is small. The interest rate does play an important role in determining the *form* of savings but has little effect on how much. Savings can be held in a variety of assets. Savings accounts pay moderate interest, but the principal is very safe. Stocks of high-tech firms have the potential for rapid appreciation of value but also for the loss of the principal. Agricultural land generates a stream of revenues, and its value depends primarily on the net income from the crops that can be grown on it. The choice among the alternate assets is strongly influenced by the interest rate and other factors, such as risk. But the influence of the interest rate on the savings level is small.

Aggregate Investment

Investment is another component of aggregate demand. Investors use the output made available by savings to build new factories, install new equipment, buy new farm machinery, and start new firms. Investors are willing to borrow from savers and make interest (or dividend) payments if they expect sufficient profit from an investment. The interest rate is a key determinant of the profitability of an investment. As the interest rate increases, profitable investment opportunities are harder to find because the cost of borrowed funds goes up. Hence, investment is related negatively to the interest rate. Because an investment has a productive life of many years, the profitability of an investment will depend on what happens in the future. Hence, expectations about future performance of the economy are more important than current performance of the economy. Thus investment is a function of interest rates and expected returns, not present income. As a result we can add investment to our diagram of the economy as a horizontal line above autonomous consumption, as in Figure 12-5. The distance from a to I gives the amount of investment. If investment increases, the line would shift upward. Comparing to Figure 12-4 note that the addition of investment shifts the AD upward by the amount of I. This shifts the GDP_{eq} to the right. At GDP_{eq} aggregate demand (the sum of

FIGURE 12-5   Production, consumption, and investment.

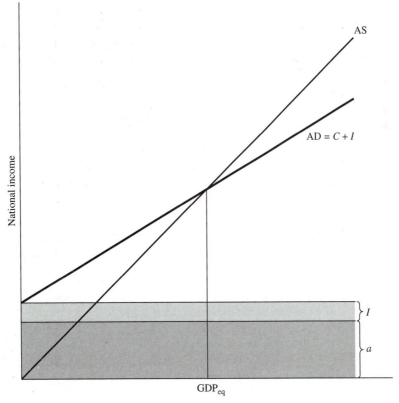

consumption and investment) equals aggregate supply. The economy is now devoting goods and services to investment as well as to consumption. This means that households consume most of aggregate production and all remaining production is invested in new productive capacity. We will show below that savings equals investment in this economy without government or trade.

INVESTMENT

When talking about the national economy, economists use the term investment differently than its everyday use. Economists define investment to mean adding *new* equipment to the stock of capital by adding plant, equipment, and facilities to the economy's productivity capacity. The purchase of an already completed building only transfers ownership from one person to another, it does not add to the capital stock.

Savers could invest themselves (and some do), but as an economy becomes increasingly complex, more specialization happens. Some individuals specialize in identifying profitable investment opportunities. A complex financial industry develops to provide a market that links savers to investors. Therefore, saving and investing activities are undertaken by different individuals. Financial intermediaries (such as banks, insurance companies, and brokerage houses) channel funds from savers to investors.

Algebraically, we can represent this economy with several equations. Aggregate demand is equal to consumption plus investment and is also equal to national income and gross domestic product. Gross domestic product is used for either consumption or investment.

$$GDP = C + I$$

In terms of national income, in an economy with no government and no trade, income must be either consumed or saved:

$$GDP = C + S$$

Therefore we can combine these two GDP functions we get:

$$GDP = C + I = C + S$$

And by elimination:

$$I = S$$

The implication is that in a simple economy, with no government or trade, saving (S) is equal to investment (I). The way the economy adjusts to equate savings and investment, that are activities of different persons and for different reasons, is one of the keys to understanding macroeconomic performance.

Suppose that the total income expected by households is less than equilibrium GDP, and therefore planned saving is less than investors plan to borrow. Investors act on their plans and order more investment goods than producers had planned to produce. These orders cause businesses to reduce inventories and produce more goods, hiring more factor services. The new employment generates additional income and results in house-holds increasing consumption and savings. Hence, planned saving increases until even-tually it is equal to planned investment.

Suppose instead that planned saving is greater than investment. Initially, this leads to too many goods available for investment. Business managers who produced, expecting a certain level of demand, have excess stocks and reduce production. Aggregate supply falls as does employment and income. With lower income, planned saving falls until it is equal to planned investment. These economic decisions lead to an important result: *an economy tends to move toward equilibrium GDP, although that level may be below full-employment GDP*.

Next, we add government expenditures and taxes to our model of the economy.

Government Expenditures and Taxes

Government expenditures G and **taxes** T are determined in the political arena and thus are not directly determined by GDP or national income. In Figure 12-6 we show how G and T are added to our graphical representation of the economy. Consider first

FIGURE 12-6   Government expenditures and taxes.

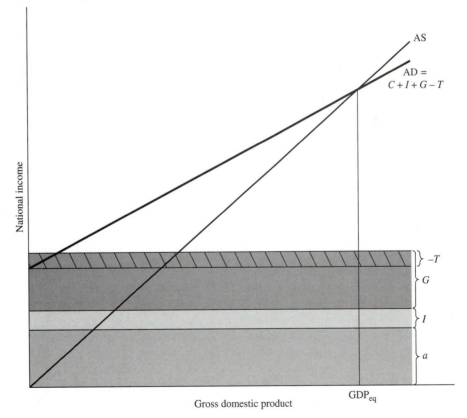

government expenditures. Government purchases of goods are an addition to the consumer and investment components of aggregate demand. The addition of G in Figure 12-6 is shown as an addition of G to a and I. This additional spending by government also shifts AD upward by an amount equal to G.

Taxes have an effect opposite that of government expenditures because taxes reduce the amount of income consumers have to spend. If we assume that taxes do not change with the level of national income (an assumption we relax below) we can add taxes to Figure 12-6 as shown. Taxes reduce rather than add to autonomous consumption, investment, and government expenditures. Therefore in Figure 12-6 T is negative. Taxes thus also shift the AD line downward.

In Figure 12-6 the addition of G is much larger than the reduction by T. As a result of G being larger than T, the combined effect is to shift AD upward and to shift GDP_{eq} to the right, as is evident if Figures 12-5 and 12-6 are closely compared.

We now turn to an algebraic model of the economy. In this model we more accurately recognize the means by which taxes affect the economy.

As you know from personal experience, taxes affect the economy by changing the amount of money available for consumption. However, because the effect of taxes on the economy occurs through an effect on consumption, we need to return to the consumption function, $C = a + mpc(Y)$ to find out how much AD is reduced. A consumer must pay taxes and may spend or save the remaining income that is called **disposable income.** Thus taxes are subtracted from Y in the consumption function giving us

$$C = a + mpc(Y - T).$$

This means that taxes have the effect of reducing aggregate demand by the amount of $-mpc(T)$. If the mpc is 0.6, a one-dollar tax increase reduces consumption and thus aggregate demand by $0.60.

We now have all the pieces in place to find the macroeconomic equilibrium. (Remember we are not considering trade in this chapter because it is not needed to understand fiscal policy.) The economy can be represented graphically as in Figure 12-6 and algebraically as follows:

$$GDP = Y = C + I + G$$

And we have modified this to reflect the consumption function and the effect of taxes to

$$GDP = [a + mpc(Y - T)] + I + G$$

Before we turn to an analysis of fiscal policy we point out one additional relationship. Remember our earlier observation that S must equal I; which was true because if:

$$GDP = C + I = C + S$$

Then

$$I = S.$$

Now we consider what happens when we add government. If the government budget is in balance, taxes and government expenditures will be equal. (Later we will consider what happens if it is out of balance, as is typically the case.) The final expenditure equation is:

$$GDP = C + I + G$$

And as we showed in Chapter 11 the uses of GDP equation is:

$$GDP = C + S + T$$

Therefore,

$$C + I + G = C + S + T$$

And canceling the Cs,

$$S + T = I + G$$

That is, savings plus taxes $(S + T)$ must equal private investment plus government expenditures $(I + G)$.
   And if:

$$S = I$$

Then

$$G = T$$

must also be true. Therefore, if G is greater than T (the government operates with a budget deficit) then private investment will be less than savings.

EQUILIBRIUM GDP

In Chapter 5, we found that if a product market is not in equilibrium, economic forces shift the market toward equilibrium. For example, if producers charge a price above the market level, a surplus will develop and the producers would have to lower the price in order to sell the product. The same forces push the economy toward equilibrium. But a crucial difference exists. When the market for a single good reaches equilibrium, there is no surplus or shortage of the good. At the macroeconomic level, however, it is possible to reach equilibrium with some unemployed resources that could be used to produce more output. That is, the economic forces that push the economy toward equilibrium do not necessarily push it toward full employment.
   The **equilibrium level of GDP** occurs where aggregate demand equals aggregate supply, which happens where the aggregate demand curve cuts the aggregate supply curve. This is GDP_{eq} in Figure 12-6. This is the only level of output (GDP) where the levels of aggregate demand and aggregate supply are equal. At that level, investment plus government expenditures equals savings plus taxes (remember that savings are indicated by the vertical distance between the aggregate supply and the aggregate demand functions).
   The forces that move an economy toward equilibrium, at both the microeconomic and the macroeconomic level, are the decisions made by consumers, producers, and investors based on their expectations about the performance of the economy.

EXPECTATIONS

The role of "expectations," what market participants expect to happen in the future, is central to economics.

An important area of research in economics and agricultural economics is how people form expectations. One school of thought is that decision makers use "adaptive expectations"; that is, information about previous prices and sales trends is used to predict future prices and sales. When actual prices and sales differ from previous expectations, the expected future outcomes are adapted to take this into account.

The new classical economists believe that decision makers have "rational expectations." Proponents of this view argue that expectations are formed with much more information than just previous prices. In the extreme, this view argues that decision makers use an economic model that considers all the market forces at work and includes predictions of the effects of changes in government policies.

Recessionary and Inflationary Gaps

Now we turn to the implications of equilibrium output that is not at full-employment GDP level. In Figure 12-7, the economy would be using all of its resources if it produced at an aggregate supply (AS) level equal to GDP_{fe}. However, if the economy operates with aggregate supply equal to aggregate demand (AD_1) at GDP_{eq1}, then not all of the economy's resources are used. The vertical distance between the aggregate demand and aggregate supply curves at full-employment GDP is a **recessionary gap.** This gap shows how much aggregate demand would have to shift to be in equilibrium at full-employment GDP.

Now consider what would happen at GDP_{eq2}. In this case there is "too much" aggregate demand. As suggested in Figure 12-7 the aggregate demand line (AD_2) crosses the aggregate supply line to the right of full-employment GDP. In reality this is not possible, because it would mean an aggregate supply (AS) larger than the economy can produce. Aggregate demand at GDP_{eq2} would be larger than the productive capacity of the economy. The vertical gap between AD_2 and supply at full-employment GDP is an **inflationary gap** which is a measure of how much aggregate demand would have to fall so that GDP_{eq} would equal GDP_{fe}.

When a model gives impossible outcomes, the assumptions of the model are wrong. For a single good, when supply is unchanged and demand shifts out, the price of the good increases; and when demand shifts in, the price falls. What is different at the macroeconomic level? Our assumption that aggregate supply responds to shifts in aggregate demand without changes in the price level gives us an inaccurate result. We can return to the economy-as-factory metaphor to explore what is happening. With unemployed resources, it is easy for the manager of an assembly line to hire more workers or buy more inputs. Thus it is possible for the economy to move from GDP_{eq1} to GDP_{fe} by using previously unemployed resources to move up the AS line. However, if all resources are fully employed and one assembly line wants to hire more services, the only source is other assembly lines. But this would reduce output from the other lines. It would also bid up the prices of the resources and, in turn, the prices of the goods produced. Thus, if all

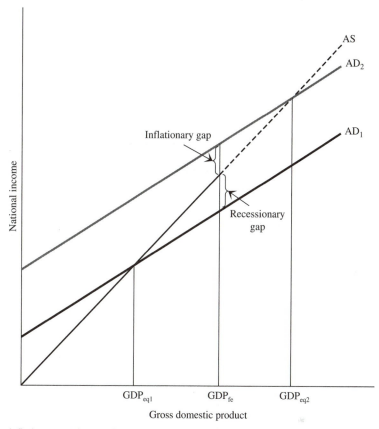

FIGURE 12-7   Inflationary and recessionary gaps.

resources are already employed, output cannot increase, but the effort to employ more resources will lead to price increases. An effort to increase output beyond GDP_{fe} would instead result in inflation due to an increase in the price of inputs. Therefore, our assumption that aggregate output responds to an outward shift of aggregate demand without changing prices is wrong, at least when the economy is at or near full employment. Next, we consider the relationship between aggregate prices and GDP.

PRICE RIGIDITY AND AGGREGATE SUPPLY

An important characteristic of the operation of an economy is that prices rise when the economy is operating near capacity but usually do not fall when equilibrium GDP is less than full-employment GDP (a notable exception occurred during the Great Depression when aggregate prices did fall). When the economy is below full employment some resources remain unemployed, in part because of an expectation of improvement. Workers tend to seek employment at their previous rate of pay rather than accept a job at lower pay or part-time employment with lower income. Property owners tend to allow

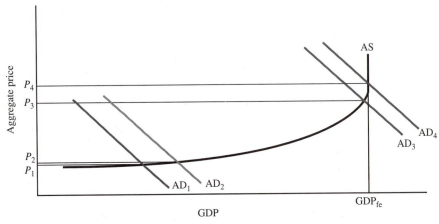

FIGURE 12-8   Aggregate demand and supply curves, and the aggregate price level.

some rental space to remain vacant rather than to reduce rents in order to attract additional tenants.

We can illustrate this effect by considering the relationship between the price level and GDP. In Figure 12-8, the horizontal axis is GDP; the vertical axis is the aggregate price level. Over a wide range of GDP-levels the aggregate supply curve, AS, is nearly horizontal which means that a change of GDP has little effect on price. This region, called the **Keynesian region,** represents output levels at which some, and even large, portions of an economy's resources are unemployed. Since in this region prices do not change as GDP changes, firms can meet an increase in aggregate demand by hiring more workers and using idle capacity to increase production. This is illustrated in Figure 12-8 by the shift of aggregate demand from AD_1 to AD_2 with essentially no change in prices.

As the economy nears full employment levels the situation is quite different. Here increases in aggregate supply, AS, are accompanied by larger and larger increases in the price level. As the economy approaches full-employment GDP, factory managers find that they do not have capacity to meet the increased demand. Businesses find it increasingly difficult to hire workers, even with higher wages, and bottlenecks become more common in industries. This is illustrated in Figure 12-8 by the shift of aggregate demand from AD_3 to AD_4. This outward shift of demand generates significant increases in the price level with little change in output.

Now we consider why the aggregate price level almost never falls but quite often increases. For two examples of this, return to Figures 11-6, 11-7, and 12-1. In spite of considerable variation in unemployment and inflation rates over more than 40 years, U.S. prices rose continuously.

A major source of **downward price rigidity** is in the labor market where demand and supply for labor services determine wage rates and unemployment. In many labor markets there are long-term contracts (either explicitly or by custom) or salary scales (such as that for civil service) that do not allow wage reductions. In these contracts wage rates, with scheduled increases, are set for a year or more into the future. However in some of these contracts, firms negotiate for the right to lay off workers when sales fall. Furthermore, in many developed economies, the "social safety net" provides benefits for unemployed workers, reducing the urgency of finding a new job. For all these reasons, a worker might choose to be temporarily unemployed rather than to take a lower paying job or a cut in pay that might become permanent. Therefore wages do not fall readily.

The pricing practices of many firms also contribute to downward price rigidity. There are costs associated with cutting prices. Products must be repriced, new price information must be disseminated, and there are other problems, including the possibility that retail firms may not pass along the reductions to consumers and that competitors may retaliate to protect their market share. In either case, the benefit of lowering prices is lost. So, it is often more profitable to reduce volume, lay off workers, and let some resources be idle than to lower prices. Also, firms that own and rent space may choose to allow some space to remain vacant rather than to reduce rents. Finally, if the expectation is that prices will rise in the near future, firms may not make adjustments that would be made if the reductions were viewed as permanent.

DOWNWARD PRICE RIGIDITY AND AGRICULTURE

Unless the government intervenes to limit price variability, prices for the major agricultural commodities such as corn, wheat, and soybeans are much less likely to be rigid downward than prices in other sectors of the economy, because their markets are among the most competitive that exist. This fact can lead to large swings in commodity prices. For example, a shift inward in aggregate demand might reduce demand for chicken. If fast-food stores keep the chicken price constant this will cause a larger decline in the corn price because the derived demand for the corn used in animal feed will shift inward more than if the fast-food stores had lowered the chicken price somewhat and thereby increased the quantity demanded.

A consequence of high agricultural price flexibility is that farmers tend to have more variable incomes but smaller fluctuations in employment than other members of an economy.

The Keynesian region of the aggregate supply curve, with unemployed workers and underutilized equipment, provides a potential role for macroeconomic policy. With some combination of fiscal and monetary policy, aggregate demand can be shifted outward, which can increase employment and output with little effect on prices and inflation.

THE ROLE OF FISCAL POLICY

As the previous section showed, an economy can be in equilibrium below full employment. An increase in demand can shift equilibrium closer to full-employment GDP, but at the risk of causing prices to rise. Two different classes of macroeconomic policy—fiscal and monetary—can be used to shift aggregate consumption, but they work in quite different ways. In the remainder of this chapter, the focus is on the effects of fiscal policy, the choice of the level of taxes, and government expenditures. The next chapter deals with money and monetary policy, how much money is available, and the effect of money supply on the performance of the economy.

Fiscal Policy and Aggregate Consumption

All governments tax and spend. Few, if any, national governments (in contrast to state and local governments) require a balanced budget. Some years, a government's expenditures might exceed spending, resulting in a deficit; in other years there may be a surplus. **Fiscal policy** refers to the use of expenditures, taxation, and budget deficit or surplus to influence aggregate demand.

Both the national political process and the state of the economy determine fiscal policy. Some government expenditures are determined by national income and other variables. For example, welfare payments rise when national income falls. Government payments to farmers generally vary inversely with farm prices. The annual budget-making process seldom alters these expenditures. Other government expenditures are determined annually. These "discretionary expenditures" can be changed from year to year to influence the performance of the economy. Examples include defense spending, research contracts, support of the arts, and federal funding for interstate highway repairs. As politicians decide on expenditures, they are influenced by domestic social concerns, foreign policy considerations, and a wide range of other factors. One important factor is, or should be, the effect on GDP.

As with expenditures, some tax revenues change automatically with GDP; others do not. For example, income tax revenues change automatically with income levels and gasoline tax revenues change with GDP. Property and cigarette tax revenues change little with year-to-year changes in economic activity. In addition the government can increase or decrease taxes.

We focus on discretionary expenditures and taxes, because these can be used by policy makers to influence aggregate demand.

An Example   To show how fiscal policy can change the performance of an economy we will work through a hypothetical example. Suppose that you are a macroeconomist asked to advise the government of a country with an economy (with no imports and exports) that is experiencing a recession. You have quantitative estimates of the consumption function, GDP_{fe}, and investment, and government revenues and expenditures; you can estimate equilibrium GDP. And if equilibrium is not at full employment, you can estimate how much government expenditures or taxes need to be changed to reach equilibrium at full employment.

Suppose you find that if the economy employs all of its resources, it can produce $750 billion worth of output. (All of the data in this example are in billions.) That is, $GDP_{fe} = \$750$ billion. Your research on the relationship between income and expenditures indicates that the consumption function is as follows:

$$C = 150 + 0.6(Y - T)$$

Investment (I) is $50, government expenditure (G) is $110, and taxes ($T$) are $40. Thus the government is operating with a deficit of $70 per year and there is concern about unemployment. You want to reduce unemployment while avoiding an inflationary surge in prices. The first thing you might do is calculate the equilibrium GDP given current conditions.

To estimate GDP_{eq} you would start with the aggregate demand equation:

$$GDP_{eq} = Y = C + I + G$$

Substituting the consumption function into this equation gives:

$$GDP_{eq} = Y = 150 + 0.6(Y - T) + I + G$$

Replacing T, I, and G by their respective values you get:

$$GDP_{eq} = Y = 150 + 0.6(Y - 40) + 50 + 110$$

Rearranging and solving this equation Y, you get:

$$GDP_{eq} = Y(0.4) = 150 - 24 + 50 + 110 = 286$$

$$GDP_{eq} = Y = 286/0.4 = 715$$

This tells you that the equilibrium level of GDP_{eq} is $35 billion below the $750 billion level of output estimated to be attainable. On this basis you report to the decision makers that the economy is operating with a recessionary gap, which is causing the unemployment. This is shown in Figure 12-9 where the GDP_{fe} is well above GDP_{eq}. To get the desired equilibrium, the level of the aggregate demand curve must be increased.

You know that fiscal policy can be used to change aggregate demand. In this case aggregate demand must be increased. This can be done with some combination of higher government expenditures and lower taxes. Suppose the decision makers ask you to estimate how much to reduce taxes to equate GDP_{eq} and GDP_{fe}. To answer this, you need to solve for the tax level (T) that sets (Y or GDP_{eq}) equal to $750 billion. You start with the same equation as above:

$$GDP_{fe} = Y = 150 + 0.6(Y - T) + I + G$$

In this case instead of solving for Y, you solve for T. You begin by substituting the values for Y, I, and G, giving you:

$$750 = 150 + 0.6(750 - T) + 50 + 110$$

collecting terms:

$$750 = 760 - 0.6T$$

and solving for T:

$$0.6T = 10$$

you find that:

$$T = 16.66$$

Thus you find that taxes, which are currently $40 billion need to be reduced by $23.34 billion to $16.66 billion. To achieve equilibrium at full employment, the government

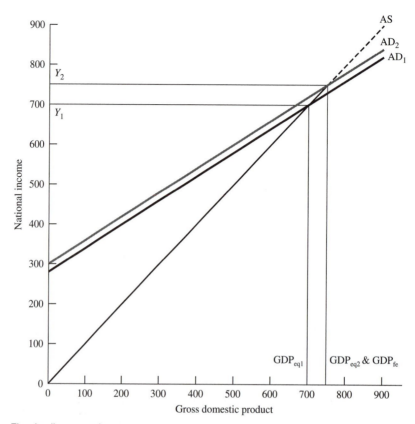

FIGURE 12-9   Fiscal policy example.

should operate with a deficit of $93.34, up from $70 billion dollars. This change is shown in Figure 12-9 as the shift of the aggregate demand curve from AD_1 to AD_2 where GDP_{eq2} and GDP_{fe} are at the same level.

You might also want to indicate how much the government would have to change its expenditure to bring the economy into equilibrium. To do this you would use the same equation, but solve for G. In this case:

$$750 = 150 + 0.6(750 - 40) + 50 + G$$

collecting terms:

$$750 = 626 + G$$

and solving for G:

$$0.6G = 750 - 626$$

you find that:

$$G = 124$$

Thus you find that government expenditures, which are currently $110 billion need to be increased by $14 billion to $124 billion. To achieve equilibrium at full employment, it should operate with a deficit of $84 billion, up from $70 billion.

Recall from our discussion of price levels and inflation that as GDP_{eq} approaches GDP_{fe} inflationary pressure builds. Thus the government should not attempt to shift AD all the way to the GDP_{fe} level.

Next we turn to an examination of several multiplier effects that provide an explanation of why a change in taxes must be larger than a change in government expenditures to have the same effect on the economy.

Multiplier Effects

A small change in taxes or government expenditures produces a large change in equilibrium GDP. This phenomenon is known as the multiplier effect. We can use the same equations as above to explain the multiplier effect:

$$GDP = C + I + G$$

$$C = a + mpc \times (Y - T)$$

And since

$$GDP = Y$$

We can substitute Y for GDP

$$Y = C + I + G$$

We want to know how changes of taxes (T) and government expenditures (G) affect the economy (Y). To do this we substitute for C to get:

$$Y = a + mpc \times (Y - T) + I + G$$

Next we bring the Y's to the left side:

$$Y(1 - mpc) = a + I + G - mpc \times T$$

$$Y = \frac{a + I + G - mpc \times T}{1 - mpc}$$

Note that this gives us another way of calculating the equilibrium level of GDP for an economy. Substituting the values from our example above we would get:

$$Y = \frac{150 + 50 + 110 - 0.6(40)}{1 - 0.6}$$

$$Y = 715$$

To find the multipliers we break our equation into three terms:

$$Y = \frac{a + I}{1 - \text{mpc}} + \frac{G}{1 - \text{mpc}} - \frac{\text{mpc} \times T}{1 - \text{mpc}}$$

We can now see the magnitude of the effect on the economy of a change of G or T. Looking first at the effect of a change in G; we want to know how much GDP will change if G changes by one dollar *ceteris paribus*. If G changes by \$1, GDP changes by:

$$\text{Expenditure multiplier} = \frac{1}{1 - \text{mpc}}$$

This is called the **expenditure multiplier.** Since mpc is less than 1, the expenditure multiplier is greater than 1. For example, if mpc = 0.65, the expenditure multiplier is 2.86. For every dollar change in government expenditure, GDP changes by \$2.86. This means that if the marginal propensity to consume is 0.65, an increase of \$1 billion in government spending (with no change in taxes) will increase the gross domestic product by \$2.86 billion. (Note that an increase in investment would also have the same multiplier effect on GDP.)

A change in taxes does not change GDP by the same amount as the expenditure multiplier. The last term in our equation gives us the **tax multiplier.**

$$\text{Tax multiplier} = -\frac{\text{mpc}}{1 - \text{mpc}}$$

If the mpc is 0.65, a billion dollar *reduction* in tax collections increases GDP by \$1.86 billion [−0.65 ÷ (1 −0.65)]. This means that if the marginal propensity to consume is 0.65, a reduction of \$1 billion in taxes (with no change in government spending) will increase the gross domestic product by \$1.86 billion.

Thus far, we have examined the multiplier effects of a change either in government expenditures or in taxes. If taxes and expenditures go up by the same amount so that the budget surplus ($T - G$) does not change, there is still a multiplier effect, called the **balanced budget multiplier** (BBM). In this case, the multiplier is equal to the effect from a change in expenditure less the effect from the change in taxes. Thus we combine the equations for the expenditure and tax multipliers:

$$\text{Balanced Budget Multiplier} = \frac{1}{1 - \text{mpc}} - \frac{\text{mpc}}{1 - \text{mpc}}$$

Since we want to know the effect of a change of both G and T by one unit, both G and T equal one. This means that:

$$\text{BBM} = \frac{1}{1 - \text{mpc}} - \frac{\text{mpc}}{1 - \text{mpc}} = \frac{1 - \text{mpc}}{1 - \text{mpc}}$$

Thus:

$$\text{BBM} = 1$$

The balanced budget multiplier is equal to one. This means that if both government expenditures and taxes increase by \$1 billion, equilibrium GDP also increases by \$1 billion.

This result explains the outcome in our example; that a larger change in taxes than change in spending was required to achieve the same equilibrium level of GDP. This is because aggregate demand increases by the full amount of the increase in spending. However, an increase in taxes has the effect of reducing consumer demand by a smaller amount. It is the tax increase adjusted by the marginal propensity to consume. If you normally spend 70 percent of your after-tax income, each additional dollar you pay in taxes means that you spend \$0.70 less.

A Cautionary Note   The discussion in the last few paragraphs might give you the mistaken impression that fiscal policy making is straightforward and that the effect of changes in taxes and expenditures on equilibrium is clear. This is not the case. We have ignored several confounding factors. The most important of these is the effect of fiscal policy on prices. Changes in fiscal policy can also affect expectations about the future and thereby change economic decisions. One of the principal contributions of the new classical macroeconomists has been to point out the dangers of ignoring the effect of macroeconomic policies on prices and expectations, especially when the economy is not in the Keynesian region of the aggregate supply curve.

It is not easy to tell if an economy is operating in the Keynesian region of the aggregate supply curve, the vertical portion where the economy is at full employment, or the intermediate region. Typically, there is at least a one-month lag in the statistics on employment and output, and they are often revised as new data arrive at government statistical offices. Sometimes the statistics give inconsistent signals. For example, the labor unemployment rate and the consumer price index might increase in the same month. Unless the economy is clearly operating in the Keynesian region and unemployment is high, it is difficult to predict how much the outward shift in aggregate demand will contribute to increased output instead of higher prices. Policy makers must weigh the potential benefits of added employment and output against the risk that prices will increase with little change in production.

Role of Prices and Interest Rates

Thus far we have ignored the role of the aggregate price level on consumption. For an individual consumer, if the price of a good increases, the individual's consumption falls, *ceteris paribus,* and consumption of other goods is likely to go up. Substitution is not a possibility for aggregate consumption. However, prices can affect consumption through a wealth effect. The **wealth effect** is the change in the real value of an individual's assets as a result of a change of the asset's price. For example, in the late 1990s there was a significant increase in stock prices with little change in the aggregate prices level. Owners of stocks were wealthier and some increased their spending from current income. It was estimated that the stock market increase from 1996 to 1999 generated a "wealth effect" that increased the marginal propensity to consume by one percent.

An example of a negative wealth effect is if the aggregate price level in the economy increases but the price of the land or other assets owned by a farm family do not increase by the same amount. The real value of the assets, the family's wealth, has declined. The family might reduce its spending and increase savings to build up its wealth. (You will see in chapter 14 that the aggregate price level also has a negative effect on net exports, although for different reasons.)

SUMMARY

In this chapter, we analyzed macroeconomic performance, specifically the determinants of equilibrium GDP. We found that an economy can be in equilibrium even with some of its resources unemployed. It sometimes pays for owners of those resources to keep them idle for a time, instead of accepting reduced payments. When an economy is not at full-employment GDP, fiscal (and monetary) policy can be used to change aggregate demand and shift the economy toward full employment. However, unless unemployment is high, it is difficult to determine the extent to which an adjustment in fiscal policy will change equilibrium GDP instead of the price level.

LOOKING AHEAD

This chapter has shown why an economy might not be at full-employment GDP and how fiscal policy can shift an economy toward full employment. In the next chapter, we examine money: what it is, what it is used for, how it is created, and how monetary policy can shift aggregate demand.

IMPORTANT TERMS AND CONCEPTS

aggregate demand   293
aggregate supply   292
balanced budget multiplier   310
capacity utilization   290
consumption   293
consumption function   293
equilibrium level of GDP   301
fiscal policy   306

frictional unemployment   290
inflationary gap   302
investment   297
Keynesian region   304
recessionary gap   302
structural unemployment   290
taxes   299
wealth effect   311

QUESTIONS AND EXERCISES

Name That Term

Read the following sentences carefully and fill in the missing term or terms.

1 The amount by which the aggregate expenditure curve must shift downward to decrease the nominal GDP to the full employment noninflationary level is the _____.

2 _____ involves changes in government spending or tax collections for the purpose of moving toward a full employment and noninflationary domestic output.

3 _____ unemployment results from workers voluntarily changing jobs or from temporary layoffs.

4 At the _____ GDP the aggregate production is equal to aggregate consumption.

5 The _____ indicates the effect on GDP of an equal increase or decrease in government spending and taxes.

6 The horizontal segment of the aggregate supply curve along which the price level is constant as real domestic output changes is the _____.

7 The relationship between income and consumption is called the _____.

8 _____ is the amount by which the aggregate expenditure curve must shift upward to increase the real GDP to the full-employment, noninflationary level.

9 Personal income minus personal taxes is _____.

10 _____ consumption is the level of consumption a consumer would make out of savings or from borrowed funds if income fell to zero.

11 The fraction of any change in disposable income that is spent for consumer goods is the _____.

12 _____ unemployment involves workers who are unemployed either because employers do not demand their skills or because they lack sufficient skills to obtain employment.

13 _____ refers to the percent of buildings, machinery, and similar resources being used in production.

True/False

Read the following sentences, then decide whether each statement is true (T) or false (F) and mark it accordingly.

T F 1 Fiscal policy is set through the political process.

T F 2 The marginal propensity to consume in most countries is between 0.6 and 0.8.

T F 3 At the equilibrium GDP there is only frictional employment in the economy.

T F 4 As real national income increases, other things equal, savings decrease.

T F 5 Other things equal, a decrease in taxes and an increase in spending of the same amount will cause GDP to increase.

T F 6 The aggregate demand curve shows the total quantity of goods and services that will be purchased at different levels of GDP.

T F 7 The interest rate has a major effect on how much is saved.

T F 8 From the perspective of a national economy, only adding new plant and equipment to the nation's capital stock counts as investment.

T F 9 An inflationary gap occurs if aggregate demand exceeds aggregate production.

T F 10 Other things equal, an increase in taxes and a decrease in expenditures will cause GDP to increase.

Multiple-Choice Questions

Circle the letter of the response that best answers the question or completes the statement?

1 Under the final expenditure approach, which of the following are sources of demand for an economy's output?

a consumption

b investment

    c government

    d net exports

    e all of the above

2 If the mpc = 0.8, the expenditure multiplier is

    a 0.2.

    b 0.8.

    c 1.

    d 1.8.

    e 5.

3 If the mpc = 0.8, the tax multiplier is

    a −4.

    b −1.

    c −.8.

    d −.2.

    e 0.

4 If the mpc = 0.8, the balanced-budget multiplier is

    a −.8.

    b −.2.

    c 0.

    d .8.

    e 1.

5 Suppose the mpc is 0.7, $Y = \$100$ million, government expenditures increase $1 million, and nothing else changes. What does the appropriate multiplier indicate that Y would be after this change in fiscal policy?

    a $70 million

    b $100 million

    c $103.33 million

    d $107 million

    e $170 million

6 Aggregate production is

    a the same as GDP.

    b the sum of all goods and services produced in an economy.

    c the final output of an economy.

    d at a maximum when all resources are fully employed.

    e all of the above.

7 A $1 increase in taxes will reduce aggregate demand by

    a GDP times the BBM.

    b mpc times national income.

    c mpc divided by one minus mpc.

    d GDP times the tax multiplier.

    e none of the above.

8 When in equilibrium, the economy

    a will use all resources to produce goods and services.

    b may have significant levels of unemployed resources.

    c will have a balanced budget.

    d all of the above.

9 To reduce or eliminate a recessionary gap, the most effective fiscal policy is to

    a increase government expenditures or reduce taxes.

b reduce government expenditures or increase taxes.
c reduce government expenditures and reduce taxes.
d increase government expenditures and increase taxes.
e none of the above.
10 The aggregate price level generally does not fall because of
  a long term contracts.
  b civil service salary scales.
  c unemployment programs.
  d the expense of re-pricing consumer items.
  e all of the above.

Technical Training

Use the following hypothetical data to answer the questions below. All numbers are in billions of dollars.

$$C = 140 + .7(Y - T); \quad G = 210; \quad I = 60; \quad T = 200; \quad Y_{fe} = 960$$

1 What is the autonomous consumption?
2 What is the mpc?
3 What is the government deficit?
4 What is the equilibrium Y?
5 What is the equilibrium level of consumption?
6 What is the equilibrium level of savings?
7 Other things equal, by how much would equilibrium GDP increase in response to an exogenous increase in I from $60 to $70 billion?
8 Other things equal, what would be the equilibrium GDP if taxes decreased $10 billion?
9 What would be the equilibrium Y if the government undertook $12 billion in new projects and raised $12 billion in new taxes?
10 Other things equal, what would be equilibrium Y if the government "balanced the budget" by increasing taxes?
11 Other things equal, what would be the equilibrium Y if the government "balanced the budget" by decreasing expenditures?
12 To what would you change G to assure Y_{fe}?
13 To what would you change T to assure Y_{fe}?

Questions for Thought

1 What is the single largest component of aggregate demand?
2 Does average aggregate demand decline as aggregate income increases? Why, or why not?
3 Where do the goods associated with household savings end up being used?
4 Savers and investors are usually different people. In a simple economy with no government or foreign transactions, why does equilibrium savings equal equilibrium investment?

5　Given the following values for an economy, what are the equilibrium levels of consumption, savings, equilibrium GDP, and the government deficit $(G - T)$?

$$C = 120 + 0.6(Y - T); \quad G = 200; \quad I = 40; \quad T = 180$$

What happens to GDP if T increases to 200?

6　Suppose full-employment GDP in the economy of question 5 is 1000. If taxes remain at 200, what level of government spending would raise equilibrium GDP to full employment?

7　Working from the example in this chapter, calculate equilibrium GDP if government expenditures are reduced to $50 billion with taxes at $50 billion.

8　Some individuals argue that we should eliminate or significantly reduce the national debt.

　　a　How would this be done?

　　b　Other things equal, what would be the effects on Y?

　　c　Other things equal, what would be the effects on unemployment rate?

9　Explain why an economy can not operate at a true full-employment level.

10　What is your role is determining fiscal policy?

11　Can you influence macroeconomic performance in a meaningful way?

12　Do you have more or less control of microeconomic than macroeconomic decisions?

ANSWERS AND HINTS

Name That Term　**1.** inflationary gap; **2.** Fiscal policy; **3.** Frictional; **4.** equilibrium; **5.** balanced-budget multiplier; **6.** Keynesian region; **7.** consumption function; **8.** recessionary gap; **9.** disposable income; **10.** Autonomous; **11.** marginal propensity to consume; **12.** Structural; **13.** Capacity utilization

True/False　**1.** T; **2.** T; **3.** F; **4.** F; **5.** T; **6.** T; **7.** F; **8.** T; **9.** T; **10.** F

Multiple Choice　**1.** e; **2.** e; **3.** a; **4.** e; **5.** c; **6.** e; **7.** c; **8.** b; **9.** a; **10.** e

Technical Training　**1.** $140 billion; **2.** 0.7; **3.** $10 billion; **4.** $900 billion; **5.** $630 billion; **6.** $70 billion; **7.** $33.3 billion; **8.** $923.4 billion; **9.** $912 billion; **10.** $876.7 billion; **11.** $866.7 billion; **12.** $228 billion; **13.** $174.3 billion

13

GROSS DOMESTIC PRODUCT AND MONETARY POLICY

CHAPTER OUTLINE

OVERVIEW

This chapter examines the role of money and shows how monetary policy influences macroeconomic performance. Money is demanded because it meets several economic needs. The supply of money is controlled by government policy. In the United States, the Federal Reserve System (the Fed) controls the money supply. Money demand and supply equilibrate in the money market, where the interest rate is the equilibrium price that balances quantities supplied and demanded. As discussed in Chapter 12, the interest rate influences investment and, therefore, demand for the economy's output. Hence, by changing the money supply, the Fed can influence private investment. If the economy is in recession, equilibrium GDP can be increased by growth in the money supply. If the economy is suffering from inflation, decreases in the money supply can reduce inflationary pressures.

This chapter continues the discussion of how macroeconomic policies are used to change equilibrium GDP. As mentioned in the last chapter, government's decisions on expenditures and taxes, or fiscal policy, are also used to meet a wide variety of other goals. The goal of **monetary policy** is improving macroeconomic performance by promoting economic growth while controlling inflation. It is even used to offset political choices with poor fiscal policy implications.

We start with the history of money and the services money provides, and then discuss the mechanisms for controlling the supply of money. Finally, we examine the chain of events linking changes in the money supply, planned investment, equilibrium GDP, and the price level.

LEARNING OBJECTIVES

This chapter will help you learn:

- What money is, how it is created, and why it is useful
- What determines the demand for money
- How a government controls the money supply
- How money supply and demand determine interest rates
- How changes in the supply of money affect the interest rate, investment, and therefore equilibrium GDP

A BRIEF HISTORY OF MONEY

The evolution of civilization and the growth of the human population required the development of farming techniques that allowed one person to feed many. As an agricultural surplus became available, specialization of economic activity became possible. Instead of producing all of one's food, clothing, and shelter, an individual could develop skills to produce a good or service to exchange with other individuals. Specialization brought with it the need to develop a means of exchanging goods and services produced by different individuals.

Barter

The first form of exchange, and one that still finds some use today, was **barter,** the exchange of goods for other goods. For example, a farm family exchanges loaves of bread for a shirt produced by a tailor. There is a major problem with barter. It requires a **double coincidence of wants** in which each party to a successful transaction must have what the other wants. In addition, they must reach agreement on a rate of exchange for the two goods (a price of one good in terms of the other).

A simple economy, with few goods and little specialization, can be successfully based on barter transactions. However, as economies become more complex many parties are needed to complete a set of transactions, and the inefficiencies of barter quickly become a serious problem. As the number of traders grows larger, the cost in time and effort of completing transactions increases. For example, suppose a tailor needs a pair of shoes. The shoemaker might not want the shirt the tailor made but does need several loaves of bread. The tailor and shoemaker must find a baker who wants the shirt the tailor has made. Even this simple three-way transaction involving a shirt, shoes, and bread can be difficult to arrange. Clearly, as the economy produces an ever-wider array of goods and services, barter becomes inefficient and eventually unmanageable.

Commodity as Money

Money is the innovation that eliminated the need for barter and the associated double coincidence of wants. The first form of money was **commodity money.** By tradition, consensus, or law, one commodity is accepted as the **numeraire,** or medium of exchange. Examples of goods that have been used as money include cowry shells, bullet casings, water buffalo, cigarettes, furs, gold, and silver.

The essential idea of commodity money is that the price of every good is expressed in terms of the numeraire commodity. For example, in a society where fur pelts are used as money, the price of a shirt might be a half pelt, and the price of a cow might be ten pelts. People accept pelts in payment because they can use pelts to buy other goods and services.

With commodity money, the double coincidence of wants is no longer necessary. However, commodity money creates other problems. A major problem is agreement on an appropriate commodity. For example, John Maynard Keynes in his *Treatise on Money* (1930) describes how, when Uganda was still a colony of Great Britain, goats were a numeraire commodity. Part of the responsibilities of the district commissioner was to decide in cases of dispute whether a particular goat was too old or too scrawny to be a "standard" goat. If it could not be a standard goat, it could not be used in trade. Other problems are that commodity money can be bulky (the water buffalo), can deteriorate (goats and water buffalo grow old), or is costly to produce (the Yap Islanders of the South Pacific used giant carved stone wheels as money until World War II). Modern money, currency, was developed to address problems such as these.

Modern Money

In every economy today, the principal medium of exchange is a standardized unit of money (e.g., dollar, yen, or peso) with a physical expression in paper, coins, checks, money market accounts, and electronic transfers. The standardized units usually have no inherent value. The cost to produce them is very small. Unlike commodity money that has intrinsic value (a pelt can be used for warmth in winter), modern money has value *only* because it serves the functions of money.

THE GOLD STANDARD

For a country to be on the gold standard, it must define its currency in terms of a certain quantity of gold and be willing to exchange gold for its currency. The gold standard began to develop in 1865, when the U.S. Treasury issued the first gold certificates. These certificates could be used in place of gold coins for money, because the holder could exchange them at the Treasury for gold coins or bullion deposits. Silver certificates followed in 1878. By 1879, many of the industrial nations had converted their monetary systems to the gold standard. Gradually gold and silver certificates replaced the use of gold and silver coins as money.

Initially, the Treasury held enough gold and silver to redeem all paper money on demand. Over time, however, the value of paper money issued was greater than the Treasury reserves. In the 1930s, the Treasury quit redeeming paper money for gold for individuals, but the United States remained ready to convert, buy, or sell gold for $35 per ounce from foreign authorities. In 1967, Congress authorized the Treasury to quit exchanging silver for silver certificates, and in 1971, the United States abandoned the gold standard completely.

SEIGNIORAGE

Although the cost of producing today's money is small, its value can be large. For example, it might cost only 1 cent to produce a new $100 bill. Whoever produces new money captures that value. The difference between the value of new money and its cost is called **seigniorage.** When governments replace worn-out currency, no seigniorage is created because an equal value of money is created and destroyed.

FUNCTIONS OF MONEY

Money serves three economic functions. First, money is a **medium of exchange** used for payments and thus ending the need for a double coincidence of wants. Second, money serves as a **unit of account** because it provides a way to quantify the economic value of a good. In this function, money solves problems such as the "standard goat" problem in colonial Uganda. Third, money serves as a **store of value.** It is much cheaper to store money, at home or in a bank, than to store most goods of equivalent value and it does not deteriorate physically as do most goods.

To serve these three economic functions, *money must be scarce*. If there is too much money or if anyone can create it, then it cannot serve any of the three functions. The challenge for macroeconomists and policy makers is to determine the "right" amount of money. All governments with the power to create money must deal with this question. If the money supply is too large, its value falls and inflation results. There have been times in which a government has increased the money supply so much that it did not serve any of the three functions. When this happened, people had to rely on barter, commodity money, or foreign money to make transactions. If the money supply is too small, the economy will slow or even falter due to insufficient money to allow efficient transactions.

Money Supply Statistics

In the United States we divide financial instruments used as money into three categories: $M1$, $M2$, and $M3$. These categories classify instruments by the degree to which they have the characteristics of money. Items in the $M1$ category are the most money-like. They are circulating coins and paper money, traveler's checks, money orders, bank drafts, demand deposits, negotiable order of withdrawal (NOW) accounts (a form of a checking account) and automatic transfer service (ATS) accounts at depository institutions, credit union share draft accounts and demand deposits at thrift institutions.

Coins and paper money in the hands of the public (i.e., not in bank vaults), also called **currency in circulation,** are readily accepted as a medium of exchange. Coins and paper money are also universally accepted as a unit of account and they function well as a store of value when inflation is low. Traveler's checks, money orders, bank drafts and similar instruments are almost interchangeable with paper money, although their use as a medium of exchange is somewhat restricted. Checking accounts, technically called demand deposits, have most of the characteristics of money. However, their use as a medium of exchange is somewhat more restricted than that of coins, paper money, or traveler's checks. Some businesses and even individuals do not accept checks as payment. NOW accounts and other interest-bearing checking accounts have characteristics similar to demand deposits. To summarize:

M1 includes the various forms of currency in circulation, plus bank accounts from which the owner can withdraw money with a check at any time (demand deposits).

PAPER MONEY IN THE UNITED STATES

The early history of paper money in the United States is checkered. The Massachusetts colony was the first to issue paper money to pay for the costs of feeding its army after an unsuccessful siege of Quebec in 1690. Shortly thereafter other colonies printed their own paper money during this period.

In 1775, facing huge expenses without adequate taxing power, the Continental Congress authorized the use of paper money, called *continentals*. However, too much reliance on this funding method led to rapid depreciation in its value. As George Washington commented: "A wagon-load of money will scarcely purchase a wagon-load of provisions." In 1790, Congress authorized the Treasury to accept continentals at the rate of 100 continentals to $1. The feeling toward paper money was so bitter that more than seventy years passed before the federal government again issued it.

Source: Adapted from "Coins and Currency," Public Information Department, Federal Reserve Bank of New York, April 1985.

M2 includes $M1$ plus items not as moneylike, such as savings accounts, time deposits (including money market deposit accounts), small-denomination time deposits (time deposits—including retail repurchase liabilities (RPs)—in amounts of less than $100,000), and balances in retail money market mutual funds.

M3 includes $M1$ and $M2$, plus time deposits of $100,000 or more, balances in institutional money funds, RPs (overnight and term) issued by all depository institutions,

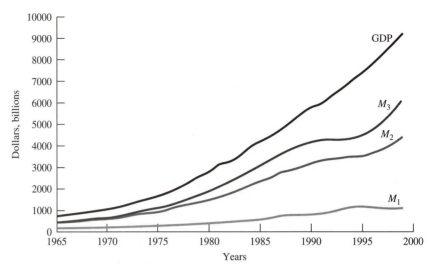

FIGURE 13-1   Money supply and GDP, United States, 1965–1999.

and Eurodollars held by U.S. residents at foreign branches of U.S. banks worldwide and at all banking offices in the United Kingdom and Canada. These deposits and assets have even fewer of the characteristics of money.

CREDIT AND DEBIT CARDS

Credit cards are not included in the various categories of money. When you use a credit card, you are borrowing money from the issuing institution (hence, the word *credit*). However, when individuals use credit cards, they need less cash and the demand for money shifts inward.

Debit cards look like credit cards but they are very different. Using a debit card is equivalent writing a personal check. Money is moved from the debit card holder's account to the account of the person being paid. Thus the transaction has no effect on the supply of money.

Figure 13-1 shows how the amount of money (*M*1 and *M*2) has changed over time and its relationship to GDP in the United States. For much of the recent past there has been about $1 of *M*1 for every $7 of GDP.

MONEY SUPPLY

In most countries the central government controls the money supply. A national institution, typically called the **central bank,** manages the supply of money. The central bank has the authority to create and destroy money. However, this is not its only means of influencing the money supply. The central bank typically supervises the banking system and uses this authority to adjust the money supply. In this section, we study the U.S. system of managing the money supply and the role of the central bank of the United States—the **Federal Reserve System**—commonly known as "the Fed."

The Banking System and Money Creation

At the start of the twentieth century, paper money was issued by national banks in proportion to the amount of U.S. Treasury bills purchased. As the federal debt was retired after the Civil War, the amount of money was automatically reduced. There was no mechanism to change the supply of money to meet the needs of a growing economy, and several "money panics" ensued. In 1913, the Congress established the Federal Reserve System. A Board of Governors runs "the Fed" and decides the nation's monetary policy. Its decisions are carried out by twelve regional Federal Reserve Banks.

The Board of Governors of the Fed also regulates the Federal Reserve Banks and all commercial banks. The regional Fed banks serve as the banks' bank. That is, commercial banks deposit their reserves with the Federal Reserve Banks and may borrow from them to cover unusual needs. Thrift institutions, such as savings and loans and mutual savings banks, are regulated by separate government agencies. However, all parts of the banking system are subject to Fed regulations that control the money supply.

It is important to understand that the Fed is an independent institution. The president of the United States appoints (with senate approval) its governors for fourteen-year terms, staggered so that there is continuity in decision making. Neither the Congress nor the president has direct control over the decisions made by the Fed. Unlike **fiscal policy**—for which the president and Congress must agree on a budget and political considerations are always paramount—the Fed's monetary policy decisions are much less affected by political pressures. As we discuss later, this independence plays an important role in macroeconomic policy making.

Central Bank Control of the Money Supply and Credit Conditions

The primary task of the Fed or any central bank is to control the money supply. It has three main policy instruments: *open-market operations* (purchases or sales in the U.S. government securities market) changes in *reserve requirements* of commercial banks, and changes in the *discount rate* (the rate at which banks borrow from the Federal Reserve). We discuss each of these in turn.

Open-Market Operations   In the United States, as in most developed economies, open-market operations are the most frequently used mechanism for changing the money supply. To understand how open-market operations affect the money supply, we start with the development of the open market. To finance public sector deficits, the U.S. Treasury Department borrows money by selling government securities (Treasury bonds) to the public. The Treasury periodically auctions bonds to replace maturing bonds and to generate cash needed to purchase goods. When the funds generated by the sale of bonds are used to purchase goods, the amount of money held by the private sector does not change. The money used by the private sector to pay for the bonds returns to the private sector when the government purchases goods.

A market in which outstanding treasury bonds held by individuals and institutions are traded exists. It is called the *open market*. The Fed uses this market as one means of controlling the money supply. The purchases and sales of government securities by the Fed in this market are called **open-market operations.**

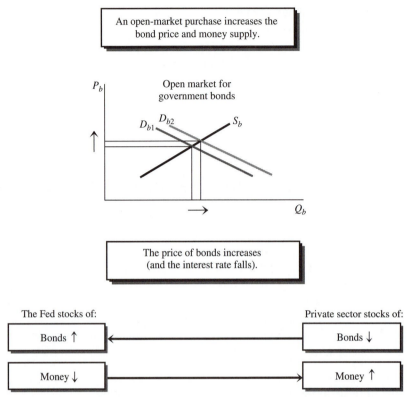

FIGURE 13-2   The effect of an open-market purchase.

If the Federal Reserve decides to increase the money supply, it buys government securities in the open market. The money paid by the Fed for these securities increases the money supply by adding to the amount of money in the private sector (see Figure 13-2). If, on the other hand, the Fed decides to reduce the money supply, it sells government securities to the public. The money received by the Fed goes into its reserves, reducing the money in circulation. The Fed and other central banks in developed countries use open-market operations to manage the money supply on a day-to-day basis.

GOVERNMENT DEBT AND THE MONEY SUPPLY IN DEVELOPING ECONOMIES

Open-market operations are the primary mechanism for controlling the money supply in most developed economies. Some less-developed countries do not have a market for government securities. When the government runs a deficit in those countries, it raises the needed funds by selling securities to the central bank, which the central bank pays for by printing new money. In this case, there is a direct relationship between the national government deficit and the money supply.

Bank Reserve Requirements   The second means of controlling the money supply is by changing commercial bank **reserve requirements**—the share of checking and savings deposits the Fed requires a bank to hold in reserve for possible withdrawals. This form of regulation is called a **fractional reserve** banking system.

Bank demand deposits make up a substantial share of the money supply. The institutions that hold these deposits—commercial banks, savings and loans, and other deposit-taking institutions—play an important role in the creation of money. To understand how changes in reserve requirements affect the money supply, we need to understand what a commercial bank does with demand deposits.

A person with a demand deposit (checking account) is assured of instantaneous access to the funds, and they are kept in a safe place.* Some of the money deposited is locked in the bank's vault, some may be placed in a reserve account in a Fed bank, but the bank will loan most to commercial customers. If the bank holds deposits in the bank vault it does not earn any return. Funds in a reserve account with a regional reserve bank earn low returns. The funds the bank lends earn interest and are the bank's primary source of revenue. Thus the bank seeks to maximize its lending. However, the bank must keep enough money in reserve to satisfy depositors who wish to withdraw funds from their accounts. In normal times a bank would need to keep only a small fraction of its deposits for this purpose. The Federal Reserve requires banks to keep reserve deposits in a regional Federal Reserve Bank higher than needed to meet safety considerations. The amount required is calculated on the basis of a **reserve ratio.** The Fed can control the money supply by changing the reserve ratio it requires banks to maintain.

Table 13-1 gives an example of how reserve requirements create a *money multiplier effect.* New money supplied by the Fed in an open-market operation is multiplied by the

---

*Savings accounts differ in that they are subject to limits on withdrawals. These limits might be on the amount that can be withdrawn at any time or on the length of time a deposit must remain in the account. Therefore, the reserve requirements are lower for savings accounts than for demand accounts.

TABLE 13-1   THE CREATION OF MONEY IN A FRACTIONAL RESERVE BANKING SYSTEM

	Maximum new deposit	Required reserve (20%)	Funds loaned	ΔMoney supply
Round 1	1000	200	800	800
Round 2	800	160	640	1440
Round 3	640	128	512	1952
Round 4	512	102	409	2392
. . .	. . .	. . .	. . .	. . .
Final round	0	0	0	4000

Notes: This table assumes that banks lend out the maximum amount possible and that all cash is put into demand deposits.

The change in the money supply is the new money created by the deposit of $1000. The fractional reserve system multiplies the original amount by $1 \div RR$ ($1000 × 1/0.2) = $1000 + $4000 = $5000.

fractional reserve system. If the Fed withdraws money in an open-market operation, the decline in the money supply is greater than the amount withdrawn. Here is an example that shows how the fractional reserve system can change the money supply.

An Example   Suppose you find $1000 under a mattress. You immediately take the money to the bank and put it in a demand deposit (checking) account. You now own a demand deposit of $1000, and the bank has $1000 of cash in its vault. Since the cash in the vault is not available to the economy, it is not counted as part of the money supply (see the definition above for *M*1). But your demand deposit is counted as part of the money supply, as was the cash under the mattress. So, $1000 cash has been turned into a $1000 demand deposit. If the Fed has set the reserve requirement on demand deposits at 20 percent, your bank must keep at least 20 percent of its demand deposits in its reserve account with the district branch of the Fed. This means that the bank must keep at least $200 (0.2 × $1000) of your demand deposit as reserves and may lend up to $800 of your deposit, as shown in Round 1 in Table 13-1. If the bank lends out the maximum, the borrower has $800. This $800 is an addition to the money supply because it is in the hands of the private sector. In this first round, the money supply has increased by $800.

ACTUAL RESERVE REQUIREMENTS

The Fed sets reserve requirements on all types of bank deposits. The reserve ratios for selected categories of deposits in August of 2000 were as follows:

Checkable deposits	Reserve requirements
$0 to $41.5 million	3%
Over $41.5 million	10%

Noncheckable deposits	
Nonpersonal savings and time deposits	0%

But that is not the end of the multiplier effect of the fractional reserve requirement. Suppose that the borrower of the $800 generated by your demand deposit makes a purchase worth $800, and the seller deposits the $800 in a bank. The bank must also keep 20 percent ($160) in reserve and may lend up to $640 to other borrowers. If it lends out the maximum, the money supply increases by another $640. If this process of borrowing, lending, and money creation continues, the bank will hold an amount equal the initial $1000 deposit as required reserves. With a reserve ratio of 0.20, an additional $4000 can be created from an initial deposit of $1000. The total amount of money in circulation as a result of the original $1000 deposit is 1 ÷ 0.20 times $1000 or $5000. The **money multiplier (MM),** the maximum amount by which the fractional reserve banking system can multiply an increase in currency, is as follows:

$$MM = \frac{1}{\text{reserve ratio}}$$

CURRENCY DRAIN

Because some currency stays in the hands of the public, leakage into hand-to-hand circulation (or "currency drain") reduces the multiplier effect of the fractional reserve. In addition, banks can choose not to lend out the maximum. Therefore, $\dfrac{1}{\text{reserve ratio}}$ is the maximum MM.

Now we can show how the Fed (or any other central bank) uses the reserve ratio to change the money supply. If the Fed raises the reserve ratio, the money multiplier falls and less money is created. If the Fed lowers the reserve ratio, the money multiplier increases and more money is created.

Suppose, in our example, that the Fed lowers the ratio from 20 to 15 percent. The money multiplier increases from 5 to 6.67. The initial deposit of $1000 (and all other money in the system) is multiplied a maximum of 6.67 times ($1 \div 0.15$), that is, to $6667. This means an increase in the money supply created by the $1000 deposit from $5000 to $6667, or $1667.

Small changes in the reserve ratio can have a large effect on money supply. A change in reserve ratios indicates the Fed has determined that a major adjustment is needed, and this occurs infrequently.

Discount Rate   The third method used by the Fed to manage the money supply is to change the discount rate. Commercial banks can borrow from the Fed to cover shortfalls in their reserve accounts and thereby increase their ability to lend. The Fed sets the interest rate, called the **discount rate,** on its loans to banks. If the Fed lowers the discount rate (an "easier money" policy), the cost of covering reserve shortfalls is reduced. Some banks can afford to reduce their reserve holding and lend more money at lower interest rates. The result is an increase in the money supply. On the other hand, if the Fed increases the discount rate, it becomes more costly for banks to have shortages. They will reduce their lending activities and increase their reserve holdings. Therefore, an increase in the discount rate reduces the money supply.

Changes in discount rates are more common than changes in reserve ratios, but less common than open-market operations. The Fed undertakes open-market operations daily to fine-tune the money supply. Since many very large private-sector transactions take place in the open market, Fed transactions cause little disruption. Furthermore, small open-market operations have larger effects on the money supply because of the money multiplier. Changes in the discount rate or reserve rate signal that the Fed believes the economy is not performing as desired. It is changing the money supply to stimulate more economic activity or to reduce inflationary pressures. We explain how in the following sections.

MONEY DEMAND

The decision of how much money an individual will hold is part of a broader issue of what an individual does with savings. Savings are held as money for *transactions, precautionary,* and *asset (or speculative)* purposes. In addition to cash, savings can be in the less-liquid form of certificates of deposit, mutual funds, or stocks in a corporation.

The **transactions demand** for money arises because income typically comes in lumps, while expenditures are made almost every day. For example, salaried employees are paid once every two or four weeks and a farmer might sell all of a crop at harvest. But production and living expenses happen throughout the year. For these needs, savings are in the form of cash, demand deposits or other liquid assets.

The **precautionary demand** for money refers to money held to meet unexpected expenses. If a family has an unexpected illness or a business has an unexpected investment opportunity, having money available means that the cost of borrowing can be avoided. Savings for these needs are generally in savings accounts, certificates of deposit, or other assets that earn interest but can be liquidated if a need arises.

The third category of money demand, **asset demand,** is more complicated. It is related to the expected return on money and other assets. Since most economies typically experience some degree of inflation, the value of money is decreasing. Thus, savings held in the form of cash lose purchasing power. Savings invested in other assets such as bonds, stocks, and real estate earn interest and may appreciate or depreciate in value. If the price of an asset is expected to rise in the future, an individual might buy the asset. This purchase would reduce the amount of savings held as cash, CDs or savings accounts (the asset reserve). Conversely, if the price of an asset is expected to fall, individuals might sell the asset and hold more money in cash, CDs or savings accounts. Hence, expectations about asset values affect the asset demand for money.

The determinants of the amount of funds held for transactions, precautionary purposes, and asset demand are not the same, but in each case there is a negative relationship between the quantity demanded and the interest rate. The interest rate is the price of money and represents the opportunity cost of holding money idle. For all three categories, the higher the interest rate, the greater the opportunity cost of holding money and thus the less money is demanded.

For both transactions and precautionary purposes, if individuals do not have sufficient money to meet current needs, they must borrow money to cover the shortfall. The cost of borrowing is determined by the interest rate. The lower the interest rate, the less costly it is to borrow to meet such needs.

For the third category, asset demand for money, the effect of the interest rate occurs through its relationship to asset prices. The price of a risk free financial asset is equal to the net return divided by the interest rate. Consider the following example using the data in Table 13-2. Suppose current interest rates are 10 percent and the government issues a bond that will pay the purchaser $110 after one year. At this interest rate, the investor should be willing to pay $100 for the bond because the alternative of putting the $100 in

TABLE 13-2 THE PRICE OF A BOND TODAY IF IT IS WORTH $110.00 ONE YEAR LATER AT VARIOUS INTEREST RATES

	5%	6%	7%	8%	9%	10%
Today	104.76	103.77	102.8	101.85	100.92	100
Next year	110	110	110	110	110	110

an interest-bearing savings account would also earn $10. However, how much should an investor be willing to pay for this same bond if the interest rate drops to 5 percent ($100 placed in a savings account will be worth only $105 next year)? If the $110 bond can be bought for $100, the $10 dollars of earnings is twice as high as the returns in a savings account. This higher interest rate will encourage many individuals to take money out of savings accounts to buy the bonds. The increased demand for bonds pushes up the bond price until the return to the bond and to the savings account investments is the same. If the investor pays $104.76 for the bond, the effective interest yield will be 5 percent ($110 ÷ $104.76 = 1.05). Hence, a drop in the interest rate causes the price of current bonds to rise, in this example from $100 to $104.76.

Using this inverse relationship between interest rates and the quantity of money demanded, we can trace the links from the interest rate to the asset demand for money. If the interest rate is high, the price of financial assets is low. This encourages individuals to hold more assets and less money for two reasons. First, if the interest rate falls, the assets increase in value as in the example in Table 13-2. Second, many financial assets pay interest, and so the earnings from the interest payments will be large. On the other hand, if the interest rate is low, the price of financial assets rises. The return from holding financial assets falls, causing individuals to hold more money and fewer assets. Thus, as with the transactions and the precautionary demands for money, the asset demand for money is related negatively to the interest rate.

EQUILIBRIUM IN THE MONEY MARKET

We can use the demand and supply relationships for money to find the equilibrium interest rate and quantity of money (Figure 13-3). All three categories of demand for money have a negative relationship with the interest rate. The supply of money, on the

FIGURE 13-3   The interest rate and equilibrium in the money market.

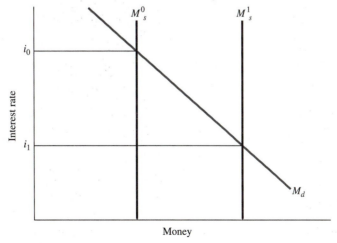

other hand, is not determined directly by the market interest rate. A central bank such as the Fed determines the money supply. The Fed might change the money supply in response to economic pressures indicated by the interest rate, but there is not a causal relationship between the interest rate and the money supply. Hence, a vertical line can represent the money supply curve. Equilibrium in the money market occurs where the money supply and demand curves cross. This point identifies the equilibrium interest rate. If the Fed decides that the current equilibrium interest rate is too high and shifts the money supply outward (from M_s^0 to M_s^1 in Figure 13-3), perhaps by buying Treasury bonds in an open-market operation, the prices of bonds and other assets rise and the equilibrium interest rate falls.

MONETARY POLICY AND EQUILIBRIUM GDP

Remember that the purpose of this chapter is to show how monetary policy (control over the money supply by a central bank such as the Fed) can change equilibrium GDP. Thus far, we have shown how the three policy instruments (reserve requirements, market operations, and the discount rate) are used by the Fed to change the money supply and how those changes affect the interest rate. Also recall from Chapter 12 that investment, one of the four components of aggregate demand, has a negative relationship to the interest rate. Using this information, we can show how monetary policy affects aggregate demand.

Remember that the final-expenditure and national-income equation for measuring GDP is as follows:

$$GDP = C + I + G + (X - M)$$

From this discussion we now know that the interest rate has a direct influence on investment similar to the effect of income and taxes on consumption. G and T are not changed by other economic variables; they are changed by public policy.

Suppose that a hypothetical economy is in equilibrium at GDP_0 (aggregate demand is equal to aggregate supply), but equilibrium GDP is below full employment as shown in Figure 13-4. The economy is in the Keynesian region of the aggregate supply curve; there is a recessionary gap. The initial level of investment in the economy is I_0. The Fed decides that economic performance is weak and expands the money supply to lower the interest rate. Some investment projects now become profitable, and investment increases to I_1. The added demand caused by this new investment is shown by an upward shift of aggregate demand. The effects of the increase in investment demand are identical to an increase in government spending described in detail in Chapter 12. Since there is unemployment in the economy (the economy is operating at less than full-employment GDP), producers can expand output to meet this new demand and equilibrium GDP increases from GDP_0 to GDP_1. This change in monetary policy leaves the economy below full employment (GDP_{fe}), which may be acceptable to the FED as a means of avoiding inflation.

Of course if the economy has an inflationary gap, the Fed could reduce the money supply to raise the interest rate and reduce aggregate demand.

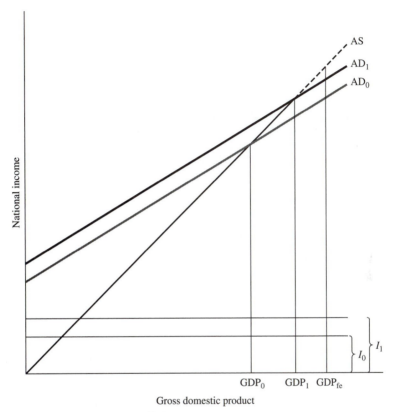

FIGURE 13-4   Changing investment and equilibrium output.

MONETARY POLICY AND CONTROL OF INFLATION

Through its control of the money supply, a central bank has control over an important lever on the engine of economic performance. By changing the money supply, a central bank can increase or decrease aggregate demand. However, the primary goal of most central banks is the control of inflation, not GDP. The reasons for this apparent paradox can be found in the roles of inflation and politics in determining macroeconomic performance. We start with inflation.

DEMAND-PULL VERSUS COST-PUSH INFLATION

Demand-pull inflation is inflation caused by excess aggregate demand, as discussed above. The Fed can act to address this type of inflation. **Cost-push inflation** is an increase in prices caused by rising costs of production. Developments in overseas markets (such as when a war reduces oil supplies) or the exploitation of domestic monopoly power can increase prices, leading to cost-push inflation. The Fed has few means of dealing with cost-push inflation.

The Costs of Inflation

As defined in Chapter 11, inflation is a rising price level for goods. A simple explanation of the cause of inflation is that "too much money is chasing too few goods." To understand what this means, look at Figure 13-5 (similar to Figure 12-7 in Chapter 12), which has GDP on the horizontal axis and the price level on the vertical axis. Remember that the aggregate supply curve is flat (the Keynesian region) when there is significant unemployment, but it is steep if most resources are employed. If the economy is operating at GDP_1 and the central bank increases the money supply, causing aggregate demand to shift outward, the result is a large increase in the price level and a small increase in GDP.

In the extreme case of a government financing its operations mainly by printing money (and collecting seigniorage), prices increase very rapidly. This condition is known as **hyperinflation.**

GERMAN HYPERINFLATION

One of the most well-known cases of hyperinflation occurred in Germany after World War I. The economy was weak, and few firms could pay taxes. The government still had to pay its employees and war reparations, and so it financed its operations by printing money. Prices skyrocketed!

"By October 1923, the postage on the lightest letter sent from Germany to the U.S. was 200,000 marks. Prices increased so rapidly that waiters changed the prices on the menu several times during the course of a lunch. Photographs of the period show a German housewife starting the fire in her kitchen stove with paper money." At the height of this inflationary period it was cheaper to burn money than wood!

Source: Adapted from Rayburn M. Williams, *Inflation! Money, Jobs, and Politicians* (Arlington Heights, IL, AHM Publishing, p. 2).

FIGURE 13-5   Aggregate supply, aggregate demand, and the price level.

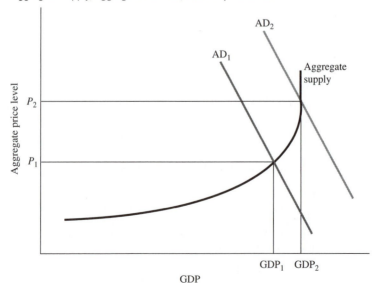

Small amounts of inflation can be stimulating for an economy, but inflation also has many negative effects. For example, it reduces the value of fixed-yield investments and the purchasing power of people who are on a fixed income. Inflation reduces the value of a loan principal to the advantage of borrowers. It causes some prices to increase much faster than others do, and this can reduce the efficiency of resource use in the economy. All such changes adversely affect some people while enriching others and reduce the efficiency of markets in allocating resources to their most productive uses. To compensate partly for the negative effects of inflation, many transactions that involve future purchases or payments are adjusted for inflation. Examples include social security payments that rise with the consumer price index and mortgages that have a variable interest rate (adjustable rate mortgages). For all these reasons, avoiding high rates of inflation is highly desirable.

With very high rates of inflation, or hyperinflation, money no longer facilitates exchange or serves as a unit of measurement or store of value. People are forced to turn to other means of completing transactions, such as barter or use of another country's currency. This has severe negative effects on an economy, reducing employment and output.

USE OF ANOTHER COUNTRY'S CURRENCY

A few countries use the currencies of other countries in domestic transactions. For example, Panama and Ecuador use U.S. dollars as their currency. In 2000, when Ecuador was considering dollarization, the adoption of dollars as its currency, the "pros" were that inflation would be stabilized at U.S. levels, foreign investors would be attracted, and long-term planning would be more feasible. The "cons" were that their government would lose the income from seigniorage, that it would not fix their underlying problems such as energy prices, infrastructure, external debt and corruption. Also their central bank would lose much of its ability to make monetary decisions.

Other countries are considering taking this action to avoid inflation problems.

Role of the Independent Central Bank

In Chapter 12 we showed that fiscal policy can be used to stimulate a recessionary economy by lowering taxes or increasing spending or to slow an inflationary economy by increasing taxes or reducing spending. A central bank can use monetary policy to affect the performance of the economy. Increasing the money supply will reduce interest rates and stimulate the economy by increasing aggregate demand. Reducing the money supply will increase interest rates and slow the economy and thereby reduce the inflation rate. However, in most countries the central banks have as their primary goal controlling inflation. The reason for this emphasis lies in the decision-making processes in fiscal and monetary policy.

The tax and expenditure decisions that make up fiscal policy are made in the political arena, where considerations other than controlling inflation are important. In the United States and in most countries, the legislative and executive branches determine fiscal policy. They face a multitude of issues and achieving a macroeconomic equilibrium is one of many concerns. It is relatively easy to reach a political decision to increase government spending or reduce taxes (i.e., expansionary fiscal policy) to stimulate an economy suffering from unemployment. It is more difficult politically to reduce government expenditures or increase taxes to control inflation in an economy that is near or at full employment.

Because the political system has difficulty controlling inflation, in most countries, a politically independent central bank is empowered to make monetary policy. Central bankers, such as the Board of Governors of the Federal Reserve and the European Central Bank, can determine the money supply without fear of being voted out of office. Therefore, most central banks see control of inflation as their primary goal.

There is another reason for central banks to concentrate on control of inflation instead of stimulating the economy. Reducing the money supply has a direct effect on the money supply, but increasing the money supply requires cooperation from the banking system. That is, a central bank can increase the money supply to banks, but the banks must lend it in order for the multiplier mechanism to stimulate the increase in the money supply. When the economy is weak, banks might choose not to increase loans, perhaps because they perceive the risk of default as too high or because there are too few qualified borrowers. In such an environment, fiscal policy can be much more effective in stimulating an economy.

SUMMARY

In Chapter 12, we examined how the government can influence economic activity with fiscal policy. Expansionary fiscal policy shifts outward aggregate demand. In this chapter, we discussed the role of monetary policy, that is, control over the money supply. Changes in the money supply affect the interest rate, which affects investment. *Monetary policies affect aggregate demand indirectly.* The effect of both fiscal and monetary policies depends on how close the economy is to full employment. When an economy is close to full employment, government intervention is more likely to cause inflation than increase output. Because the political process determines fiscal policy, it can be difficult to use fiscal policy to reduce aggregate demand. Monetary policymaking is often left to an independent central bank, and so it can make difficult economic decisions less burdened by political interference.

LOOKING AHEAD

In the next and final chapter on macroeconomics we examine the role of an economy's international transactions. We introduce the fourth category of aggregate demand—net exports—and examine the factors that influence its level. We find that the international economy can have an important effect on the performance of the domestic economy and on the role of macroeconomic policy.

IMPORTANT TERMS AND CONCEPTS

asset demand   328
barter   318
central bank   322

commodity money   319
cost-push inflation   331
currency in circulation   321

QUESTIONS AND EXERCISES

Name That Term

Read the following sentences carefully and fill in the missing term or terms.

1. _____ is a very rapid rise in the price level of a nation.
2. _____ inflation is the result of an increase in aggregate demand.
3. An economy's _____ is what sellers generally accept and buyers generally use to pay for a good or service.
4. The _____ is the specified minimum percentage of its deposit liabilities that a member bank must keep on deposit at the Federal Reserve Bank or as in-vault cash.
5. _____ inflation is caused by increases in the price of inputs.
6. _____ is the exchange of one good or service for another with no money used in the exchange.
7. _____ involves changing the money supply to assist the economy in seeking a full-employment, noninflationary level of total output.
8. The _____ is the government agency responsible for managing the nation's money supply and credit conditions.
9. The coins and paper money in the hands of the public are the _____ in circulation.
10. _____ is the difference between the value of money and the cost of producing it.
11. The _____ is the interest rate that the Federal Reserve Banks charge on the loans they make to depository institutions.
12. A _____ is the commodity that serves as a measure of value or unit of account in a commodity money economy.

True/False

Read the following sentences, then decide whether each statement is true (T) or false (F) and mark it accordingly.

T F   1 Changes in reserve ratios are more common than changes in discount rates, but less common then open-market operations.

T F   2 The U.S. currency has value today because each dollar can be redeemed for gold, the international currency.

T F   **3** Because of leakages in hand-to-hand circulation, the money multiplier represents the minimum amount by which the fractional reserve banking system can multiply an increase in currency.

T F   **4** When it comes to money, "more is better" for a nation because it makes each unit of money stronger.

T F   **5** If the economy is suffering from inflation, an increase in the money supply is an appropriate prescription.

T F   **6** If the economy needs an increase in the money supply, the Fed may reduce the discount rate, the interest rate it charges on its loans to banks.

T F   **7** The reserve ratio under a fractional reserve banking system is usually over 50 percent.

T F   **8** If an economy is in the Keynesian region of the aggregate supply curve, the Fed can expand the money supply, which lowers the interest rate, increases investment, and thereby causes aggregate demand to increase.

T F   **9** Other things equal, an increase in the money supply increases the interest rate.

T F   **10** By changing the money supply, the Fed can influence the interest rate, the amount of private investment, and, if the economy is in the Keynesian region of the aggregate supply curve, equilibrium GDP.

T F   **11** Because the governors of the Fed are appointed by the President for fourteen-year terms, monetary policy is less affected by political pressures than fiscal policy.

T F   **12** An open-market operation involves the purchase or sale of government securities by the Fed from or to the private sector.

T F   **13** To decrease the money supply, the Fed would buy government securities in the open market.

T F   **14** The price of financial assets and the interest rate are inversely related.

Multiple-Choice Questions

Circle the letter of the response that best answers the question or completes the statement.

1 Which of the following is not an example of money in the *M3* category?
  a paper money
  b savings accounts
  c shares in a mutual fund held by a household
  d a $250,000 time deposit held by a large corporation
  e none of the above

2 If the reserve ratio in a fractional reserve banking system is 11 percent, the maximum that each $1000 in new money supplied by the Fed in an open-market operation can increase the supply of currency is
  a $1000.
  b $1100.
  c $2100.
  d $9000.
  e $9091.

3 About how many dollars in GDP are there in the United States for every $1 of *M1*?
  a .5
  b 1

   c 2
   d 7
   e 15
4 Which of the following is not an example of money in the *M2* category?
   a coins
   b paper money
   c savings accounts
   d a $250,000 time deposit held by a large corporation
   e none of the above
5 Which of the following is not a cost of inflation?
   a reduction in the value of fixed-yield investments
   b reduction in the purchasing power of people of fixed investments
   c reduction in the wealth of owners of land and other real assets
   d increase in the transaction cost associated with making deals by requiring
      adjustments for anticipated inflation
   e none of the above
6 In what year was the Federal Reserve System established in the United States?
   a 1776
   b 1865
   c 1913
   d 1932
   e 1968
7 Which of the following is not a function of money?
   a a medium of exchange to end the need for barter
   b a unit of account used to quantify the economic value of goods and services
   c a cheap way to store value
   d a way to encourage barter
   e all of the above
8 Why are savings held as money?
   a because expenditures are made almost every day while income typically comes
      in lumps
   b to meet unexpected expenses without borrowing
   c to be able to purchase assets in the future that are expected to decline in price
   d all of the above
   e none of the above
9 Which of the following is not an example of money in the *M1* category?
   a coins
   b paper money
   c checking accounts
   d savings accounts
   e none of the above
10 Which of the following is not one of the main policy instruments used by the Fed to
   control money supply?
   a changing the reserve requirement of commercial banks
   b making purchases or sales in the U.S. government securities market
   c changing the rate at which banks borrow from the Federal Reserve
   d running or shutting down the printing presses
   e all of the above

Questions for Thought

1 What are the three mechanisms the Fed can use to control the money supply? Describe how they work. Which is most commonly used in developed countries?

2 What are the three categories of demand for money?

3 How does money eliminate the need for a double coincidence of wants?

4 What are the three economic functions of money? Can you think of economic conditions that might make commodity money more effective than paper money in meeting these economic functions?

5 Furs, gold, and goats are examples of goods that have been used as commodity money. Discuss the pros and cons of these commodities as money.

6 If the reserve requirement is 0.15, what is the maximum amount of money created by a new dollar?

7 Land can be considered an asset. The return on that asset is the net profit earned from growing the most profitable crop. Suppose that an acre of land is expected to produce a net profit of $10 and the market interest rate is 10 percent. What should an investor be willing to pay for that land? Suppose that the government raises the support price for the crop grown on an acre of land so that net profit rises to $20. What is the new market price of that acre?

8 What's happened to interest rates in the last few months? Are they declining, stable, or increasing? What does that trend indicate about the Fed's sense of the difference between equilibrium and full-employment GDP? What does that trend indicate about the Fed's sense of the danger of inflation increasing?

9 Are you more likely to borrow money to buy a car when interest rates are high or low? Why? Do you think other people considering investing in capital assets feel the same way? How do lower interest rates "stimulate" the economy?

10 Suppose your bank was offering a loan at 8 percent annual interest. If you anticipated that the inflation rate in the next year was going to be 10 percent, would you be likely to borrow money? Why?

ANSWERS AND HINTS

Name That Term   1. Hyperinflation; 2. Demand-pull; 3. medium of exchange; 4. reserve ratio; 5. Cost-push; 6. Barter; 7. Monetary policy; 8. Federal Reserve System; 9. currency; 10. Seigniorage; 11. discount rate; 12. numeraire

True/False   1. F; 2. F; 3. F; 4. F; 5. F; 6. T; 7. F; 8. T; 9. F; 10. T; 11. T; 12. T; 13. F; 14. T

Multiple Choice   1. e; 2. e; 3. d; 4. d; 5. e; 6. c; 7. d; 8. d; 9. d; 10. d

14

GROSS DOMESTIC PRODUCT, INTERNATIONAL TRANSACTIONS, AND MACROECONOMIC POLICY

CHAPTER OUTLINE

OVERVIEW

The previous chapters in this section on macroeconomics have ignored how a country's domestic economy interacts with the world economy. This simplification was needed because the interactions within a national economy are very complex. We needed to develop first an understanding of the workings of the domestic economy. However, every country in the world has important dealings with the global economy. Indeed, a country's interactions with the rest of the world can have a profound effect on the performance of the domestic economy. In many of the smaller nations of the world, external economic decisions may be more important than any internal policy decisions the government can make.

This chapter adds international transactions to our model of the domestic economy. These transactions include exports, imports, foreign loans and investments, and loans and investments at home by foreigners. In addition, we consider several new economic variables: the exchange rate, foreign prices, foreign interest rates, and foreign GDP. It is not possible to discuss the interactions among these items in detail. Whole books are devoted to that topic! Instead, we focus on the U.S. economy because of its importance to the rest of the world economy. And we examine in some detail the dramatic changes in the relationship of the U.S. economy to the rest of the world that occurred near the end of the 20th and the beginning of the 21st centuries.

LEARNING OBJECTIVES

This chapter will help you understand:

- How to measure U.S. economic interactions with the rest of the world
- How these interactions affect the U.S. economy
- How fiscal and monetary policy affect international trade

AN INTERNATIONAL ECONOMY

In previous chapters, we largely ignored the effects of the international economy on the domestic economy. But even a casual observer of current economic affairs is aware that international transactions have important consequences for the domestic economy. It is not uncommon for Americans to drive a Japanese car, wear jogging shoes made in Korea, and drink French wine. Germans are likely to eat French cheese, use American software on their Taiwanese computers, and listen to Italian compact discs on Japanese-made CD players. Japanese consumers eat meat produced with American-grown soybeans, wear clothing made in China, and take vacations in Bali. While the implied flows of goods and services are impressive, the international flow of investment funds is much larger. Decisions by investors to move funds among countries can dramatically affect a country's economic performance.

Economic developments in the last several decades dramatically illustrated the links between the United States and world economies. Large trade deficits and disputes over trade agreements are regular topics of discussion in the news media. Agriculture was also buffeted by international developments. The booming export years of the 1970s were followed by a crash in the early 1980s that caused many farmers to declare bankruptcy. Late in the 1980s and early 1990s there was a rebound in agriculture, stimulated in large measure by foreign demand, but a decline again at the end of the decade and into the 21st century was the result, in part, of the collapse of international markets.

BALANCE OF PAYMENTS

Just as the national income accounts discussed in Chapter 11 measure domestic economic transactions, the **balance of payments** records a nation's international economic transactions. A nation's balance of payments (BOP) is an account of all *international* transactions made by individuals, firms, and governments of that country. Examples of international transactions include U.S. exports of corn to Russia, French imports of oil from Saudi Arabia, Japanese purchases of U.S. government bonds, and a flight taken by a Swiss citizen on a German airline. Each transaction is recorded as a **credit** to the BOP of one nation and a **debit** to the BOP of the other nation. An export sale by a Canadian firm to a Mexican customer generates a flow of foreign exchange from Mexico into Canada. This flow of funds is a credit for Canada; it increases the Canadian BOP. It is also a debit for Mexico; it decreases the Mexican BOP.

Current and Capital Accounts

All international transactions are recorded as current-account or capital-account transactions. In a current-account transaction, the money pays for goods and services received at the time of the sale. In a capital-account transaction, a claim is received on goods and services in the future (e.g., a government bond that matures in the future or a factory and its future output).

Current-account transactions include the merchandise and the nonmerchandise trade balance. The **merchandise trade balance** includes all payments for goods such as U.S. corn exports (sometimes called trade in **visibles**). The **nonmerchandise trade balance** includes all payments for services such as a U.S. musician appearing in Japan (sometimes called trade in **invisibles**). Figure 14-1 illustrates how the sale of a good from one country to another affects the current accounts, and therefore the BOP, of the two countries. Suppose that an Argentinian grain firm sells a ton of wheat to a Japanese importing firm for $95 (most grain sold in international markets is priced in U.S. dollars). To pay the Argentinian firm, the Japanese firm buys $95 at the current exchange rate of 110 yen per U.S. dollar. The firm pays 10,450 yen ($95 \times 110 = 10,450$) for the $95.

TRANSACTIONS INVOLVING THE UNITED STATES

If the wheat transaction had involved a U.S. exporter, the United States would have seen an inflow of U.S. dollars, not a foreign currency. The U.S. dollar is the most important **reserve currency** for the world. Most countries keep a large share of their foreign exchange reserves in U.S. dollars or dollar-denominated assets. The United States is thus unique in that changes in its BOP do not necessarily affect its foreign exchange reserves, because most of its international transactions are made in dollars, not a foreign currency. However, these transactions do affect the dollar reserves in the U.S. banking system and therefore the money supply.

In recent years, other currencies, especially the Japanese yen and the German mark, have become more widely used as reserve currencies. It is likely that the Euro, the currency of the European Union, will replace the mark as a reserve currency.

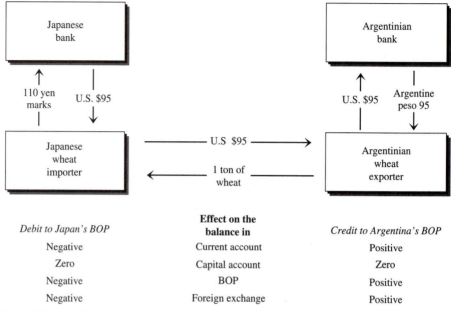

Debit to Japan's BOP	**Effect on the balance in**	*Credit to Argentina's BOP*
Negative	Current account	Positive
Zero	Capital account	Zero
Negative	BOP	Positive
Negative	Foreign exchange	Positive

FIGURE 14-1   Financial transactions associated with a grain trade.

The Argentinian firm exchanges the $95 for Argentinian pesos at a rate of $1.00 peso ÷ $1.00 and deposits the resulting $95.00 pesos in its bank account. The transaction in Japanese is counted as a debit, Japanese reserves of U.S. dollars have fallen, and the Japanese current-account balance falls. In Argentina, the reverse is true. The export is a credit in the Argentinian balance of payments, and Argentina's U.S. dollar reserves have increased, as has its current-account balance.

Capital-account transactions give the purchaser a claim on future output. Money from one country is used to make a **foreign investment** by purchasing an asset within another country. The person or firm spending the money receives some form of claim on future output from an asset of the recipient country. There is no movement of goods from the recipient country to the originating country at the time of the purchase.

Again, an example is useful to understand a capital-account transaction. Figure 14-2 illustrates what happens if a Japanese firm purchases a soybean farm in Brazil. The Japanese firm buys dollars from its bank and uses them to pay for the farm. This transaction is a debit to the Japanese BOP (foreign exchange flows out), and the Japanese capital account and Japanese dollar reserves fall. In exchange, the Japanese firm has a claim on the future output of the farm and assumes the risk associated with production on that farm. This foreign investment has increased Brazil's reserves and its current BOP has increased, but no Brazilian has a claim on the output of the farm.

Capital-account transactions can also occur as loans. For example, a Japanese bank lends yen to a Brazilian who purchased the soybean farm. The effect on the capital account (and therefore the BOP) would have been the same as in the example of a purchase. The Japanese bank would receive future payments from the Brazilian farm

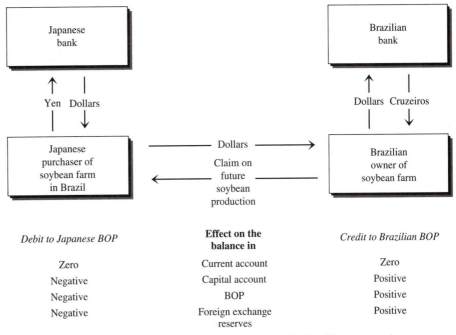

Debit to Japanese BOP	Effect on the balance in	Credit to Brazilian BOP
Zero	Current account	Zero
Negative	Capital account	Positive
Negative	BOP	Positive
Negative	Foreign exchange reserves	Positive

FIGURE 14-2   Financial transactions associated with a Japanese purchase of a Brazilian soybean farm.

owner to retire the loan. The new owner of the farm would have assumed the risk that the farm would produce enough to repay the Japanese loan.

OFFICIAL SETTLEMENTS

The final category of the BOP is called **official settlements.** The BOP is a double-entry book-keeping system. For every transaction in either the current or the capital account, there is an equal and opposite transaction in the official settlements account. Hence the sum of the current and capital accounts and the official settlements is, by definition, zero.

The Changing U.S. Balance of Payments

The volume of both current- and capital-account transactions expanded dramatically in the 1970s. Since then U.S. merchandise exports have grown dramatically, but imports have grown even more rapidly. The United States has a positive trade balance in the nonmerchandise or services account. However, this positive balance is not large enough to balance the deficit in the visibles account. The result is a trade deficit (exports greater than imports) in the current account. Interestingly, the agricultural component of the merchandise trade balance has generally been positive. We export more (primarily grain and soybeans) than we import (primarily fruits, vegetables, and processed products).

The picture for the capital account is quite different. This account has generally been positive. This reflects a net in-flow of funds from other countries to the United States for the purpose of purchasing assets such as treasury notes, stocks, businesses, and even farmland. There is a relationship between the capital and current accounts. If funds flow out of a country due to a current account deficit, at some point there will be a counter flow of funds into the country which will result in a positive balance in the capital account.

DETERMINANTS OF INTERNATIONAL TRANSACTIONS

In this section, we examine the underlying links between the rest of the world and the domestic economy.

Determinants of the Current-Account Balance

Recall that the final-expenditure equation (including all components of aggregate demand) is as follows:

$$GDP = C + I + G + (X - M)$$

In the previous two chapters we have discussed the determinants of consumption, investment, and government. Now we turn to the explanation of imports and exports. We first consider the current-account balance, exports and imports of goods and services. We start with exports.

Exports are purchases of domestically produced goods and services by foreigners. They add to aggregate demand for the output of the exporting nation. At the macroeconomic level, two important factors influence exports. The first is economic conditions in the importing countries. The demand for imports is stronger from large, growing economies and weaker from small or depressed economies. Thus the health of the economies of the rest of the world is important for U.S. exports. This effect is especially important for agricultural exports and is one of the major factors causing periods of economic prosperity and hardship in this sector.

The second important factor affecting exports is relative prices in the trading nations, adjusted for the exchange rate. If the domestic price level is low relative to the foreign price level, exports are encouraged. This is because to compete in a foreign country, firms must sell their products at the market prices of that country. To understand how both relative prices (prices at home compared with prices abroad) and the exchange rate affect export competitiveness, consider the following example. Suppose the U.S. price of a ton of corn is $90. If the exchange rate is 150 Japanese yen per dollar, the cost to a Japanese buyer is ¥13,500 (90 × 150). If the U.S. price rises, say to $100, the cost to the Japanese buyer also rises, to ¥15,000. Hence, a rise in the U.S. price will reduce the quantity demanded in Japan, *ceteris paribus*.

CURRENCY APPRECIATION, DEPRECIATION, DEVALUATION, AND REVALUATION

Several terms are used to describe the change in the exchange rate between currencies. When the amount of a foreign currency that can be purchased with the domestic currency increases (decreases) as a result of market forces, the domestic currency is said to appreciate (depreciate). The terms *revaluation* and *devaluation* are used when a government declares an increase or a decrease in the amount of a foreign currency that can be exchanged for a domestic currency.

Exchange-rate changes also affect the cost of U.S. goods abroad. If the U.S. dollar depreciates relative to the yen (a dollar buys fewer yen), say from 150 yen per dollar to 100 yen per dollar, the cost of a ton of corn priced at $90 to the Japanese buyer drops from ¥13,500 to ¥9,000. Thus **depreciation** of the domestic currency makes exports more competitive. An **appreciation** of the domestic currency (the domestic currency is worth more units of the foreign currency), *ceteris paribus,* makes exports less competitive abroad. Clearly, the exchange rate is an important economic variable, and its determination is explored below.

Domestic policy has little if any influence over foreign prices, but it can influence domestic prices. We saw in Chapters 12 and 13 that excess aggregate demand can cause the price level to increase. We show below that macroeconomic policies do influence the exchange rate. Further, domestic macroeconomic policies can change exports through their influence on domestic prices and the exchange rate.

Imports are purchases of foreign goods and services for domestic consumption. They reduce aggregate demand for domestic output because some income earned producing goods at home is spent on goods produced abroad. Both national income (GDP) and prices affect imports.

• Imports are directly related to national income. As national income goes up, a part of the increase is spent on imports, and total imports rise. If national income falls, imports will also fall.

• The relative price of foreign goods to domestic goods is an important determinant of imports. Just as the volume of exports depends on how expensive they are abroad relative to competing goods, so the volume of imports depends on how expensive they are relative to domestic goods.

• As in the case of exports, the exchange rate has an important effect on the competitiveness of imports because of its effect on relative prices.

To see the importance of the exchange rate on imports, suppose that the price of a CD player in yen is ¥25,000. If the exchange rate is 150 yen per U.S. dollar, the price is $166.67. If it is 100 yen per dollar, the price is $250, *ceteris paribus.*

From the discussion above, we see that exports and imports are both functions of the exchange rate and the ratio of domestic to foreign prices. Imports are also a function of national income. Also, we can see that U.S. importers (consumers) prefer a strong dollar because the dollar price of imports is low. However, U.S. exporters (businesses that rely on exports) prefer a weak dollar because the foreign price of the exported goods is low.

Determinants of the Exchange Rate

Since the exchange rate plays such an important role in determining net exports, we look now at how it is determined. The **exchange rate** is the price of domestic currency in terms of a foreign currency. The exchange rate is determined in the foreign exchange market, which equates the demand for domestic currency by foreigners and the supply of domestic currency for sale to foreigners, as shown in Figure 14-3. Individuals and firms with foreign currencies who buy U.S. goods, who make investments in U.S. assets, or who make loans to U.S. borrowers create the demand for dollars. The higher the price is for dollars the fewer dollars are demanded.

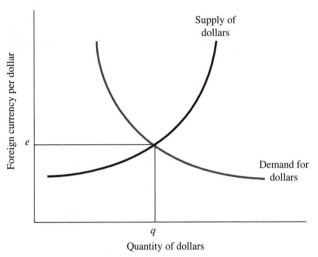

FIGURE 14-3   The foreign exchange market.

FLOATING AND FIXED EXCHANGE RATES

Before 1971, the exchange rate of the U.S. dollar with other currencies was fixed under the gold standard (described in Chapter 15). Any nation was free to buy gold from (or sell it to) the U.S. Treasury. In 1971, the United States broke the link between gold and the U.S. dollar and switched from a fixed to a floating exchange rate. Most other industrialized countries followed suit.

American individuals and firms who buy foreign goods or invest abroad provide the supply of dollars. To buy or invest they need foreign currencies as indicated in Figures 14-1 and 14-2. The higher the price of dollars (e.g., the more yen a dollar will buy), the cheaper Japanese goods are in the United States, the greater will be imports and thus the more dollars will be supplied. The exchange rate is the equilibrium price for dollars in terms of a foreign currency. When the exchange rate increases, the dollar purchases more yen; the dollar is said to get stronger. When the dollar buys fewer yen, or other currencies, it is said to be weaker.

Many economic factors determine the demand and supply curves for dollars, including perceptions of relative safety of investment. Three sets of factors are key: foreign and domestic GDPs, foreign and domestic aggregate price levels, and foreign and domestic interest rates. We focus on the relationship between the exchange rate and interest rates because the U.S. interest rate (relative to foreign rates) is a major determinant of foreign willingness to buy U.S. assets and to lend to U.S. citizens. To do this, we need to examine the determinants of the capital-account balance and the effect on the exchange rate.

EXCHANGE RATE QUOTES

Most exchange rates involving the U.S. dollar are quoted as units of foreign currency per dollar. The only exception is the U.S. dollar/British pound exchange rate, which is quoted as dollars per pound ($/£). Exchange rates that don't involve the dollar are usually written so that the rate of exchange is larger than 1.

Determinants of Capital-Account Balance

This account measures the flow of investment funds in to and out of a country. A surplus in the U.S. capital account occurs if there is more foreign investment in U.S. assets and loans to the U.S. from abroad than U.S. investments and loans abroad. There are three reasons why foreign investors invest their money in the U.S. economy (or more generally why any investor would choose to invest abroad): rate of return, safety, and diversification.

The first consideration is the rate of return on investments and loans in the U.S. economy relative to returns in other countries. The higher the interest rate is in the United States, the more attractive it is to lend money to U.S. borrowers, to purchase U.S. treasuries.

Second, the United States has effective legal safeguards to protect investors from unscrupulous behavior, and nationalization of an industry is extremely unlikely. The United States is also politically more stable than many other countries.

Third, a general rule of investing is to diversify investments to reduce the overall risk in an investment portfolio. For foreign investors, the U.S. economy provides a market with different risks than exist in their home market.

The safety and diversification reasons for investing in U.S. assets are little influenced by macroeconomic policy. General characteristics of the U.S. economy such as its size and legal institutions are important but not affected by economic policy.

The rate of return, on the other hand, can be changed with macroeconomic policies, especially monetary policies that change the money supply. If the Fed decreases the supply of money, the interest rate rises. If interest rates are higher in the United States than in other markets, the attractiveness of lending money to American borrowers goes up and the demand for dollars in the international market increases. This shifts the demand for dollars outward, like the shift from D_1 to D_2 in Figure 14-4. As loans and foreign investments flowing to the United States increase, foreign exchange flows into the United States raising the capital-account balance and increasing the value of the dollar.

Note that a change in the capital account can affect the current account. Figure 14-4 indicates that an increase in the capital-account balance causes the exchange rate to increase. This makes prices of exports rise in foreign markets and prices of imports fall at home. Therefore, changes in the capital account affect the exchange rate which in turn affects exports and imports which determine the current-account balance. An improvement in the capital account will lead to a stronger currency and exchange rate meaning increased imports and reduced exports and thus a fall in the current-account balance.

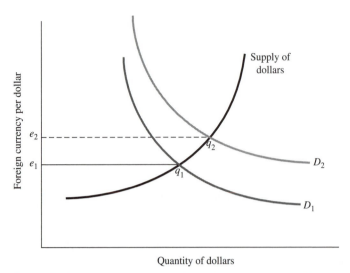

FIGURE 14-4 The effect of interest rate changes in the foreign exchange market.

Note: D_1 denotes relatively low interest rates in the United States. *D_2* denotes relatively
high interest rates in the United States.

MACROECONOMIC POLICY AND THE BALANCE OF PAYMENTS

Fiscal Policy Let's consider the effects of an expansionary fiscal policy, for
example, increasing governmental expenditures or reducing taxes to increase aggregate
demand. As national income rises, some of the increase is spent on imports. The
increase in imports due to the change in fiscal policy causes a decrease in the current
account and thus the BOP. In addition, if the expansionary fiscal policy causes an
increase in the price level (inflation), foreign goods become cheaper in the U.S. market
and American goods become more costly abroad. As a result, this also causes exports to
fall and imports to increase. Therefore, there is a negative relationship between
expansionary fiscal policy and the current-account balance.

Monetary Policy Monetary policy affects both the current and the capital
accounts. If the central bank (Fed) decides to reduce the money supply and raise the
interest rate, this increases the attractiveness of foreign lending to U.S. borrowers.
Hence, foreigners lend more money to U.S. borrowers, and the capital-account balance
increases. Simultaneously, the exchange rate rises, as shown in Figure 14-4. This
makes domestic goods more expensive and imported goods cheaper. The U.S. demand
for imports rises, and foreign demand for U.S. products falls. Hence, the effect of a
contraction in the money supply is to raise the capital-account balance and lower the
current-account balance.

EQUILIBRIUM GDP IN AN OPEN ECONOMY

We can now add the final component of aggregate demand, **international trade,** to our
model of equilibrium GDP. In Figure 14-5 we add imports and exports to the fiscal policy example in Figure 12-9. Clearly imports and exports can increase or decrease

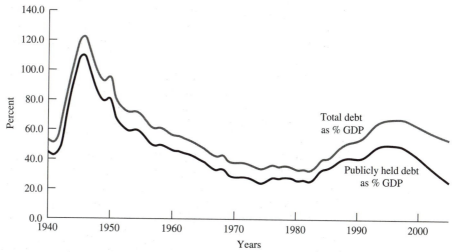

FIGURE 14-8   The U.S. debt as a percent of GDP, 1960–2005.
Note: Data for 2000–2005 are estimated.
Source: Office of Management and Budget estimates based on data from Bureau of Economic Analysis.

government began to operate with a budget surplus, which means the debt as a percent of GDP fell. Early in 2000 president Clinton proposed that the government adopt a plan that would pay off the national debt, excluding Social Security debt, by 2013.

Is there an appropriate ratio of public debt to GDP? How much is too much? How little is too little? These questions have no simple answers. If the debt is so high that the interest payments make it impossible for a government to finance basic public services, the ratio is too high. On the other hand, it is not difficult to find public sector investments—such as roads, education, and defense—for which it is prudent for a government to borrow and repay over several years instead of raising taxes. There is a wide range of debt to GDP ratios between these two extremes. The determination of the best ratio for a given country and a given time is both a political and an economic question that is beyond the scope of this book. However, the rate of increase in the debt experienced from 1980 to the early 1990s cannot be sustained without adverse economic consequences.

SUMMARY

In this chapter, we have completed our analysis of macroeconomics by examining how international transactions affect equilibrium GDP. We showed that by changing the money supply and therefore the interest rate, the government could control the exchange rate. The exchange rate has an important influence on net exports, the final component of aggregate demand. Since the exchange rate is influenced by the interest rate, monetary policy can be used to change two components of aggregate demand: investment and net exports.

LOOKING AHEAD

Having worked our way through microeconomics and macroeconomics, and having considered the role of agriculture and agribusiness, we next analyze the role of economics in several important policy arenas. We will first consider environmental and resource policy issues. Then we will turn to economic development and price and income policies.

IMPORTANT TERMS AND CONCEPTS

balance of payments   341
capital-account transactions   342
credit   341
currency appreciation   345
currency depreciation   345
current-account transactions   341
debit   341
exchange rate   345
exports   344

foreign investment   342
imports   345
international trade   348
invisibles   341
merchandise trade balance   341
nonmerchandise trade balance   341
official settlements   343
reserve currency   341
visibles   341

QUESTIONS AND EXERCISES

Name That Term

Read the following sentences carefully and fill in the missing term or terms.

1 Domestically produced goods and services sold in the market of a foreign country are called _____.
2 _____ are goods and services purchased by businesses, individuals, or governments from a foreign country.
3 The government-approved transactions that settle the outstanding financial accounts in international trade are the _____.
4 _____ transactions include the merchandise and the nonmerchandise trade balance.
5 Expenditures by a government, firm, or individual to purchase assets in another country is called _____.
6 _____ is the record of a nation's international economic transactions.
7 The _____ indicates the amount of a foreign currency that can be purchased with a unit of domestic currency.
8 A country that sells more tangible goods than it buys has a positive _____ trade balance.
9 In a _____ transaction a claim is received on goods or services in the future.
10 Currency _____ involves an increase in the number of foreign currency units that can be purchased with the domestic country's currency.

True/False

Read the following sentences, then decide whether each statement is true (T) or false (F) and mark it accordingly.

T F   1 A debit in the U.S. balance of payments results in the flow of foreign exchange into a country and an increase in the U.S. balance of payments.

T F   2 Current-account transactions include the merchandise and nonmerchandise trade balances.

T F   3 The purchase of Japanese cars by a U.S. dealer would generate a debit in the U.S. balance of payments.

T F   4 When the British pound depreciates relative to the dollar, a dollar buys more pounds.

T F   5 Safety, rates of return, and investment diversity are the primary reasons individuals in other countries invest in the United States.

T F   6 The agricultural trade balance was negative throughout the 1980s and thus contributed to the overall trade deficit of the United States during these years.

T F   7 A sale of farm machinery by John Deere to Russia would generate credit in the U.S. balance of payments.

T F   8 If the dollar appreciates relative to the yen, a dollar buys fewer yen.

T F   9 The share of trade in GDP leveled off and even declined in recent years as the United States closed foreign markets and restricted foreign imports.

T F 10 Capital-account transactions involve claims on goods and services in the future, such as bonds or ownership of a factory.

Multiple-Choice Questions

Circle the letter of the response that best answers the question or completes the statement.

1 Which of the following is not an international transaction that would appear in the U.S. balance of payments?

a U.S. firm exports soybeans to Mexico

b U.S. firm purchases oil from Saudi Arabia

c You spend the night in a French hotel

d You purchase a Japanese car from your local dealer

e A Japanese citizen buys U.S. government bonds

2 Which is the most important reserve currency in the world?

a the U.S. dollar

b the Japanese yen

c the Euro

d the British pound

e the Canadian dollar

3 Which of the following does not encourage increased exports from the United States to Japan?

a a depreciation of the dollar

b an increase in Japanese income

c an appreciation of the yen

d an appreciation of the dollar

e an increased appreciation of U.S. goods by the Japanese

4 If the U.S. exchange rate is 110 yen per dollar, 2 dollars per pound, and 1.8 dollars per mark, what is $10,000 worth?

a 1,110,000 yen or 20,000 pounds or 18,000 marks

b 90.91 yen or 5,000 pounds or 18,000 marks

c 1,110,000 yen or 5,000 pounds or 5,555.56 marks

d 90.91 yen or 20,000 pounds or 18,000 marks

e 1,100,000 yen or 20,000 pounds or 5,555.56 marks

5 Which of the following will decrease the current-account deficit?

  a an increase in government spending

  b a decrease in taxes

  c contracting the money supply

  d an appreciation in the dollar

  e none of the above

6 The exchange rate is

  a is the price of domestic currency in terms of a foreign currency.

  b is determined in a market in which international investors trade.

  c can be fixed by a government.

  d is a major determinant of imports and exports.

  e all of the above.

7 Which combination of these economic factors is the worst for U.S. agriculture?

  a high interest rates and a weak dollar

  b high interest rates and a strong dollar

  c low interest rates and a weak dollar

  d low interest rates and a strong dollar

8 The total federal debt currently is approximately what proportion of GDP?

  a 10 percent

  b 30 percent

  c 60 percent

  d 110 percent

  e 135 percent

9 If the Fed reduces the money supply it will tend to

  a reduce the current-account and increase the capital-account balances.

  b increase the current-account and reduce the capital-account balances.

  c increase the current-account and the capital-account balances.

  d reduce the current-account and the capital-account balances.

  e none of the above.

10 A nation's budget deficit can be "financed" by

  a domestic savings greater than domestic investment or imports less than exports.

  b domestic savings less than domestic investment or imports greater than exports.

  c domestic savings greater than domestic investment or imports greater than exports.

  d aggregate consumption greater than aggregate demand.

  e none of the above.

Technical Training

Suppose you are the purchasing agent for Oz, charged with purchasing iglets, your people's primary food. Iglets are only produced in three countries: Amer, Bost, and Cran. The iglets are of identical quality in each country and each is a reliable supplier. Your budget is 1000 wizs per month.

1 In April, you call the iglet sales representatives in each country and learn that their current prices are:

Amer	120 alms per ton of iglets
Bost	540 blogs per ton of iglet
Cran	6,000 cews per ton of iglet

When you look in your Yellow Brick Road Review, you learn that the exchange rates for the three currencies are as follows:

2 alms per wiz
10 blogs per wiz
120 cews per wiz

From whom do you purchase your monthly iglets? How much do you buy? What is your total payment in terms of the currency of the providing country?

2 In May, you learn that your three suppliers are offering iglets at the same price as in April. The relative values of the alm, the blog, and the cew are unchanged too. However, a recent devaluation of the wiz means that each wiz is only worth half as much as each of the other currencies. From whom do you purchase your monthly iglets? How much do you buy? What is your total payment in terms of the currency of the providing country?

3 In June, you find that your three suppliers still are offering iglets at the same price as in April. The exchange rate for alms and cews is the same as it was in May, but depreciation in blogs by the Bost government means that the new exchange rate is 11 blogs per wiz. From whom do you purchase your monthly iglets? How much do you buy? What is your total payment in terms of the currency of the providing country?

Questions for Thought

1 How do the above transactions affect the current/capital-account balance of Oz?

2 What is the principal difference between a foreign loan and a foreign investment?

3 How does the health of the world economy affect U.S. exports?

4 Explain how the foreign demand for U.S. dollars influences the exchange rate.

5 How does the exchange rate affect the relative competitiveness of U.S. exports and imports?

6 In late 1992, the central bank of Germany (the Bundesbank) raised the German interest rates to attract capital to Germany to fund German reunification. What effect do you expect this action would have had on:
   a U.S. interest rates?
   b the dollar/mark exchange rate?
   c the German trade balance?

7 Make a list of the things that you own that were produced in other countries. Do you know where everything you own was produced? Do you care?

8 The "Buy American" slogan encourages you to buy goods produced in the United States to save domestic jobs and increase domestic income. Have you ever purchased something produced abroad when a similar good produced domestically was available? Why did you ignore the slogan's advice?

9 From the late 1970s to the late 1980s, the United States ran a large deficit in its current account because it imported more goods and services than it exported. During the same period, the country ran a surplus in its capital account because of the large net inflow of foreign capital into the country. Some have suggested that this combination is analogous to using a credit card to go on a spending spree. Do you think this analogy is valid? What happens to an individual who uses a spending card to go on a shopping

spree? What happens to a nation that imports goods and services in exchange for claims of future outputs?

10 Look in the *Wall Street Journal* and find what the current exchange rate is for the dollar versus the yen, the mark, and the pound.

ANSWERS AND HINTS

Name That Term **1.** exports; **2.** Imports; **3.** official settlements; **4.** Current account; **5.** foreign investment; **6.** Balance of payments; **7.** exchange rate; **8.** merchandise; **9.** capital-account; **10.** appreciation

True/False **1.** F; **2.** T; **3.** T; **4.** T; **5.** T; **6.** F; **7.** T; **8.** F; **9.** F; **10.** T

Multiple Choice **1.** d; **2.** a; **3.** d; **4.** c; **5.** d; **6.** e; **7.** b; **8.** c; **9.** a; **10.** c

Technical Training **1.** Cran, 20 tons, 120,000 cews; **2.** Cran, 10 tons, 60,000 cews; **3.** Bost, 20.37 tons, 11,016 blogs

PART FIVE

PUBLIC POLICY
AND AGRICULTURE

15

RESOURCE AND ENVIRONMENTAL MANAGEMENT

CHAPTER OUTLINE

OVERVIEW

So far, we have described how producers and consumers come together in a market to allocate society's resources, and how the larger macro economy functions. In the next three chapters we turn to the economic consequences and implications of the decisions made by governments and governmental agencies. The decisions made in governments are popularly called "public policy."

Public policy involves the plans and decisions that people make in creating and guiding governments in solving problems that require a public decision. It refers to the problems that can be approached and solved only through decisions and actions by government. A public policy issue arises when there are differing opinions as to the best approach to a public problem. Decision making in public policy differs from decision making within a private firm operated by a single management. In the public policy arena the values and goals of different groups of people conflict, and what may be politically acceptable to one group may be opposed strongly by another group. Public policy development involves many people, with different ideas about what to do, interacting in a public institution. In all forms of representative government (the subject of our concern), compromises must be made among differing groups and there must be a system or process for reaching a conclusion. While the same can be said about decision making within a firm, the participants are limited to those in the firm and the focus is on the firm's performance.

Three sources of conflict may arise. First, there may be disagreement about the facts at issue, or beliefs about what is true; the costs and benefits that will result or who will gain or lose. These disagreements can be resolved through further research or investigation. In economics, this research is called positive analysis, as discussed in Chapter 1.

A second source of conflict arises out of the way people value the possible outcomes of a governmental action; whether it is desirable for them, the country, or for an interest group important to them. That is, often people agree about the facts, but still disagree about what should be done. In economics, such disagreements are called normative issues.

Third, people may agree on the basic facts and what to do, but may disagree about the appropriate way to implement the action. The economic analysis of these questions, or studies about what has worked in the past and the costs and benefits of alternative approaches are called prescriptive analysis.

Economists learn how to:

• discover the important facts in a case
• analyze these facts in ways that will lead to meaningful generalizations and conclusions and
• use these conclusions in generating alternate approaches, with different costs and benefits.

Economists have little or no special ability in making decisions concerning the values people hold about a particular policy. Economists contribute to public policy formulation by identifying major categories of policy, and by refining policy alternatives and analyzing their effects. For example, there are important alternatives in a national policy for economic development, income stabilization, conservation of resources, and protection of the environment. These are convenient categories for studying public policy related to the economics of resources, agriculture, and food.

This chapter begins by indicating some of the conditions under which the market does not operate in an acceptable fashion and governmental action is appropriate. An important role of government is to establish market rules and regulations under which individuals, firms, and industries have the greatest opportunities to prosper. This chapter shows that in some situations the government must go beyond this role in:

- controlling the costs to society of the nonmarket effects of firms' activities, called *externalities;*
- establishing organizations that can manage resources on a group basis, called *common property management;* and
- developing policies to deal with resources, goods, and services that are difficult to buy and sell in a market (such as national defense), called *public goods.*

We then turn to public policy issues that relate to resource management problems and environmental quality. We also discuss the natural resources important to agriculture (focusing on land and energy), examine the types of environmental problems important to the sector (using water and solid wastes as examples), and consider how public institutions influence the use of soil, oil, forests, fish, and other naturally occurring resources.

In many cases there is overlap between resource and environmental management issues. Often resource management decisions—for example, those focusing on areas such as mining, logging, fishing, and even recreational uses of forests—can have environmental implications. Likewise, most environmental questions focus on the effects of pollutants on the quality of the environment in terms of health and welfare of people— but often there are also resource management implications of the choices made.

LEARNING OBJECTIVES

This chapter will help you learn:

- The roles for government in a competitive market economy
- When market failure occurs
- What externalities are and the alternatives for dealing with them
- The difference between common property and open access resources
- About public goods, and their creation and management
- That there are resource management problems related to agriculture
- About selected environmental problems related to agriculture
- The role of governmental agencies in managing resource and environmental problems

THE ROLES OF GOVERNMENT

This chapter discusses conditions under which a government might want to modify or participate in markets. The specific topics are the control of negative externalities, management of common property resources, the creation of public goods, and actions to deal with open access resources. Before a government can deal with these issues, however, it must have laid the groundwork. A key element is the existence of enforced property rights. The development of an approach to property rights and the institutions to deal with enforcement of property rights is a basic role of government.

Brief reference to American history will help explain why development and enforcement of property rights is key to the functioning of a market economy. When the American continent had only a few inhabitants, it was possible for everyone to use the land as he or she wished. But once land became scarce, a system of land allocation and property rights was essential. The transition of the western frontier from a lawless region (at least as the European settlers saw it), with gunfights over land and its use, to a highly productive agriculture required the development of property rights, institutions, and government enforcement. On the open range, anyone could hunt, graze, or grow crops. In a situation such as this, conflicts among users were sometimes resolved with force. Battles between cattle raisers and farmers over rights to graze were sometimes fought in court and at other times decided with guns. The struggles between Native Americans and European settlers were also largely over control of land.

The government established a set of property rights and committed the resources necessary to enforce those rights (including a police and court system). Much of the land was assigned to individuals. Corporations were also given significant amounts to help finance and build transcontinental railroads. The owner, whether an individual or a firm, had more or less unfettered rights to use the land. Other lands were kept in public hands and managed by federal, state, or local government agencies, or leased to individuals for specific uses. A small amount of land was set aside for Native Americans (the Indian reservations), to be managed by the tribes. Public lands and the reservations are examples of common property. The land is owned collectively by a group of people. In the case of federal lands, ownership includes all Americans. In the case of the reservations a tribe is the owner.

The development of property rights and their enforcement provided the security necessary for the owners to invest in the development of the resources. With secured property rights, individuals could make long-term capital investments, assured that other individuals, firms, or the government would not take the land without agreement and compensation.

Even when markets are well established, the development of new technology can require the government to modify existing property rights and institutions for the market to operate effectively. For example, the biotechnological advances in agriculture have generated intense debates on property rights involving such issues as patentability of new life forms, the profit rights from the sale of the new technology, and the controls placed on the release of new materials. The government must resolve these issues as products of a new technology enters the market. For biotechnology, American patent laws have been interpreted as protecting the property rights of the inventor of new crop and livestock hybrids for seventeen years. How broad that patent protection will be is not yet fully resolved. The decisions will determine the major gainers and losers from the development of the technology.

Another way of illustrating a fundamental role of government is to observe the different ways nations resolve questions of resource allocation. Capitalist nations rely mainly on private ownership to allocate natural resources. In socialist nations, most factors of production are owned by the state. But no nation is strictly capitalistic or socialistic, and so the mixture of the two systems is important. In some countries, including several capitalistic nations, the transporting and exporting activities for agricultural products are done by a governmental agency. In others, including the United States, the private sector provides these services. Although the United States is among the most capitalistic of the major nations, the U.S. government manages many important fishery and forest resources. Many less-developed nations leave control of these resources to private or local management.

INTERNET REGULATIONS

The Internet provides an example of both the need for market regulation and of the global scale of some issues. It is generally assumed that businesses oppose regulation of their industry. In late 1999 CEOs of numerous major Internet corporations from around the world met to attempt to agree upon a set of regulations they would ask all nations to impose on their industry. The basic reason is that they realized the need for a commonly accepted set of rules under which they could conduct business. This is often characterized as having a "level playing field."

It is clear that in socialistic countries the government is involved in markets. In capitalistic countries, the government is also involved in markets, albeit to a lesser extent. One might ask whether a country could rely entirely on markets to allocate resources. Because some market failures are inevitable, the answer is that such a pure system would not work satisfactorily. That is, there are instances in which unregulated markets do not allocate goods and services in an optimal manner. We now turn to an analysis of this misallocation and the available remedies.

MARKET FAILURE

Market failure occurs when the price mechanism does not allocate goods or services in a socially acceptable pattern. The basic reason for market failure is the lack of a complete set of exclusive property rights. With a complete set of property rights, all resources, goods, and services are owned by an individual, or individuals, who have the right to determine how that property is used. Individuals would pay for all goods and services they consume as well as for any damages they impose on others. But such a complete specification of property rights is not feasible or practical.

When a market failure occurs there is no way for producers to interact with consumers in determining the amount of a good to produce. Too much (as in the case of a pollutant) or too little (as in the case of recreational areas) of a good or service is produced. Consumers might be denied a product or service that would provide utility if it were available in the market. In this case, producers would not be using resources to produce and market this good, and too many resources would be used to produce other goods and services. Although it is relatively easy to establish property rights to land and most physical things, it is difficult to establish property rights to the ocean, flowing waters in a stream, the fish in the stream, or the exhaust of an automobile. Externalities, such as damage caused by automobile exhaust, are the most common cause of market failure.

EXTERNALITIES

An **externality** is a positive or negative effect of the actions of one individual, firm, or nation on another without compensation. A classic example of a positive externality is the honey bee pollinating a neighbor's crop as it collects nectar to produce honey for the bee owner. If compensation is not paid for the pollination service, it is an externality. Other examples of positive externalities include enjoying another's flower garden or appreciating the scenery during a drive in the country.

An example of a negative externality is passive smoke. Some nonsmokers are annoyed by or may develop health problems because of another person's cigarette smoke. Under normal circumstances the smoker does not compensate the nonsmoker; thus it qualifies as an externality. Air and water pollution due to discharges by industries, automobiles, and many other sources are typical examples of negative externalities.

There are numerous examples of negative externalities involving agriculture. Farms and agribusinesses are often the source of pollutants, and both farm and nonfarm families can be affected adversely by these pollutants. Examples include:

- pesticides drifting to a neighbor's field, entering the groundwater or a lake or stream;
- livestock wastes entering surface water or groundwater;
- livestock odors drifting to a neighbor's property;
- eroded soil covering a neighbor's crops, filling drainage ditches or lakes; and
- food processing wastes contaminating waterways or generating odors that annoy neighbors.

Clearly, there are costs associated with negative externalities. Just as clearly, these costs fall if the externality is reduced, although it also costs money to reduce the externality. Two questions come to mind. First, by how much should an externality be reduced? Second, should the polluter pay for the reduction of a negative externality, or should those who are negatively affected pay the polluters to stop polluting? Surprisingly, the answer to the first question does not necessarily depend on the answer to the second. The optimal level of pollution control does not depend on whether the polluter, or the party damaged by the pollution, pays.

Optimum Levels of Pollution

Figure 15-1 provides a conceptual basis for determining the socially optimum level of pollution. The marginal damage (MD) curve indicates the additional damage caused by each unit of pollution, moving from left to right, or the reduction of damages by reducing a pollutant moving from right to left in the figure. The marginal cost (MC) curve indicates the marginal cost of reducing the pollutant by each unit, moving right to left. As the amount of a pollutant released into the environment increases, the damage increases. Therefore, as the amount of pollution is reduced, damage falls. Similarly, the cost of controlling pollution increases as the extent of control increases.

The optimal level of pollution, and thus pollution control, is where the marginal cost of pollution control equals the reduction of marginal damages from pollution. This occurs at point *a*, where MC = MD. At this point, the increase in pollution control cost from a reduction of one unit in pollution (MC) equals the reduction in damage (MD) realized by reducing pollution by one unit. Notice this suggests that it is seldom efficient to eliminate pollution entirely. If the pollutant were completely eliminated, we would have used too many resources for pollution control. More resources would be spent on controlling pollution than the value of the reduced damages. In some places, stopping all pollution would preclude agricultural production.

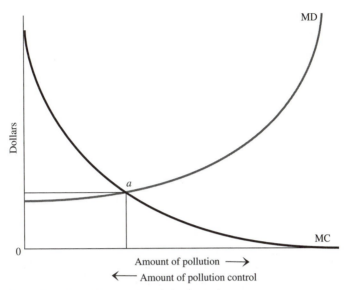

FIGURE 15-1   Marginal costs and benefits of pollution control.

Pollution problems are complex. Usually, there are numerous sources of pollutants, and numerous individuals and firms are adversely affected. Often a private firm, or firms cause the pollution, and the general public incurs the damages. In a major metropolitan area there may be thousands of sources of pollutants and millions of individuals affected. Other pollution problems, such as acid rain and ozone depletion, have global consequences. For many types of pollutants, including those from agriculture, it is difficult (impossible at reasonable cost) to identify the sources of the pollutant. Such pollutants are called **nonpoint pollutants.** Because it is difficult to identify the sources of nonpoint pollutants from agriculture, the costs of measuring and addressing this problem can be substantial. For example, the water quality problems associated with agricultural chemicals occur as water leaches through the soil and runs off the soil into streams. Measuring the quantity of a chemical lost to the environment, and the quantity originating from any one farm, is clearly difficult. Despite these problems, it is important to try to control pollution to the point where MC = MD. What options do we have for doing this? If a market for pollution control existed, an equilibrium price and quantity for pollution would exist, as suggested in Figure 15-1. (Such markets do exist in some instances, and more are being developed.) In the absence of a market, the government might choose regulation to force producers to an acceptable level of pollution, or it might establish pollution taxes or subsidies.

Policies for Dealing with Externalities

Externalities occur when one individual or firm produces a pollutant that adversely affects another individual or firm, and no payment is made to compensate for the damage done by the pollutant. Several strategies can be used to address externality problems.

Do Nothing This response is usually chosen for positive externalities. The fact that one person enjoys unintended benefits from the actions of another is seldom a public policy problem. Thus there is little incentive to incur any public expense to deal with it. There may, however, be a response in the private sector. For example, the owner of a beautiful garden might charge admission, and there is a sizable bee-keeping industry with bee owners receiving payment for bringing hives to orchards and fields to provide pollination services.

With negative externalities, the do-nothing response is also sometimes chosen because even the least-cost means of dealing with the externality is more costly than the benefits achieved. The benefits of adopting a policy of control must be greater than the costs that would be imposed on the individuals or firms discharging the pollutant, plus the administrative costs of implementing the control program. When no action is taken, the polluting firms are implicitly given the right to pollute.

Internalization This alternative can be appreciated by considering a simple example. If the polluter and the party damaged by the pollution were the same individual or firm, the amount of the pollutant would be controlled to the appropriate level, as indicated in Figure 15-1. Also, if two firms in this situation were to merge, the incentive to limit pollution to the appropriate level would be realized. In effect, there is no longer an externality because the polluting and receiving parties are now one. The limitation to this alternative is that most environmental problems, including those associated with agricultural pollution, involve many sources of pollution and many individuals and firms are affected. In most cases, the damages caused by pollution cannot be internalized by the firm acting alone or with a few other firms. Some other way of addressing the problem must be found. This usually involves government action such as suggested in the following five alternatives.

Education The society might support an educational program to inform polluters and the public of the damages and possible means of control. For example, American farmers have been the focus of a long-term and intensive education program about controlling erosion to protect the soil and avoid the negative effects of sediment. If erosion control improves profitability by increasing productivity, farmers will adopt conservation techniques. If, however, the benefits of reduced externalities are not realized on the farm, it is unlikely that the farmer will voluntarily incur costs of pollution control. When control of the externality is not in the economic interest of the decision maker, there is sometimes an appeal to a social responsibility or moral obligation to control the pollution. "Give a hoot, don't pollute" and "Pitch in America" are examples. Generally, educational programs alone are not effective in dealing with major environmental problems. They might work well when there has previously been a lack of information, but this is not typical.

Investment A government might allocate resources to control or neutralize pollutants after their discharge instead of attempting to prevent them from getting into the environment. For example, aeration equipment can be installed in a lake or stream to replace the oxygen depleted by organic wastes. Governments might share in the expenses of erosion control facilities in part to reduce the amount of sediment entering waterways; or they might dredge sediment from streams, lakes, and harbors, rather than attempt to control erosion completely.

Subsidies   The government can offer the polluter a payment for each unit of reduced pollutant. For example, a firm that has been discharging 100 units of a pollutant might be offered $5 per unit to reduce the level of pollutant discharged. If the firm reduced emissions to 50 units, it would receive a payment of $250. In Figure 15-1, the appropriate subsidy would be at level *a*, where marginal benefits and costs are equal. Then the polluter would maximize profits by reducing pollution to that point. This option is not often chosen, however. Documenting the amount of pollution that would have occurred without the subsidy (this information is needed to calculate the subsidy) can be difficult, and many people object to paying polluters not to pollute.

Not all subsidies are made on the basis of a payment per unit of reduced pollution. Sometimes the subsidy is paid to encourage operators to invest in pollution control practices. For example, the agricultural sector has received substantial subsidies to encourage farmers to reduce erosion. Most of these subsidies have been contributions to the cost of planning and constructing erosion control facilities such as terraces and grass waterways.

Taxes or Charges   An alternative to paying polluters to reduce their discharge is to require them to pay for each unit of pollutant discharged. If a polluting firm is charged a tax of $5 for each unit of pollution discharged, it might decide to reduce the discharge from 100 to 50 units to reduce the tax liability from $500 to $250. Again, the appropriate tax is at point *a* in Figure 15-1, where MD = MC. This approach also encourages the firm to seek new pollution control technologies to reduce the pollution fees. Although the difficulties of implementing such a program are less than with subsidies (because the base level of pollution does not have to be established), the appropriate level of tax must be determined, discharge levels must be monitored, and the taxes must be collected. Of course many firms object to paying such taxes.

Regulations   A common approach to controlling environmental externalities is to limit the pollution an individual or firm can produce, or to regulate the method of production to limit pollution. Many types of regulations have been used to address the diversity of environmental problems. From the economic perspective regardless of the type of regulation, the goal should be to approximate point *a* in Figure 15-1.

Firms and industries faced with the possibility of either pollution taxes or regulations will probably oppose them because they must bear the costs of pollution control. Some polluters argue that their property rights include the right to use the environment as a sink for waste discharges.

Many environmentalists prefer regulations to taxes and fees. They feel that regulations assure a fixed limit on the amount of pollution. Economists generally favor discharge taxes and fees over regulations. Taxes allow firms to choose the least-cost method of achieving a given level of pollution.

If taxes or regulations are not enforced uniformly, they place some firms at a competitive disadvantage. For example, if farmers in one state were forced to incur costs of pollution control, they might be placed at a disadvantage to farmers in other states not required to control pollution. Some firm owners object to pollution control because producers in other nations are not held to the same standard.

ALTERNATIVE OWNERSHIP SYSTEMS

Open Access Resources

Property rights are not designated for **open access resources.** An individual or a group does not own these resources, and thus no one controls the use of the resources. Thus, anyone can use them. In the long sweep of history, all resources were initially open access resources. As long as the resource is not scarce, open access does not result in problems. This is the case of a free good. For almost everyone, sunshine is a free good. We do not pay anyone for the sunshine we enjoy and depend on for all plants to grow and provide our food supply. We do not need a set of property rights because we can all enjoy as much sunshine as we want.

Open access problems arise when potential uses of open access resources are incompatible, or when others judge the rate of use by one group of users excessive. Examples include overgrazing to the point of causing erosion; too many hikers in an area, causing congestion or damage to the habitat; and excessive hunting or fishing that depletes the stock of game or fish. When each user makes independent decisions about how to use the resource, overuse or congestion usually occurs. If there is no private ownership, there is no way for the market to equilibrate the marginal costs and returns of the several uses, and a market failure occurs. Some refer to this problem as the **tragedy of the commons,** although it should be called the "tragedy of open access."

When excessive or incompatible use occurs, it is appropriate for the government to take action. For example, at different times, ranchers, timber companies, miners, hunters, campers, and hikers have all pressed their claims for exclusive use of forestland. But who resolves these conflicts? More generally, how does a society decide how resources are to be allocated and controlled? The answer in the United States is through the government, be it local, state, or federal. In other countries, the government, or equivalent socially accepted body, allocates the resources.

Granting exclusive property rights to one firm or industry could eliminate many open access problems. But often it is not practical to assign exclusive property rights, as in the case of oceans. In many situations where it would be possible to assign property rights to a person or firm, as with public lands, it is not politically acceptable to do so. People who want to make various uses of the land feel that, as taxpayers and citizens, they should have access to it. The challenging goal in such cases is determining how to manage resources for use by numerous individuals wishing to use the resource for competing purposes.

Common Property and Public Resources

A group, rather than an individual or a firm, owns common property. Decisions about how to use common property must be made by the whole group, or a part of the group that has been delegated decision-making authority.

Governmental Management   As the pressures on resources increase and concerns about the quality of the environment grow, the instances of public policy action to convert open access resources to common (or private) property increases. Often the resource is managed as a **common property** by an agency of the government. Here the government assumes the property rights to the resources and makes the decisions about resource use. National, state, and local parks are examples of the management of common property resources by government.

In centrally planned economies, most resources, including land, are owned and managed by agencies of the government. Generally, some land is designated for public recreational uses and other land is assigned to collective farms. In addition, most countries have established, or will establish, an agency that develops policies for the management of the air, water, and other ubiquitous resources. When the government assumes control over the use of resources, it is asserting a property right to the resource and constraining the right of individuals and firms to use the resource in the discharge of pollutants.

Although most resources in the United States are privately owned, a large amount of land, especially in the West, is held by the federal government or by state governments. These are called *public lands*. **Public lands** are common property owned by all Americans (federal lands) or by state citizens (state-owned lands). Some private land is owned and managed by a group of individuals as common property.

Problems with common property management arise when members of the group do not agree about how to use a resource.

Collective Management of Common Property   Globally, this may be the most common management system for most natural resources, including important agricultural resources. Here, a group of private owners of the resource agree on a system for making decisions regarding its use.

The actual management structures used in common property management are quite varied. They usually include a democratic system of representation and direct involvement in decisions. Many housing developments include a lake or green space that is under the control of the owners of property in the subdivision. The owners designate a group of individuals who are responsible for the management of the area, including the rules for use, and who assess fees on all property owners to cover the cost of operation. This form of management was, and is, common in many parts of the world. The allocation of land and grazing rights is often managed in this way. Sometimes a dictatorial form of management is employed, with decisions made by a chief or equivalent leader. The collective farm, where a large number of individuals share equally in the management and labor, is another example of collective management.

If the common property management unit is appropriately structured, the collective management system will make the same resource allocation decisions as a single owner, although the transaction costs may be higher. If there are many common property management units and they buy and sell in open markets, the allocations will approximate those of a competitive market. If, on the other hand, the collective management unit has control over enough resources to influence market prices, the problems identified in Chapter 7 with imperfect competition will occur.

Private Property Rights

Where possible, the institution of private property rights is the most common recommendation from economic studies on the organization of agriculture and other economic activity. If control is placed in the hands of individuals, firms, or designated groups (common property) the market will allocate a resource optimally among alternative uses. Private property was widely established in North and South America in the early colonial period of settlement by European immigrants in place of the common property system used by the American Indians. Early in the development of the United States much of the land was used by hunters, trappers, and grazers on an open access basis. As use became too heavy to be sustained, conflicts among users developed. Governments usually responded by allocating the ownership of the land to individuals who would be responsible for its management and giving them exclusionary rights.

The management of fisheries also poses property rights issues. If open access is allowed, over-fishing can occur. One approach is for the government to allocate to individuals the right to catch a specified quantity or size of fish. In this way, the catch can be limited to an appropriate level, and individuals can choose when and how to fish. In some cases, they can also sell the right to catch the quantity of fish to another individual. Under this approach the market failure is addressed through the creation of private property rights to a certain amount of fish. These rights allow the market to function in a competitive fashion while also ensuring the continuing productivity of the resource over time.

In summary, there are several alternatives to private property rights in the management of resources, including creation of common property rights and management by government. Usually when management is by a government agency, the market price system is not used to guide the allocative decisions.

MANAGEMENT OF PUBLIC GOODS

A **public good** is a good or service that is available to all regardless of whether they have paid for it. Furthermore, use by one person does not affect use by another. Examples include national defense, parks, public schools, and broadcast signals. The reason the private sector doesn't provide public goods is that it can't charge individuals for their use. If the private sector tried to provide a public good, it would suffer the free-rider problem. A **free rider** is an individual who uses a good or service without paying for it. For example, it is difficult to identify people tuned to a particular radio or television station to charge them for the service. (Pay-for-view television eliminates the free-rider problem by using scramblers.) With public goods like national defense, there is no way to identify how much of the service is used by an individual. The benefits are collective.

Collecting, publishing, and broadcasting agricultural market information by the government is an important public good that benefits both producers and consumers. A government agency collects information about prices and quantities of many crops and other agricultural commodities from farmers and others in the agricultural sector. Prices and other market data are collected at the farm, wholesale, and retail levels and made available to the public for business and personal decisions. (Remember that one of the conditions for a competitive market is the availability of information to all parties.) Sometimes the government also makes projections about anticipated trends in the form of "outlook reports." Although some larger firms collect information on the same items, publicly provided information is the only comprehensive and readily available source for many small farms and other small firms. Generally, it is not practical to establish a fee-paying market for such service, although some firms collect and package public, and some private, data and sell access over the Internet or similar outlets. As with many other data series on the performance of the economy, the value to the nation in improved economic performance is probably much greater than the cost of collecting and disseminating the information.

Since most public goods are created by action of the government, a basic economic question is how to finance them. There are two basic options: use tax revenues or charge the users. In cases like national defense, the only viable option is tax revenues. No one "uses" national defense in a way that makes a market for the service possible. Sometimes, a public good can be sold like a private good. For example, in some cases users are charged fees to pay some or all the costs of publicly built roads and bridges. In still other cases, tax revenues are used to provide a basic service, and fees are charged for higher levels of service. For example, those who wish special access to market information, perhaps computer-ready data, might be charged a fee, while those who rely on the same information carried in the media pay nothing. In determining whether to provide additional services, the government must determine whether sufficient revenues can be collected to cover the cost of administration of the sale of information. Also, the government should consider who would be helped and who would be harmed by the policy. If the provision of the information on a fee basis would give some market participants an unfair advantage, the government should decide not to provide the information.

Agriculture depends on public goods (roads, bridges, waterways, irrigation systems, information services, grades and standards, and inspection services) often supplied by the government. The general test of the appropriate level of public goods is whether their marginal contribution to economic development and growth is greater than the marginal cost. The same economic criterion used in the private sector is appropriate in determining whether and how much public goods should be provided.

NATURAL RESOURCE MANAGEMENT

Natural resources are products of nature, rather than products of industry, although it often takes human effort and industry to make them useful. Examples include ores, fossil fuels, land, and water. These are used in combination with human and capital resources to produce goods demanded by consumers. The primary concern in agriculture is with resources that support the growth of flora (crops, forests, wildlife habitat) and fauna (livestock, fish, wildlife). Agriculture is an industry that converts stocks of these natural resources into flows of commodities used by the economy.

The decisions of farmers can have an effect on the quality of natural resources. For example, erosion can reduce or even destroy the productivity of the soil, although in most of the United States this is of far less economic importance than the off-farm effects of soil sediment on the environment. Farmers are guided in their use of agricultural land and other resources by the prices of inputs and outputs, as described in Chapters 4 and 5. However, depending on the governmental system, there will be different degrees of regulation concerning land use. The quality of natural resources in agricultural areas is also affected by activities in the nonfarm sector, for example, by the location of landfills in rural areas, acid rain deposition, and the growth of urban areas.

Resource Renewability

One of the most important characteristics of natural resources is whether they are renewable or nonrenewable. For a **nonrenewable resource,** there is a finite stock, and when it is gone, society must exist without it. Fixed quantities of ores such as copper, iron, phosphate, potassium, and bauxite (aluminum) make up a certain percentage of the earth's crust. Many of our energy sources are also nonrenewable, such as coal, oil, and gas. Before most of these nonrenewable resources are exhausted or exploited to the point where further use is not economical, society must conserve them or find substitutes for them—or suffer the consequences. Fortunately, so far, scientists have found good substitutes for resources as they became scarce, and economic growth has continued. When we exhaust the supply of oil, ending the petroleum age, society will face a major challenge.

Renewable resources such as forests, water, fish, wildlife, and the sun can provide inputs to the economic system on a continuous basis indefinitely, providing they are managed properly. Some resources can be regenerated, as in the replanting of forests. Some can be used continually because they are not exhausted by use, such as the sun. With improper management, however, it is possible to use some of these renewable resources to the point of extinction or exhaustion, and then they are no longer renewable! For example, a species of wildlife that becomes extinct is not renewable, and forests can be completely destroyed.

Fish are an example of a renewable resource, and Figure 15-2 suggests the desired economic fishery management strategy. Suppose that one firm owns the rights to all the fish in this lake. If the fish are sold in a competitive market so that all fish sell at the same price, the total revenue curve reflects the quantity of fish caught. Initially, as additional

FIGURE 15-2   Optimal use rate in a fishery.

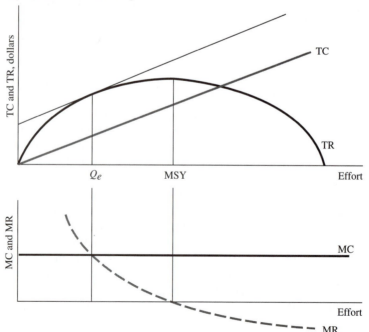

effort is devoted to fishing, the catch increases, resulting in an increase in revenue. With further increases in effort, a **maximum sustainable yield (MSY)** of fish is reached. Here the rate of reproduction and growth is equal to the rate at which fish are being caught. Additional fishing effort by competing firms will result in catching fewer fish in the long run, which is analogous to operating in Stage III of the production function. In Figure 15-2 the cost of fishing rises in a linear fashion as more effort is devoted. The firm will operate at the point where the increase in costs of additional effort is equal to the increase in total revenue. Note that the firm is realizing a profit, the difference between TR and TC.

To understand the management challenge associated with open access resources, we return to the example of determining how many fish can be taken from a lake in a year, as suggested in Figure 15-2. In Figure 15-3 the total cost line (TC) suggests that as more effort is expended, the costs of fishing rise. Also, as more fishing effort is expended, the catch increases rapidly at first and then more slowly because there are fewer fish in the lake, as indicated by the TR curve. With even more effort, the catch begins to fall because fewer fish remain. At some point, reproduction is not possible, and then the catch falls to nothing. In Figure 15-3 the possibility of operating at excessive levels of effort is suggested. If one firm had all the fishing rights and that firm had perfect information, it would operate at Q_e, But if there are numerous firms competing for fish caught from the lake, the result is quite different. Without property rights (assigning property rights is an appealing solution conceptually but difficult to implement in practice), fishing firms compete to get the largest possible share of the fish. As explained in the discussion of Figure 15-2, a few aggressive firms will probably increase their efforts to catch more fish. As other firms respond, the total effort can easily reach Q_{ex}, where again MC = MR. There is overfishing by the industry. If firms have invested in the equipment required to catch Q_{ex}, none of the firms will find it in their interest to reduce their effort. If a firm reduces the effort expended, other firms will catch more fish. Because the size of the fish population is unknown, there is a possibility of destroying the fishery by unintentionally going well beyond Q_{ex}. The stock of fish may fall to the point that reproduction ceases and the species becomes extinct.

The alternative to allowing overfishing, and possibly destroying the fishery, is for a management agency, generally an agency established by the government, to control the

FIGURE 15-3   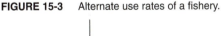 Alternate use rates of a fishery.

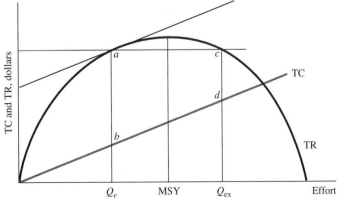

fishing effort. The socially optimal level of fish taken is at Q_e, which is also the quantity that would be taken if one firm owned the fishery. A commonly advocated goal is to maximize the quantity of fish taken from the lake on a continuous basis, the maximum sustainable yield (MSY). Here the rate of fish reproduction and growth is equal to the quantity of fish taken. Operating at MSY is analogous to a firm operating at the maximum point on the TPP curve; it would be correct only if fishing were costless.

Management could limit the number or weight of fish taken, the number or size of fishing boats, the technology allowed, and so on. A management agency could set limits by issuing fishing permits and allowing the firms to buy and sell them. The limits can be set at a level that is too low, below Q_e. Here too few fish are caught to maximize profits or to optimally allocate resources. Limits can also be set at levels too high, resulting in lower profits and, at levels above MSY, the possible destruction of the fishery. Since the management agency can only estimate the number of fish in the lake, the reproduction rates, and the rates of growth, determining the optimal catch and fishing limit is often a difficult challenge indeed. The management agency's goal is to find Q_e, where the marginal cost of additional fishing effort is equal to the marginal revenue of the fish caught, and the fishery will remain productive for a sustained period.

The optimal level of fishing effort occurs where the MC of fishing effort is equal to the MR of that effort, which occurs where slopes of the effort and revenue curves are equal, as indicated in Figure 15-2. And while fishing at the MSY level will maintain the viability of the resource, it is not the profit-maximizing level of effort.

The possibility of overfishing to the point of extinction and the desirability of controlling fishing efforts to the less aggressive level where MC = MR are the factors that make intervention appropriate. This situation is especially challenging in international waters where it may be profitable for the firms from several nations to exploit an ocean species beyond what would be profitable if one firm or nation owned the ocean or the species. Thus these renewable resources are sustainable only if nations agree to require firms to limit fishing to a suitable level.

Nonrenewable Resources

For nonrenewable resources, inappropriate decisions can adversely affect society permanently. A resource could be exhausted, instead of being used at low rates, and thus made unavailable to future generations. Many believe that private owners will not make socially correct management decisions and argue for policies that constrain the options.

Think about exhaustible resources. Those who believe that the market should function without government regulation argue that the market-based free enterprise system correctly allocates exhaustible resources. Several factors that work to control the rate of use are cited.

• First, resource owners have a choice of extracting either now or in the future. If the resource is extracted now, the owner can invest the income and earn interest. If the resource is extracted later, it will be sold at the price at that time. Thus, if the price is expected to increase faster than the interest rate, it will be in the owner's economic interest to delay extraction. Delaying extraction means conserving more of the resource for later use. That is, resource owners, expecting that future prices may be higher, will offer less for sale currently and thus force up the prices of the resources and the products

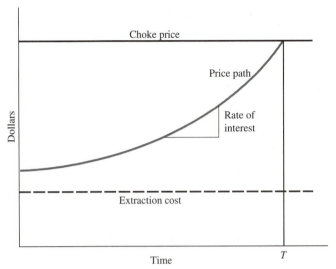

FIGURE 15-4   The price of a resource over time.

produced from them. This will reduce the quantities demanded and thus conserve resources for the future. The decisions of resource owners can be expected to result in
price increases at least equal to the rate of interest.

• Second, as a resource becomes scarcer and the price rises, the resource users will
find substitutes. If the price reaches a level so high that the resource is no longer used in
the economy, this is called the **choke price,** because use is "choked off" by the high price.

• Third, as scarcity increases and expected prices rise, induced innovations will
occur. Inventors will focus their energy on finding substitutes because they can expect
substantial rewards from a new development that addresses scarcity problems. For example, the fact that the supply of oil for gasoline will eventually be depleted stimulates
the interest in developing ethanol, solar-powered automobiles, energy conservation, nuclear energy, alternative ways of transporting goods, and so on.

Taken together, these three market forces provide a powerful system for dealing with
scarce nonrenewable resources.

Figure 15-4 indicates a simple model of expected price changes, over time, as a nonrenewable resource is used to economic exhaustion. Given a constant real cost of
extraction, the price will rise at a percentage rate at least equal to the rate of interest. This
is caused by the decisions of resource owners to adjust their rate of extraction based on
the expected price of the resource as compared with the interest rate. The choke price is
reached at the point where the resource is exhausted or other resources are substituted.
Of course, changing extraction costs, exploration for new reserves, and changing use
patterns make the decisions more complex than suggested in this basic model.

Economic Exhaustion

A nonrenewable resource reaches economic exhaustion when the cost of mining and
processing it into a usable product is higher than its choke price. However it is possible

for a resource that was previously uneconomical to become economically viable with advances in extraction technology or higher prices for the products produced from the resource.

Many people, including some notable economists, are not confident that a competitive market system can conserve and distribute resources among this and future generations equitably and efficiently. They suggest we should not rely solely on the development of new technology or on markets to determine the optimum rate of resource use. They fear that excessive use of a resource may occur because private entrepreneurs may not be able to capture all the benefits of developing new technologies or substituting alternative resources. Also, some are concerned that the interest rate used by market participants may not be suitable for a decision as important to society as the exhaustion of a resource. The private operator discounts future earnings at the current market rate of interest. When interest rates are high, operators discount future earnings heavily. For example, a farmer might allow excessive soil erosion to occur since significant reductions in productivity will not happen until ten or more years later. Similarly, some energy and mineral resources can be *economically exhausted* over relatively short time horizons. As a result of such concerns there is considerable interest in achieving a sustainable agriculture for future generations. Although there are numerous definitions of sustainable agriculture, the central thrust is to maintain output levels over an indefinite period. Concerns about achieving a sustainable agriculture are amplified by soil erosion and the use of nonrenewable resources, such as fertilizers produced from fossil fuels. We will return to these topics below.

AGRICULTURE AND ENVIRONMENTAL POLICY

As indicated above, environmental quality problems resulting from externalities are important to agriculture. Some agricultural externalities affect the environment adversely, and some nonagricultural externalities affect agriculture. For example, besides the on-farm productivity effects of soil erosion, the eroded soil carried off the farm also creates problems. The sediment can adversely affect water resources by clogging drainage channels and filling lakes. Pesticides and plant nutrients can move with the sediments into water resources and adversely affect water users. The location of a sanitary landfill in an agricultural area is an example of the nonagricultural sector imposing a negative externality on rural residents. Because of heavier truck traffic, the possibilities of blowing paper and other items, odors, and even the possible contamination of groundwater used for personal consumption, nearby residents nearly always oppose the opening of a sanitary landfill.

MEGA LIVESTOCK FARMS

The growth in the number of mega-farms that produce large numbers of hogs or other animals have become controversial mostly due to the externalities associated with the waste produced. This waste can create odor problems for nearby residents and can cause adverse effects on water quality. Political pressure has resulted in the development of new regulations on livestock production intended to deal with these externalities. These regulations have resulted in many of the operations dealing effectively with the externalities.

Society responds with new policies if concerns about such environmental externality questions become important enough in the political arena. Sometimes, the policy response is one of direct intervention. Pesticide regulations and air and water pollution control standards are among the regulatory actions instituted to address environmental problems caused by agricultural activities. Taxes on pesticides, conservation subsidies, public sector investments in wastewater treatment plants, and environmental educational programs are examples of economic and educational policies implemented to reduce environmental damages.

In the sections that follow, we have selected a few major issues to give you an appreciation of some of the complexities of environmental policy problems and how an economist might approach them. While there are both resource management and environmental quality policy aspects of all the problems discussed here, we focus on the resource elements in the first two examples (land policy and energy) and on the environmental elements of the second two (water quality and solid wastes). There are, of course, many additional important issues with agricultural implications that are beyond the scope of this text—such as ozone depletion and global warming.

Land Policy Issues

Land has attributes of both a renewable and a nonrenewable resource. Improper management that causes excessive soil erosion can render it so unproductive that it is not profitable to farm. However, land that is not subject to these problems, or land that is susceptible to erosion but is well managed, can produce indefinitely. It is accurate to classify most farm and grazing land in the United States as a renewable resource. In essence, it is agriculture's ultimate renewable resource. There are numerous land resource management issues at the world, national, state, and local levels.

Land Ownership Issues   First we consider the differences among national **land ownership** and control systems that have important implications for economic performance. In the United States, about two-thirds of the land area is privately owned and operated, including most of the productive agricultural land. Approximately one-third of the nation's land is owned and managed by the federal government. Significant amounts of the federal land are leased to individuals for cattle grazing, to firms for the right to harvest timber, or to other firms for the rights to extract mineral or energy resources. Generally, public land is also available for recreational purposes. No commercial activity, except a few retail concessions, is allowed on state and national parkland.

OWNERSHIP OF MINERAL RIGHTS

The United States is one of very few countries that give the owners of the land surface the rights to the associated subsurface mineral, energy, or water resources. In most other countries, subsurface resources are controlled by the central government. The revenues generated from the sale of resources are collected as taxes.

At the other extreme, in some communistic countries and under some forms of socialism, land and other resources are public property owned and controlled by the central government as common property resources. In a purely centrally planned system, the government makes all investment, production, and distribution decisions. State farm managers and workers have little to say about how to run their farms. However, in many planned economies, individuals operate small plots on which they produce for sale, barter, or their own consumption (generally the individual has the right to use the plot but not ownership of the land).

Another alternative is feudalism, which often was operated as an oppressive form of land ownership. In this system a few individuals, perhaps royal families, own and control the land. Peasant families work the land. The peasant has few or no employment alternatives, and the landowner can keep the peasant impoverished without fear of losing the labor provided. When the landowners control the national government, the peasants have little power to effect change. Over time, as a result of education, democratization, and information, the incidence of feudalism and communism in the world has diminished.

PRIVATE VERSUS PUBLIC OWNERSHIP

In the early 1990s as the former Soviet Union and eastern European countries moved away from their planned economies, they were forced to make difficult decisions including whether and how land should be converted to private ownership status. Those favoring the shift to private control pointed to the efficiency and productivity advantages. They argued that as individuals compete, they make decisions that expand the productivity of the economic system. Effort and risk taking are encouraged by the fact that the individual receives the benefits generated. They noted that in a state-controlled system, the gains produced by the individual accrue to the society, which is a less direct incentive. Those who favored governmental ownership argued from a normative perspective. They pointed out that the private system allows some owners to acquire control over large tracts of land and to acquire considerable wealth while others remain in landless poverty. Some also took a moral position that it is wrong to buy and sell "the motherland."

The final form of land ownership we consider is common property. Here groups of individuals make the economic decisions. For example, in Africa much of the land is a resource controlled by kinship or ethnic groups. The decision-making process used to manage the land has evolved over centuries and continues to develop. In some cases a group leader, with the advice of elders, may allocate land to families and may impose restrictions on land use, such as enforcing a period of fallow when the land is held out of production. The importance of custom can make this form of management slow to adopt new technological developments. For example, restrictions on transfering land use-rights to a non kin-group individual might exclude a progressive farmer from expanding while protecting a less efficient producer.

In the relatively recently developed Kibbutz in Israel, those who live and work in the co-op make the economic decisions and determine how the earnings will be shared. Generally these farms have been more efficient and productive than the "state farms" in centrally planned economies, but less efficient than privately owned farms. The right to participate in decision making is the significant distinction between a Kibbutz and a state farm, and is a major factor explaining the difference in performance.

Land Management Issues   In the United States, policies affecting the use of land have changed over time. It is important to understand these policies and some of their implications for resource management. Initially, the nation encouraged the settlement of the frontier by offering land free (or at very low cost) to those who were willing to develop it for agricultural production. This was a major incentive for immigration to the country. Most of the land was converted from open access to private property status. Also, federal, state, and local governments have reserved control of substantial amounts of land as common property for multiple uses. Both the federal and state governments have also leased some of this land to private individuals and firms for grazing, mineral extraction, or forestry.

Although we may think of private property as being entirely under the control of the owner, federal, state, and local governments may exercise some control over private property for the general benefit of society. For example, state governments retain the power of **eminent domain** (also granted to federal and local governments in some instances). Governmental agencies use this power to purchase land for use in the public interest from a private owner at an appraised market price plus damages to the owner (if any). This power is necessary for efficient provision of public services.

EMINENT DOMAIN

Think for a moment about the difficulty in constructing a highway without this power. The government would have to pay exorbitant prices to individuals reluctant to sell or would have to build the highway around their properties. Either the costs would be extremely high or the highway would be quite crooked!

State and local governments often exert direct or indirect control over the use of private property. An example of the direct method of control is zoning. State law grants specific powers to local governments for procedures to follow in voting to establish a local planning and zoning commission. If approved by the voters, the unit of government designates areas of land as a zone for specified uses, instead of allowing the landowners free choice of how to use the land. For example, a planning commission might designate agricultural, commercial, industrial, and residential use zones to control the location of these activities and thus achieve a more desirable land-use pattern. This form of land-use control is more evident in Europe, where there is a clearer line of demarcation between agricultural land and urban zones. Also, there may be controls on the size and nature of the construction or activities allowed within these urban development zones. For example, some communities restrict the height of buildings to retain a desired skyline and to control the density of activities and the associated parking requirements.

An indirect means of influencing the pattern of land development is the provision of public services. The public sector can influence the location of private housing and commercial development by providing water, sewer, and other utilities in the areas targeted for development, and by refusing to provide it elsewhere.

Soil Conservation Issues   U.S. **soil conservation** policies are also important for land management. These policies were designed to control erosion and reduce sedimentation. We consider first the concerns about **erosion,** and then the associated **sedimentation** problems. U.S. erosion policies sprang from the public concern about conserving soil productivity for future generations. There was, and still is, a concern that farmers tend to plan for the immediate or short-term horizon and not for the long-term horizon. Therefore, future generations may be deprived of the productive land resources required to produce an adequate food supply.

The Erosion Process   Erosion is a natural geological process whereby the soil is worn away by the action of water and wind. Erosion can reduce soil productivity and is a location-specific problem. Agricultural practices increase the erosion rate by exposing the soil to the elements for part of the year. The rate of erosion depends on the properties of the soil, the length and degree of slope of the land, the nature and amount of rainfall, the crop produced, and the management practices employed. Erosion also varies greatly from year to year depending on the nature and timing of rainfall.

The federal government initiated soil conservation policies in the 1930s because of public pressure to conserve topsoil for the future. The "Dust Bowl," a major wind erosion episode that severely damaged soil in the Great Plains and spread dust as far as the East Coast, was an important factor in the adoption of the legislation. Public policies encouraging and subsidizing farmers to do a better job of controlling erosion were adopted and have been modified over the years. One of the major thrusts of soil conservation policy has been the provision of subsidies to farmers for the adoption of conservation practices. The **Natural Resource Conservation Service** (previously named the Soil Conservation Service) was established to provide education and technical assistance necessary to design the structures such as terraces and grassed waterways. Financial assistance, generally half of the improvement cost, is provided to farmers on a cost-sharing basis through the Natural Resource Conservation Service. The legislation establishing the original Soil Conservation Service also encouraged the development of soil and water conservation districts at the local level to encourage conservation and to set priorities for the distribution of the cost-sharing funds. This voluntary program includes:

- education, to make farmers aware of the erosion problem and alternative means of control;
- research, to identify better and less costly methods of control;
- technical assistance, to aid farmers in the design of a conservation program and conservation measures; and
- subsidies, to share in the cost of implementation.

This multifaceted program has been in effect since the mid 1930s. Figure 15-5 indicates the basic concepts underlying a technical assistance program. MD_f is the on-farm damage; MD_t is the total damage. The difference is the cost of the external damage. Without an assistance program, the farmer will consider only the farm-level damages (MD_f) and equate them with the unassisted marginal costs of control (MC_1). Thus, the amount of sediment leaving the farm will be S_{f1}. With a technical assistance program the farmer's marginal cost falls from MC_1 to MC_2, and sediment erosion falls to S_{f2}.

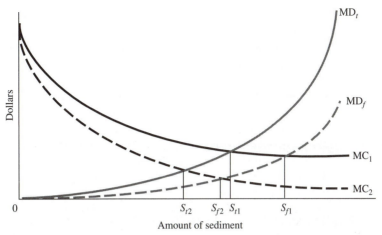

FIGURE 15-5 Optimal erosion and sedimentation rates at the farm and social levels with and without an incentive policy.

Note that forcing the farmer to incur both the off-site and on-farm damages (the total marginal damages, MD_t) would lead to controlling sediment loss to S_{t2}.

Other policies with conservation implications were adopted over this period. For example, in many years, farmers were encouraged to remove land from production to control output and increase farm incomes. These programs, called soil bank or set-aside programs, have included a requirement that land taken out of production be protected from erosion.

In the 1985 and 1991 farm bills, the federal government adopted several policies that moved away from the voluntary approach by requiring farmers to meet specified conservation requirements to qualify for other federal programs. Farmers with erosion-prone land were required to develop by 1990, and then adopt by 1995, conservation plans to continue to enjoy the benefits of a range of other farm program provisions, including commodity price supports. Also, farmers were required to preserve wetlands or prairie pasturelands under "Swampbuster" and "Sodbuster" provisions. These require cross compliance among program provisions, since the farmer must meet acceptable soil erosion standards to receive the benefits of the government programs.

SWAMPBUSTER AND SODBUSTER

The goal of the Swampbuster provision is to prevent the drainage of wetlands; the goal of the Sodbuster provision is to prevent the plowing of prairie not capable of sustained crop production. An important reason for these programs is to preserve biodiversity through habitat protection. Producers are required to comply with Swampbuster and Sodbuster provisions on their entire farm if they wish to participate in any federal support program.

The government also initiated the Conservation Reserve Program (CRP), under which the federal government purchases the cropping rights on highly erosive soils. The farmer must agree to cease production for ten years and maintain a conservation cover crop. Over 45 million acres of land were entered in this program on the basis of bids submitted by farmers. In this case, the primary objective was to take highly erosive land out of production to reduce erosion and sedimentation; the secondary objective was to enhance farm incomes.

More recently a Conservation Reserve Enhancement Program (CREP) was initiated. Under this program marginal lands are restored to wetlands. The farm operator receives payments for 15 years under this program.

The sedimentation aspect of the soil erosion problem has important environmental dimensions. Sediment is the major water pollutant, by weight, in the United States. The eroded soil is an externality that can cloud water and adversely affect aquatic life, clog waterways to the point that drainage is slowed, and build up in lakes and harbors, sometimes to the point that boat use is impossible. Also, eroded soil may be deposited in roadside ditches and on low-lying cropland. The deposited sediment reduces the carrying capacity of drainage ditches, streams, and rivers; increases the severity of flooding episodes; requires the dredging of harbors, waterways, and roadside ditches; and reduces the quality of the habitat for aquatic life. Soil erosion policies were modified to include consideration of the costs of sedimentation when the magnitude of the problem was recognized.

This sedimentation problem associated with erosion differs from most urban and industrial externality problems since the sediment does not enter the water from an easily identifiable point, such as a factory discharge pipe. For that reason, sedimentation is called nonpoint pollution. Its diffuse nature makes it harder to measure and manage than most other externalities.

After over fifty years of education and subsidy programs, erosion and sedimentation remain significant problems. Referring again to Figure 15-5, research indicated that the total damages included a substantial amount off the farm. Farmers will operate at S_{f2} because they do not suffer the off-site damages. Society must, therefore, design additional programs to encourage the movement to S_{t2}. The federal government has responded by targeting the soil erosion policies to areas where the eroded sediments caused significant water quality problems. For example, the conservation reserve program was modified and expanded to address the sedimentation problem. Bids are accepted from farmers who contract to take land along streams out of production to create filter strips, and thus reduce sediment delivery to the streams. With encouragement from the federal government, especially the Environmental Protection Agency (EPA), states began to develop a variety of different programs. Some states augment the cost-sharing subsidies provided by the federal government. Some mandate the adoption of conservation measures, although in many cases the non-compliance penalties have been weak.

TOTAL MAXIMUM DAILY LOAD (TMDL)

At the end of 1999 the Environmental Protection Agency moved toward implementing a program, the Total Maximum Daily Load, which will require all municipalities, industries, and farm operators to develop a plan for controlling the total maximum daily load of pollutants entering the streams, rivers and lakes in a watershed.

What can we conclude about soil erosion and conservation policy? The policy has evolved over a long period. It started as a voluntary resource management program that stressed erosion control to preserve the soil and maintain productivity. Over the years it has evolved to a more comprehensive policy with the addition of an environmental quality objective of controlling sedimentation to improve water quality. The policy began to move away from the voluntary approach and toward mandating farmers to comply with the program to qualify for other benefits of the federal "farm bills." However, the 1996 legislation, discussed in Chapter 17, eliminated some of these mandates.

CONSERVATION PROGRAMS

Federal programs in effect in 2001 that target improved soil and water management include:

USDA Environmental Quality Incentives Program which offers payments to farmers and ranchers for incentive, educational, and technological assistance to reduce nonpoint surface pollution and promote underwater supply cleanliness ($200 million budgeted)

USDA Conservation Reserve Program which provides rental payments to producers to retire highly erosive land from production ($1.8 billion budgeted)

USDA Conservation Technical Assistance program that helps farmers and ranchers to use better methods of soil and water quality improvement to achieve improved water quality ($540 million budgeted)

USDA Farmland Protection Program that provides easements and other types of interest in land with highly productive soils ($18 million budgeted)

USDA Wetland Reserve Program that helps farmers cost sharing to return "farmed wetlands" to their wetland state ($200 million budgeted)

USDA Emergency Conservation Program that helps farmers to recuperate from disasters and to hold on to water during droughts ($35 million budgeted)

EPA Nonpoint Source Program that provides states program direction, technical assistance, and funding to put in place nonpoint-source pollution management plans ($120 million budgeted)

EPA Coastal Zone Management Act Reauthorization Amendments require states to give EPA a coastal zone management plan to protect the offshore water supplies from nonpoint source pollutants

EPA Wellhead Protection Program that seeks to promote and make safe the groundwater used as drinking water from chemical, pesticide, and nutrient pollution ($12 million budgeted)

Agriculture and Energy

Energy use in agriculture is of concern to some, even though farming and all agribusiness-input firms use less than one-sixth of the total energy consumed in the entire U.S. food and agricultural sector. The rest is used in transporting, storing, processing, freezing, cooking, and so on. The advocates of sustainable agriculture point out that as fossil fuels are exhausted, the shortage of energy could limit the ability of agriculture to provide an adequate food supply. Proponents of reductions in agriculture's use of energy (including fuel for power and crop drying as well as the production of commercial fertilizers and pesticides) are skeptical that the price mechanism will lead to appropriate adjustments in the agriculture or energy sectors in time to avoid a crisis. They support a variety of changes designed to convert agriculture from nonrenewable to renewable energy resources. Some argue that individuals should be convinced to shift to a diet that includes less meat to reduce the demand for the grain crops fed to cattle. Others argue that farms should reduce the use of fertilizers by adopting management systems that include more reliance on nitrogen-fixing crops in the crop rotation and a combination of crops and livestock that will derive more of the required plant nutrients from natural sources. The policy response has been to encourage research and educational efforts on alternative agricultural production systems.

Some view agriculture as a potential source of energy for the larger economy. Brazil, a country with extensive land resources and limited fossil fuel deposits, attempted (with limited success) to run its domestic transportation system on fuels derived from agricultural products, primarily sugarcane and cassava, an edible root. Partly because of concerns that oil resources used to produce automobile fuel will eventually be exhausted, the U.S. government and several states have provided substantial tax incentives that encourage the use of ethanol produced from corn as a partial substitute for gasoline. Also, some governments have instituted requirements that a portion of their fleet vehicles shift to engines that burn alternative fuels including methanol, which can be produced from wood and a broad variety of crops. The goal of these policies is to encourage automobile manufacturers to design vehicles that will operate more efficiently on such fuels. The policy also seeks to develop a market for the alternative fuels as well as the production of these fuels at lower costs. It is hoped that, eventually, they may prove to be competitive with gasoline. Unfortunately, the goals of those advocating reduced reliance on fossil fuel energy, those advocating use of agricultural products as a source of energy, and those placing a high priority on environmental quality sometimes conflict.

Water Quality

Agriculture is a major source of water pollutants. This is an externality problem; most of the damages occur off the farm that is generating the pollution. The problem is also nonpoint in nature; the pollutants enter the system from diffuse sources. There are four major types of water pollutants from agriculture: sediment, plant nutrients, pesticides, and (in some cases) salt. Sediment, a product of soil erosion as discussed above, is the major pollutant in terms of quantity, although it may not cause the greatest damage.

Plant nutrients, fertilizers applied to stimulate crop growth but not used by the crop, are another important pollutant contributed by agriculture. Nitrogen and phosphorus are the main plant nutrient pollutants. When they are present in water in concentrations much higher than normal, a process called *eutrophication* occurs. This process results in the growth of algae and other oxygen-using organisms to the point that all oxygen is depleted and a die-off of fish and plants occurs. High levels of nitrates in drinking water are also hazardous to the health of baby children. Plant nutrients in manure produced from livestock production facilities have been addressed through requirements for control developed at the state level resulting from pressure from the federal government.

Because plant nutrients that enter the water system from crop production come from diffuse sources, it is difficult to develop and implement a program for control. Improved soil conservation practices generally reduce nutrients lost, and research and education efforts to address the problems have been supported.

Changes in fertilizer technology can also help to address this problem of off-farm pollution. For example, encapsulation of urea slows the release of nitrogen in the soil, allowing more to be taken up by the plant and less to be leached into the groundwater. However, this makes the fertilizer more expensive. Since the problem being addressed is an externality, most farmers will not use the more expensive fertilizer unless there are other advantages to the farmer of doing so. Alternatives include subsidizing the cost of encapsulated fertilizer, taxing nonencapsulated fertilizer, or mandating application of the desired form and at the appropriate time. Such questions can be viewed in terms of which group is given the property rights. If farmers retain the right to operate as they have in the past, the policy must compensate them for additional costs. If the property right is bestowed upon the consumer, the farmer must bear the additional cost.

Pesticides are also important water pollutants. There are many pesticides used for various purposes. The two major categories are herbicides, used to control weeds, and insecticides, used to control insects. There are several damages associated with pesticide use. One of the adverse effects concerns the handling of the pesticides by the applicator. This danger is *not* an externality because the negative effect is on the individual using the product. To address this problem, continual efforts are made to improve the packaging of the products, the techniques for handling them, and the warning labels. Also the cooperative extension service and manufacturers sponsor training schools to educate users.

Externalities associated with pesticides occur at several levels. Pesticides can drift from one farmer's field to that of another and damage a nontargeted crop. Some pesticides can be transported to streams and lakes, where they kill fish and other aquatic life and enter water supplies for human consumption. Pesticides also can leach into groundwater, which is especially problematic because some of this water may be used later for human consumption. Some pesticides cause health problems including birth defects and cancers. Finally, repeated and extensive use of a pesticide often results in the development of resistant strains of the targeted pest.

The policy response to the pesticide problem has been quite different from response to the other environmental problems. One of the primary thrusts has been based on regulatory policy instead of educational and voluntary economic incentive programs. The EPA has taken off the market several pesticides that have serious health consequences. The use of some other pesticides has been restricted. Over time the requirements for the registration of pesticides for sale have become progressively more stringent. As a result, the pesticides being placed on the market today are generally much safer than the pesticides that have been marketed for years. Also, as the regulations have become more stringent, the use of some pesticides has fallen as farmers shift to other pesticides or other means of control.

In recent years, increasing numbers of farmers have shifted from the use of chemical insecticides to **biological control** agents to protect their crops. This shift has occurred for a variety of reasons, including social, scientific, and, perhaps most important, economic concerns. Socially, farmers recognize that the public is pressuring the government for more stringent controls and that consumers prefer products grown without pesticides. They are concerned about environmental and health risks associated with repeated exposures to some chemical pesticides, and they want to protect lands and groundwater as well as workers. Scientifically, they are also aware that insects develop resistance to traditional chemical insecticides and that significant advances have occurred in the biological techniques.

Economically, many farmers have found that by using more careful management strategies—for example, integrated pest management techniques such as scouting for infestations rather than applying on a preventative basis—they can reduce their expenditures for pesticides and increase profits. Another cost of pesticide use is the administrative cost, especially in states with strict environmental legislation. In these states farmers must deal with the required paperwork when working with potentially harmful chemicals. Some farmers also face rising insurance premiums associated with the use of such chemicals.

Considerable research and education have been devoted to the development of **integrated pest management (IPM)** programs, which involve moving away from the use of pesticides on a preventative basis. Instead, there is careful monitoring of the crop during the growing season so that use of the pesticides can be limited to times when it is necessary to deal with an outbreak of a problem. There is more use of a broader range of control techniques, which may also lead to the use of less pesticide to control a pest problem. In an economic context, IPM involves the substitution of labor (scouting) and management (determining the appropriate control approach) for the preventative use of pesticides. Often the IPM technique is more profitable than applying pesticides on a preventative basis. In these cases IPM meets both the farmer's and society's objectives.

Salt is the last pollutant considered here. This is a problem primarily in agricultural areas where a dry climate necessitates irrigation. As water evaporates from the surface of the soil, the salts in the water are left on the soil surface. Over the years the level of salt in the topsoil can become excessive. This process, called **salinization,** can result in reduced crop yields and can even prevent crop production entirely. Sometimes, especially in developing countries where there is limited knowledge and resources to deal with the problem, the land is lost from production. In other areas, irrigation districts operate a program of backwashing and flushing to remove some of the salts from the soil. In this process, salt can be washed into streams and rivers, damaging the natural habitat and increasing the cost of processing this water for human use.

Agriculture and Solid Wastes

Society must find a way to deal with products after their useful life, that is, of dealing with the tons of discarded products and packaging materials called **solid waste** (trash). Some of this material is reused or recycled. In developing countries, there is considerable recycling in the private sector because labor is relatively cheap and material goods are expensive. In the United States, substantial effort is being devoted to increasing the proportion recycled; but for most of the waste, final disposal is in landfills that are generally located in a rural setting. Landfills do not use enough acreage to affect the productivity of agriculture, but any externalities (odors, blowing waste, groundwater contamination) from the landfills will affect the people living in the area. Besides the adoption of more stringent regulations concerning the construction and operation of landfills, there are several economic policy options that will reduce their use.

THE SOLID WASTE PROBLEM

The public has only recently begun to appreciate the magnitude of the solid waste problem. (New York City is creating a mountain of trash that will be one of the highest points in the eastern half of the United States and will dwarf the Egyptian pyramids in overall mass.) Media coverage of the problem has expanded greatly. Illinois, as is true in many states, is rapidly filling existing landfills to capacity and must site many new landfills or build alternative systems, such as incinerators. The result has been intense debate in the localities where facilities are proposed.

An international dimension of the problem is suggested by the experience of several ships loaded with solid waste that were denied the right to unload their cargo in nation after nation. They sailed the oceans for months in search of a disposal location.

Reuse and Recycling   Many governments provide both positive and negative incentives to encourage reuse and recycling. A number of states have adopted "bottle bills," which require merchants to pay a fixed amount for each container returned. Objections to such legislation generally come from merchants who do not want to handle the used containers and from the container industry because of reduced demand for their product.

LANDFILL SITING DECISIONS

Landfills are almost always located in a rural setting for two reasons. First, a landfill requires substantial acreage, which is less costly in a rural area. Second, the siting of a landfill is a political decision, with the "Not In My BackYard" (NIMBY) phenomenon playing a major role. Thus, when a new landfill is sited, one can expect strong opposition from those who live near the designated location because of the fear of negative externalities including odors, blowing paper, and water pollution. To minimize the number of individuals opposed, those charged with the decision seek a location with low population density, namely, a rural area.

A major hurdle to recycling is the development of a market for the recycled product. For example, newsprint recycling and processing plants must be constructed, and newspaper printers must convert to the use of recycled paper. Without buyers of newsprint it isn't profitable to recycle but without recycling there is no recycled paper to process and reuse. Numerous means of initiating a recycling market are possible. Some communities operate curbside or central drop-off recycling programs and sell the paper collected to processors. Governmental agencies can stimulate the demand for the recycled paper products by buying the products themselves. Also, governments can require newspaper publishers and others to use recycled newsprint. If these activities are successful, a market will develop for the recycled product, that will encourage recycling and reuse. When consumer recycling increases rapidly, as in the case of paper in the early 1990s, the quantity may exceed processing capacity and dramatically force down the price paid for recycled material. The lower price places considerable stress on the firms in the industry, while the reduction in input costs encourages new firms to enter the field (as discussed in Chapter 5).

Limiting Disposal of Selected Materials in Landfills   Some states have set limits on the proportion of the waste stream that may be placed in a landfill. The burden falls on local communities to devise programs to accomplish the objective. Communities can select from among several approaches. They might ban certain classes of materials, such as grass clippings and yard wastes, from landfills. Another alternative is to invest in the development of a collection system and a composting facility funded through general tax revenues or through fees paid by residents who use the service. If funded through general tax revenues, a form of public good is created because all residents may use the system without being charged. If sufficient participation could be assured to operate a system funded through user fees, a marketlike system is developed, and a private entrepreneur might provide disposal service. Another approach is to ban the material from the landfill and depend on incentives provided for the private sector to develop a means of dealing with the wastes.

Incineration and Waste to Energy Facilities   Some localities invest in the construction of incineration plants to burn the material or plants that burn the wastes to generate electricity. In both cases, there are concerns about air pollution, and there will be an ash residue (which may be toxic) that must be disposed of, generally in a landfill. Also, some environmentalists oppose these alternatives because they believe that reuse and recycling are inherently superior alternatives, regardless of the economic aspects. To date, the waste disposal problem has fallen primarily on state and local governments. However, the federal government could be involved by developing consistent requirements, developing regional programs, or providing federal funds. Consider the possibility of a federal tax on packaging materials, perhaps with the tax rate determined partly by the proportion of the material recycled or reused. Such a tax would discourage the use of excessive packaging, encourage recycling, and provide tax revenues that might augment general revenues or be earmarked for support of waste disposal.

SUMMARY

In this chapter, we recognize that markets do not always function to allocate resources as described in earlier chapters. Several reasons for market failure were discussed. When markets do fail, resources are not allocated in the best interests of the society. Governments may establish property rights and participate in markets to reduce negative externalities, manage open access resources, and provide and regulate the use of public goods. A fundamental role of government is to take actions to help markets function. It is an activity that is of great importance to agriculture: There are significant externalities associated with agricultural activities, many resources used in agriculture are managed as common property resources, and agriculture depends in important ways on public goods. This chapter has addressed only some of the fundamental issues. A more thorough analysis is beyond the scope of this book.

Perhaps the essential observation from the above set of resource and environmental issues is that the problems and the policy responses are quite diverse. Economic analysis can play a significant role in the development of cost-effective policies. The nature of the resource is important. For nonrenewable resources, the goal is to use them efficiently and search for substitutes that may be used as prices rise. For renewable resources, the challenge is to ensure that they remain productive for the indefinite future. Where private individuals own resources, we can expect the market mechanism to allocate the resource in the present and over time. When open access resources become scarce, intervention by a government to create a system of property rights is encouraged. A government may develop a common property management agent, assume the management responsibility directly, or assign private property rights. Some problems are international in scope, while others are important at the farm-to-farm level. Some involve immediate problems, while others involve concerns for future generations.

Where the problem is a negative externality, the challenge is to reduce the pollutant to acceptable levels and determine a satisfactory allocation of the costs. Sometimes private groups can take actions that will make a difference, but usually the intervention of governments—at the local, state, national, or international level—is required to produce the necessary changes. Sometimes the externality can be internalized. However, more often the policy response is one of direct intervention. The government may participate by establishing, or allowing to be established, an organization with the responsibility and authority to manage the resource in question. Besides these responses, governments provide support for research and education.

LOOKING AHEAD

In the next chapter we turn to an analysis of some of the economic aspects of economic development. We will show that the public sector plays a key role in the development process.

IMPORTANT TERMS AND CONCEPTS

choke price   377
common property   370
eminent domain   381
erosion   382
externality   365
free rider   372
integrated pest management
 (IPM)   388
internalization   368
land owership   379
market failure   365
maximum sustainable
 yield (MSY)   375
Natural Resource Conservation
 Service   382

natural resources   373
nonpoint pollutants   367
nonrenewable resource   374
open access resources   370
pesticides   387
plant nutrients   386
public good   372
public lands   371
renewable resources   374
salinization   388
sedimentation   382
soil conservation   382
solid waste   388
tragedy of the commons   370

QUESTIONS AND EXERCISES

Name That Term

Read the following sentences carefully and fill in the missing term or terms.

1 An _____ is an uncompensated (positive or negative) effect of a production or consumption activity on another producer or consumer.
2 A _____ enjoys the benefits of a group's efforts without sharing in the cost of their provision.
3 An _____ resource can be used by anyone without restrictions and without regard to the effects on other users.
4 Pollution and other negative externalities lead to market _____, the inability of the market to bring about the socially optimal allocation of resources.
5 _____ management involves establishing organizations that can manage resources fora group.
6 _____ are chemicals that are used to control insects, weeds, or fungi that damage or compete with crops.
7 Soil _____ practices are adopted by farmers to reduce erosion.
8 _____ is the right reserved by the government to take land for public use and compensate owners at fair market value.
9 Minerals are examples of _____ resources.
10 The maximum amount of fish that can be harvested on a continuous basis is the maximum _____.

True/False

Read the following sentences, then decide whether each statement is true (T) or false (F) and mark it accordingly.

T F 1 Developing and enforcing property rights is crucial to the functioning of a market economy.
T F 2 Allocations based on free markets generate the optimal level of all goods and services.
T F 3 Markets fail by producing too much of goods that generate negative externalities and too little of goods that generate positive externalities.
T F 4 Markets work best with no government involvement.
T F 5 One common cause of market failure is the lack of a complete set of exclusive property rights.
T F 6 The economically optimal level of pollution is zero, leading to the pure environment we all desire.
T F 7 Renewable resources are in unlimited supply at the current price.
T F 8 Farmers consume about two-thirds of the energy consumed in the food and agricultural industry.
T F 9 For fish, the maximum sustainable yield in the long run occurs when additional fishing will result in the catching of fewer fish with additional effort.
T F 10 Economic exhaustion occurs when the cost of processing a resource into a usable product is higher than the value of the product.
T F 11 About two-thirds of the land in the United States is privately owned and managed.

Multiple-Choice Questions

Circle the letter of the response that best answers the question or completes the statement.

The private aggregate demand (Dp) and supply (Sp) curves for good X are shown in the graph below. Each unit of good X generates $6 of negative externalities.

Use this graph to answer questions 1 to 6.

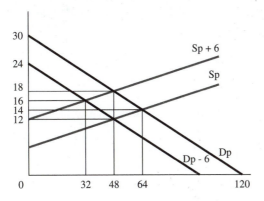

1 Without any action by the government, what will be the market clearing level of X?

   a 0

   b 32

   c 48

   d 64

   e 120

2 Considering the negative externality associated with X, what is the socially optimal level of X consumption and production?

   a 0

   b 32

   c 48

   d 64

   e 120

For questions 3–6, assume the government imposes a $6 per unit tax on the producers of X. Assume Sp is the aggregate marginal cost curve of X producers, all producers must pay the tax, and all else remains the same.

3 How much are demanders willing to pay per unit for 48 units of X?

   a $12

   b $14

   c $16

   d $18

4 Considering the tax on suppliers, what will be the new market clearing level of X?

   a 0

   b 32

   c 48

   d 64

   e 120

5 How much revenue will the $6 per unit tax on X raise for the government?

   a $6

   b $72

   c $192

   d $288

   e $864

6 After paying the tax, how much do producers have left to pay other factors of production?

   a $0

   b $512

   c $576

   d $896

   e $864

7 Which of the following is not an example of a renewable resource?

   a fish

   b forests

   c agricultural crops

   d coal

   e all of the above are renewable resources

8 Which of the following is not an example of a natural resource?

   a fossil fuels

   b land

   c water

   d all of the above are natural resources

   e none of the above

9 Which is true of eminent domain?

   a a means by which the government can take private property

   b a property owner must be compensated at a fair market price

   c the government could take your home

   d the means for government to get access to land for public works

   e all of the above

10 Which of the following is not a major type of water pollutant from agriculture?

   a sediment

   b plant nutrients

   c pesticides

   d salt

   e all of the above are types of water pollutants from agriculture

Technical Training

Assume Righteous must routinely work in a closed room with Hacker, a cigarette smoker. Righteous dislikes smoke and complains to the boss, who suggests Righteous and Hacker work it out themselves. Assume that Hacker's demand curve for cigarettes takes the form $P = 0.50 - 0.02C$, where C is number of cigarettes smoked during an eight-hour shift and P is the price per cigarette. Each cigarette costs 20 cents and causes Righteous 10 cents in discomfort.

1 Sketch Hacker's demand curve for cigarettes, the marginal cost of cigarettes to Hacker, and the number of cigarette's Hacker will choose to smoke in a shift.
2 Graph how the 10 cents per cigarette negative externality affects the social marginal cost of cigarettes. Considering the externality, what is the optimal level of cigarette consumption?

For questions 3–6, consider proposal A: that Hacker be required to contribute 10 cents to the snack fund for every cigarette consumed.

3 What is the marginal cost (including the "tax") to Hacker of each cigarette smoked?
4 Will this proposal cause Hacker to choose to consume the socially optimal level of cigarettes?
5 How much will Hacker have to contribute to the snack fund?
6 What is the value of the negative externalities that Righteous still must endure?

For questions 7–10, consider proposal B that Hacker bring exactly one pack (20 cigarettes) to work each shift and Righteous pay Hacker 10 cents for each cigarette Hacker has left in the pack at the end of the day.

7 Will this proposal cause Hacker to choose to consume the socially optimal level of cigarettes?
8 How much will Righteous have to pay Hacker at the end of the shift?
9 What's the value of the negative externalities that Righteous still must endure?
10 Which proposal works better at achieving the optimal level of cigarette consumption? Which proposal does Righteous favor? Which proposal does Hacker favor?

Questions for Thought

1 What is a market failure? What are the three major causes? What are the implications of a market failure for the economy? Give three instances of market failure in agriculture.
2 Give three examples of externalities, and indicate the alternative possibilities for control. Is it always appropriate for a government to act when an externality exists?
3 What is the MC = MD rule? Under what circumstances can it be applied to agriculture? Give an example.
4 Redraw Figure 15-1 to indicate that all of the pollutant should be eliminated.
5 What is the difference between private property, common property, and open access resources? Can an open access resource be converted to a common or private property resource?
6 What is a public good? Give two examples public goods used by the agricultural sector.
7 What is the fundamental role of a government in the development of market institutions? Give several examples of the government developing market institutions in agriculture.
8 Give examples of renewable and nonrenewable resources. Can you list any policies in effect to manage these resources?

9 What is the justification for keeping some 300 million acres of grazing land under federal or state ownership?

10 Describe the economic forces that mitigate against the exhaustion of a nonrenewable resource. Why are some people concerned that these forces might not be adequate to ensure an appropriate use rate?

11 What are the alternative policy responses to externality problems? Give an example of an appropriate response to an externality problem caused by agriculture.

12 Make a list of externalities you have experienced this week.

13 What open access resources have you seen in the last six months? In your view, has that resource been overused?

14 Suppose you live in a communistic economy that decides to switch to a market economy. Your job in the new regime is to determine how property rights to the property that is presently owned by the government should be determined. Would you place all property in individual private ownership? Why or why not?

15 When you drive your car you generate negative externalities. Rank the following proposals in terms of their effectiveness in getting you to reduce the pollution you generate by driving: (a) an ad campaign urging to you be a good citizen and drive less, (b) improved bus transportation at a lower price, (c) a substantial (say $1 per gallon) tax on gasoline, and (d) a higher registration fee (say $200 per year) for all "gas-guzzling" cars that average less than 20 miles per gallon. What other programs would you suggest to decrease the amount of negative externalities generated by cars?

16 Earlier chapters discussed how price levels reflect scarcity. If we were "running out of" a nonrenewable resource, we would expect its price to increase. Yet in recent years the prices of many nonrenewable resources have decreased in real and nominal terms. What could explain these price declines?

17 If economic growth continues and standards of living continue to increase, does it make sense to deprive the present (relatively poor) generation of the use of a nonrenewable resource today so future (relatively rich) generations can use more of that resource? Would it have made sense for our ancestors to burn less coal so we could have more today?

18 Describe an example of an experience you have had with a common property resource. Hint: Sharing a kitchen in an apartment is every bit as good an example as raising sheep in Switzerland.

19 Make a list of things you know the world needs but you would not want to have in your neighborhood. Where should these things be located? Will there be anyone who will voice objections to the location you recommend?

ANSWERS AND HINTS

Name That Term **1.** externality; **2.** free rider; **3.** open access; **4.** failure; **5.** Common property; **6.** Pesticides; **7.** conservation; **8.** Eminent domain; **9.** nonrenewable; **10.** sustainable yield

True/False **1.** T; **2.** F; **3.** T; **4.** F; **5.** T; **6.** F; **7.** F; **8.** F; **9.** T; **10.** T; **11.** T

Multiple Choice **1.** d; **2.** c; **3.** d; **4.** c; **5.** d; **6.** c; **7.** d; **8.** d; **9.** e; **10.** e

Technical Training   1.   **2.**

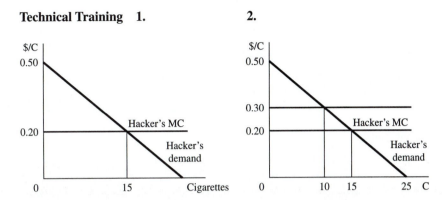

3. 30 cents; **4.** yes; **5.** $1; **6.** $1; **7.** yes; **8.** $1; **9.** $1; **10.** both proposals "work" in the sense of generating the optimal consumption of cigarettes; Righteous favors Proposal A and Hacker favors proposal B.

16

AGRICULTURE AND DEVELOPMENT

CHAPTER OUTLINE

OVERVIEW

What explains the development of civilization from traditional agrarian societies to the highly technological service and information society we now enjoy? Did it "just happen," or did human beings make decisions and form governments that implemented polices encouraging these developments? Our answer is that governments do make a difference and did influence the developments that occurred. Governments work to provide a "level playing field" or the "rules of the road" for individuals and businesses.

The study of economic development and the role of public policy in helping (and sometimes hindering) that development provides a perspective from which to view society's problems and options today. Furthermore, the lessons learned by studying the experiences of developed nations can provide helpful signposts to policy makers in the less-developed nations of the world about what to do, and what not to do.

This chapter focuses on agriculture and development. A common national policy goal of governments throughout the world is to develop and adopt policies and programs that promote economic development. Since agriculture is the dominant form of economic activity in the early economic development of most nations, agricultural policy can play a critical role in hindering or speeding the development process. Also, even in the most developed nations agricultural development is important. In this chapter, we examine some of the noteworthy U.S. agricultural development policies.

LEARNING OBJECTIVES

This chapter will help you learn:

- What is meant by economic development and what is the role of agriculture
- The basic roles of government in economic development
- The policy foundations for American agricultural development
- The role of science and technology in agricultural development
- The importance of the international context in development

THE ROLE OF AGRICULTURE IN ECONOMIC DEVELOPMENT

In the macroeconomic section of this book, we discussed two central economic goals that must be balanced by all nations: producing for consumption today while providing for the future. If an economy is performing well, the real incomes of its people are growing and more goods are available for consumption over time. With economic development, a nation moves from low incomes (one measure is GDP per capita) to higher incomes. Accompanying this change in per capita income, or levels of living, are important changes in resource availability and public policy. Because economic development is a process as well as a goal, it is useful to identify characteristics of countries at the opposite ends of this process. To do this, we use four economic variables: GDP per capita, production agriculture's contribution to GDP, share of labor force in production agriculture, and share of food in consumption expenditures. We also use two socioeconomic variables: literacy and infant mortality. To illustrate these variables, we present data on the United States and Germany, two of the richest and on Nepal and Chad, two of the poorest countries in the world (Table 16-1). In this table the GDP data are in terms of purchasing power parity, a measure of purchasing power of income adjusted for factors such as home-produced food. If compared on actual GDP basis, the differences would be even greater.

For countries at the low-income end of the scale, GDP per capita is low, most of the labor resources (and other resources for that matter) are employed in the farming and the share of production agriculture in GDP is high. Food consumption accounts for a large share of total expenditures. Many people are malnourished and some are on the brink of

TABLE 16-1   SELECTED ECONOMIC DEVELOPMENT STATISTICS FOR TWO DEVELOPED
AND TWO LESS-DEVELOPED NATIONS

	United States	Germany	Nepal	Chad
GDP per capita	$31,500	$22,100	$1,100	$1,000
Infant mortality rate (per 1000 live births)	6.33	5.41	80	115
Life expectancy at birth	76.2	77.2	58.4	48.5
Literacy	97%	99%	27.5%	48%
Production agriculture's share of GDP	2%	1.1%	41%	39%
Labor force in production agriculture	2.7% (including fishing and forestry)	2.7%	81%	85%

Source: 1999 CIA World Fact Book.

starvation. The high rates of infant mortality and the short life expectancy reflect the fact many citizens of these countries live with hunger.

The situation is very different in the developed economies. GDP per capita in the United States is more than 30 times higher than in Nepal. Most of the resources of the economy are invested and employed outside of farming, as is indicated by the relatively small share of the management and labor force working in production agriculture. In real terms, much more is spent on food, but the share of the nation's total expenditure spent on food is much smaller. The average per capita caloric intake is well beyond the minimum level required to prevent malnutrition. In fact, the negative effect on health of a too abundant diet, with too much cholesterol and sugar, is a serious nutritional health problem for many people in the developed countries. Resources are available, however, to provide adequate diets for poor people and there can be special health-care programs for pregnant women and infants. Hence, although health care may not be provided or used uniformly, malnutrition is practically nonexistent and the infant mortality rate is much lower than in the less-developed countries.

Structural transformation of an economy in the process of economic development involves a relative decline in employment in farming, in the proportion of economic activity in production agricultures, and in expenditure of food relative to nonfood goods and services. This change in expenditure is consistent with **Engel's Law** (named after Ernst Engel, a nineteenth-century Prussian statistician who first observed this relationship), which says that the share of food in consumption expenditure declines as incomes rise. See Figure 11-5 for the declining share of GDP that is received by the farm sector.

We know that the income elasticity of demand for food becomes much less than one as incomes rise above subsistence levels. Hence, as incomes rise, the share of expenditures on food rises less than proportionately. As the demand for nonagricultural goods and services grows proportionately more than the demand for agricultural commodities, fewer resources are attracted into production agriculture. In addition, agriculture's share, and especially the farm sector's share, of the GDP and of employment of the economy's resources fall. Even so, many maintain that economic development is an important goal of food and agricultural policy. Development is a means of achieving two other goals: higher food standards for consumers, and improved balances of trade.

AGRICULTURAL DEVELOPMENT AND FOREIGN ASSISTANCE

It is widely accepted that agricultural development is an essential first step toward general economic development. An important issue that is still undecided is the role of foreign aid in the developmental process. Most people are aware that a variety of sources of assistance are available to today's developing nations. International agencies such as the **World Bank** and the United Nation's Food and Agriculture Organization (FAO), private foundations, such as the Ford Foundation and the Rockefeller Foundation, and government of many countries including the United States, have devoted substantial technical and financial assistance to help the less-developed nations. In some countries, this assistance has made great contributions, while in some others, the effects are less obvious. Formidable challenges remain in identifying new ways to improve the productivity of the agricultural sectors.

The amount of U.S. foreign assistance per capita or as a percent of GDP is quite low as compared to other industrialized nations. Regardless, some farm groups (and others) in the United States oppose foreign assistance to agriculture on the grounds that the increased production means increased competition for U.S. production. Aside from humanitarian reasons for continuing foreign assistance, economic development in a developing country nearly always results in greater increases in demand for food than increases in supply. The most rapid growth in U.S. agricultural exports has come from those developing countries with the most rapidly growing incomes. Economic development for most countries has started with increases in efficiency in their agricultural sector and yet such development results in more agricultural imports. A policy of reducing agricultural assistance because of its supposed negative effects on U.S. agricultural exports is shortsighted and generally wrong.

THE ROLES OF GOVERNMENT

What can a government do to assist the developmental process of its country? There are four roles that a government can play: provide physical infrastructure, provide institutional infrastructure, perform as a market participant, and modify market outcomes with policy instruments. We briefly consider each of these roles.

Physical Infrastructure (PI)

The **physical infrastructure (PI)** most crucial to agriculture for economic development is the transportation infrastructure (roads, railroads, harbors, canals, and waterways). Development of this infrastructure allows more efficient movement of inputs to the farm and output from the farm. Losses and spoilage of food can be minimized, production can be specialized in lower-cost areas, and products can move to the most profitable market. In the United States, as in most other nations, the government provides transportation infrastructure, generally from tax revenues, although it is not uncommon for user fees to be charged (see the box on turnpikes). In the United States, a large portion of the expenditure on physical infrastructure, which involves construction and maintenance, is the responsibility of state and local governments.

TURNPIKES

When road construction began in the United States in the early 1800s, most construction was by private companies that collected tolls from users. Many roads built during this era were called "turnpikes." The name was derived from a large turnstile, made of two timbers, rotated by the toll collector to allow a wagon to pass. The ends of the timbers were usually capped with a metal point and looked similar to the pike, a favorite weapon of medieval infantryman . . ."

Source: Truman E. Moore, *The Traveling Man,* Doubleday and Company, Inc., New York, 1972, p. 8.

Indirectly a government can also provide a variety of physical infrastructure services by subsidizing the initial investment costs. Examples include the development of rail-roads, communication facilities, and utilities. For American agriculture, an important example was the Rural Electrification Administration (REA). Beginning in the 1930s during the Great Depression, the federal government subsidized REA cooperatives, mainly with low interest rates, to bring electricity and telecommunications to farms and rural areas not served by private sector power companies. Since then, the use of electricity has increased greatly on farms, and in homes and businesses in rural areas. The capacity of the private sector to deliver electricity has also expanded greatly, making the provision of governmental support less urgent, but still important, in some areas.

There are two justifications for government to sponsor or provide support for developing physical infrastructure. First, for some types of infrastructure, if a firm provided the infrastructure it would be in a monopoly position, or the industry would be so highly integrated, that either the firm or the industry would be able to charge excessive prices. The government may choose to operate a facility itself, or it may regulate the activities of a private firm, as in public utility firms. Second, for some types of physical infrastructure, it is very difficult to charge the public for its use (e.g., city streets) and the private sector might not develop such infrastructure. As explained in the next chapter, these are termed public goods. Any or all members of society may use or benefit from these goods and it is not feasible to charge for their use. National defense is a classic example.

Institutional Infrastructure (II)

In a market economy, an extremely important role of a government is to provide for the **institutional infrastructure (II).** Perhaps the most important part of this institutional infrastructure is property rights. For Americans, the right of the individual to own and manage land and other property has been established as a basic right. The right is a fundamental factor in development because it enables owners to use the property as they please, providing such use does not violate the rights of others as established by law. In other societies, property rights are either implicitly or explicitly in the hands of groups as "common property" (discussed in Chapter 15) or they belong to the government. Even in the United States, the property rights of an individual landowner are circumscribed by zoning laws and other government regulations, adopted and enforced at the local level.

Another important part of the II is the enforcement of rules and regulations governing market participation. All market participants must know and accept the rules and regulations of the market for a complex market to function efficiently. Formalizing these rules into law, and enforcing them, reduces the risk of a participant violating market rules for personal gain at the cost of other market participants.

The government establishes rules under which businesses organize as corporations, partnerships, joint ventures, and cooperatives. As discussed in the macroeconomic chapters, the government may establish a central bank or banks and set the policies for operating the banking and financial system. Governmental agencies grant and enforce patents, set grades and standards, and inspect products to ensure that they meet the established quality standards (meat inspection, tests for pesticide residues, inspections and quarantines of imports to protect domestic agriculture and consumers). The federal government facilitates and regulates the operation of futures markets, enforces antitrust regulations, and establishes rules for international trade by entering into international treaties and agreements. State governments license certain businesses, such as pesticide applicators, which may operate under either state or federal law, or both. All of these governmental actions establish the institutional infrastructure necessary for smooth operation of a market-oriented economy. In even the least regulated economies, this institutional infrastructure is needed. In most countries, however, the role of government in agricultural markets goes well beyond this basic organizational function.

Government as Market Participant (MP)

In many countries, the government is a market participant (MP) as agencies are established to produce and market many types of goods and services. In a few nations, governments still own most resources and government agencies augment or supplant activities of the private sector. In other countries, domestic markets are privately organized, but a governmental agency conducts part or all of the international trade activities of agriculture and other industries. Often, these agencies provide services such as utilities and transportation. In many developing countries, governments produce and distribute farm inputs such as fertilizers and improved seed varieties to farmers to encourage their use, and the government may operate banks or other institutions to provide credit to farmers and other local businesses.

GOVERNMENTAL PARTICIPATION IN MARKETS

One way to categorize governments is by the degree to which they become market participants; that is, whether they buy and sell goods directly and engage in production. In capitalistic economies, the role of the government as a market participant is smaller than it is under most other forms. At the other extreme would be the case where only the government participates in markets. No country currently falls into this category. Albania and China may have come the closest, but a substantial and growing share of economy activity in these two countries now occurs in markets.

It is common, especially in developed countries, for governments to conduct research activities or to support those providing these services. Also, a government may provide additional sources of capital to targeted sectors.

Market Modification (MM)

Lastly, most governments adopt market modification (MM) policies to alter market outcomes. In developed countries, these policies generally favor agriculture, although in some underdeveloped countries the policies have disadvantaged agriculture. Government policies adopted in many developed countries have set goals to raise prices of farm commodities, augment farm incomes, or restrict imports or subsidize exports to benefit domestic producers. In the 1930s the federal government subsidized cooperatives that provided electric service to rural areas. In Chapter 17, we look at a specific example, the agricultural price and income policies of the United States government. In addition to price and income policies for agriculture, governments have begun recently to adopt policies to influence the use of agricultural resources to reduce negative externalities. Natural resource policies of the federal government, and some state or local governments, are discussed in Chapter 15.

AGRICULTURAL DEVELOPMENT POLICY

A government's role in general economic development, as well as in agricultural development, is to assist the private sector in improving the contribution of the nation's resources (its land, labor, and capital) to the well-being of its citizens. This assistance can provide an institutional environment supportive of the private business sector or can provide direct assistance by creating public goods, financing technological advances, and organizing resources for development. In the following sections, we examine some key policy choices, which relate primarily to agriculture and the rural economy. We will point out whether the policies are physical infrastructure (PI), institutional infrastructure (II), market participant (MP), or market modification (MM) in nature.

PRESSURE FOR AGRICULTURAL SUPPORT

Throughout the nineteenth century, American farmers complained of lack of support from the government. The nation's conservative monetary policy strengthened the dollar and depressed the export demand for and prices of farm commodities. Many industries that served the farm sector had a monopolistic pricing structure. The government imposed high import tariffs on industrial goods, which raised the cost of production for agriculture and reduced the real value of agricultural products and farm property. In the thirty years immediately following the Civil War, these complaints resulted in a large, rural-based political movement, called "the Populist Revolt." The goal of the movement was electing candidates for Congress and president who would support such things as expansion of the currency, national regulation of all banks, and control of monopoly in all its forms. By the 1920s that many of the goals of this movement had become national policy.

Land Policy

Colonial development in North America was heavily influenced by the circumstances that the European settlers found on their arrival in the early 1600s, and later. Leaving behind a relatively heavily populated continent with a well-entrenched political and social structure, they found a vast and sparsely populated land untouched by agricultural practices with which they were familiar, and with no political or social institutions they recognized. For 300 years, from the initial colonial settlements in the early 1600s, to the early part of the twentieth century, governmental policies were dominated by issues of land settlement and problems faced by settlers in their encounter with the frontier. Underlying beliefs in the importance of democracy and individual freedom in economic matters served as an inspiration for laws to guide land settlement policy. These laws provided the institutional infrastructure (II) for the early development of agriculture and helped set a pattern for more recent development policies.

TWO TRAGEDIES

Native Americans, the original inhabitants of the land, received little attention during this development. One of the two tragedies in the development of the American nation was the destruction of the existing population. The other was the enslavement of African Americans until the Civil War and the continued exploitation of many African Americans for decades after that. Both were elements of U.S. policies affecting agricultural development.

A fundamental concern of the early colonists was the establishment of a system of property rights and the initial allocation of land to individuals. With the establishment of the United States this became an explicit element of public policy because the U.S. government claimed ownership of most of the land of the continental United States. Over the course of American history, the federal government's land policy has had four major elements: (1) the settlement of agricultural land by private farmers (II), (2) the creation of large amounts of public lands (as a MP), (3) the management of land used to improve farm income (through MM), and (4) the control of land uses to achieve environmental objectives (also MM).

Settlement Policy   In the late 1700s, when the United States was founded, the colonial governments agreed that all previously unappropriated land should be ceded to the new federal government. Alexander Hamilton, among others, argued that the vast lands lying beyond the original settlements should be sold in large blocs to private companies and individuals. Thomas Jefferson and others wanted the government to grant land to individual settlers, or to sell it to anyone at a very low price. However, it was not until 1862 that an official "homestead policy" was established to provide free land for each family of settlers. Thus, early in the nation's history a fundamental II choice was made to place agricultural land under the ownership and control of individuals as private property, instead of employing common property or governmental management of agricultural land and any associated mineral or energy resources. Simultaneously, an II policy was adopted to allocate that land not on the basis of existing market power or wealth, but to any settler whether they were poor, new immigrants or well-established citizens.

INDIAN REMOVAL

"The lands beyond the Appalachian Mountains onto which the settlers poured in flood-like proportions between 1800s and 1860 were not vacant lands. They were the ancient and established lands of many Indian tribes, some of which were highly civilized. The tribes and individuals did not have pieces of fancy paper, registered with a government somewhere, stating that they had title to their lands. So when the whites, with their insatiable demand for land, poured over the Appalachians, the Indians were pushed aside by superior military might and forced to sign treaties ceding the most promising of their lands to the whites. Later, in the 1820s, when more of the country was settled, the cry went up to remove the Indians from their homelands. A policy was formalized by the federal government to remove all Indian tribes east of the Mississippi and to settle them in an Indian Territory—a territory that later became the state of Oklahoma.

The policy of Indian removal was vigorously pursued in the 1830s, and thousands of Indians east of the Mississippi were rounded up at the point of the bayonet and driven, or herded by government contractors with the support of the army, over the "Trail of Tears." Perhaps one-quarter to one-third of those who started the trek perished. (A similar policy was followed towards Indians west of the Mississippi in later periods.) Hard and cruel as the age of pioneering and settlement was on the white settlers and their families, the same period was even harder and crueler for the Indians and their families. They lost their home and their way of life. They lost all."

Source: Adapted from *The Development of American Agriculture,* Willard Cochrane, University of Minnesota Press, Minneapolis, 1981, pp. 62–65.

The Homestead Act of 1862 was the most definitive of the numerous acts adopted to transfer land from the federal government to settlers. It provided for a grant of 160 acres to each head of a family who settled on the land and lived there for five years. When this acreage proved insufficient to support a family in most areas, there were eight more land settlement or homestead acts that increased land available to a family.

Irrigation Policy   Vast stretches of land west of the Mississippi River presented new problems for successful farming because much of it was too dry to grow good crops without irrigation. Because large-scale irrigation projects required a substantial investment, far beyond what settlers could finance, settlers began to use their political influence to get the federal government to help provide irrigation. The federal government first responded by transferring large amounts of irrigable land to the states for reclamation and development. Most states did not finance this however and, after several years of delay, the Federal Reclamation Act of 1902 authorized a MP policy of direct federal funding to build and operate dams and other facilities for developing irrigation, public power, and flood control. The 1902 act also limited the sale of water for irrigation used on federally financed projects to 160 acres for a single owner. This

MM provision, which has been the subject of much political contention, was never fully enforced. The lack of enforcement became a significant issue and after years of argument the 160-acre limitation was adjusted upward. The early and continuing lack of enforcement made it possible for a small number of very large farms to be developed based on water purchased from federally funded projects at a fraction of its cost. Continued lack of enforcement has enabled this pattern of land ownership and use to survive.

Public Land Policy   Despite the policy of distributing land free to individuals, the federal government is a MP in that it still owns substantial amounts of land, particularly in the western states. In the late 1800s a strong political movement developed to conserve natural resources. The Revision Act of 1891 gave the federal government new responsibilities for conserving the nation's natural resources, and established authority for the federal government to set aside more land for additional national forests. This policy also was extended to include public lands valuable for oil, potash, copper, phosphate, and other minerals. Finally, in the General Leasing Act of 1920 an overall leasing policy was written into federal law (II).

Most of the land that is still owned by the public is managed by public agencies in the U.S. Department of the Interior (Bureau of Land Management and the Park Service) and the U.S. Department of Agriculture (Forest Service). Most of this land is available for multiple uses. For example, a national forest might be used for forestry, wildlife habitat, leased livestock grazing, mineral extraction, and recreation. It is not uncommon for these uses to come into conflict and lead to intense public debates. For example, the U.S. Forest Service subsidizes the harvest of timber from federal lands as a means of supporting the lumber industry, providing wood products for construction, and earning international exchange by exporting lumber. Those who oppose this policy argue for the preservation of forests, care of wildlife habitat, the protection of endangered species, and protection of land against erosion. They also oppose the use of scarce fiscal resources to support private firms in the forest industry. Consequently, some land has been withdrawn from commercial use and rules and regulations have been instituted limiting the use of some land.

PUBLIC LAND USE CONFLICTS

An example of the conflict among different users of public lands was exemplified by the fight over the spotted owl in the old-growth forests of the Pacific Northwest. The owl survives only in dense forests that have not been logged. Harvesting the old-growth forests reduces the number of spotted owls. Placing the owl on the endangered species list ended logging in the areas where the owl lives.

Loggers in the region argued that their economic livelihood depends on continued logging of the old-growth forests, and should take precedence over the owl. Naturalists argued that the subsidies to logging firms are inappropriate and that the old growth timber should be protected for biodiversity, scenic, recreational, and wildlife habitat values. They used the Endangered Species Act protection of the spotted owl as the legal method of stopping the logging.

Land Retirement Policy   With roots in the 1930s, and gaining substantial influence in the 1980s and 1990s, environmental and conservation interests have secured adoption of regulations to protect resources and reduce the adverse environmental effects of agricultural production. The federal government has incorporated provisions to improve productivity into price and income support policies. These MM price and income policies are discussed at some length in Chapter 17. The use of some agricultural lands is also constrained by regulations to preserve wetlands and natural prairie (the "swampbuster" and "sodbuster" provisions of the farm legislation). These lands provide habitat for wildlife, help to maintain biodiversity and, in the case of wetlands, are the source of ground water recharge (see Chapter 15). The crucial point is that land management policy no longer focuses on a single objective of increasing production of agricultural commodities or controlling production at the lowest possible cost. Land management policy now attempts to balance multiple objectives including farm income support, increased productivity and sustainability, conservation of land resources, and protection of the environment.

Results of U.S. Land Settlement Policy   Well before the end of the 1920s, large-scale land settlement came to a close with about 1.0 billion of the 1.4 billion acres in the original forty-eight states being transferred to private ownership under the public land laws. By the middle of the 1930s, the number of farms reached a peak of nearly 6.8 million farms. About 90 percent could be classified as family farms; large enough to support a family and yet small enough so family members could do most or all the work. The remainder were either too small to support a family, or much larger than a family could operate. Since then consolidation of land holdings has created much larger farms, and there are now well below 2.0 million operating units, the vast majority of which are family owned and operated. This trend is a result of the economies of scale resulting from specialization, technical change, a more efficient marketing system, education, and other government policies that encourage agricultural development and growth.

Educational Policy

Education makes workers more productive. Note the literacy disparity in Table 16-1. Improving the education of citizens has always been a central goal of economic development. In the United States, educational policy is determined primarily at the state and local level, but the federal government has played a unique role in agricultural education. The Ordinance of 1785, under the Articles of Confederation, provided land in each township for the support of primary and secondary schools that states could use to provide both PI (buildings) and II (teacher salaries).

In 1862, the Morrill Act (commonly called the "Land-Grant College Act") provided for a publicly supported system of higher education and established three new policy goals. First, it declared that an opportunity for higher education should be available for students from all walks of life, instead of just the wealthy or the elite. Second, the teaching of science and technology should receive primary attention (in contrast to many private universities where the traditional subjects of liberal arts and literature were dominant). Third, this education should have a practical application to agriculture and industry.

The Morrill Act was followed by (1) the Hatch Act of 1887, that provided for research in the land-grant universities; (2) the Smith-Lever Act of 1914, that founded the agricultural cooperative extension system; and (3) the Smith-Hughes Act of 1917, that initiated vocational education programs. These programs have been expanded and have provided additional PI and II support in the years since.

Agricultural Credit Policy

The capital resources of a country are the third key category of resources necessary to development. A well-functioning system of credit markets is crucial to the efficient transfer of capital resources from savers to borrowers; whether the use be short term, to finance farm production operations, or long term, to finance agricultural real estate or a new factory building. Access to sufficient credit was a serious problem for early American farmers (as it is today for farmers in many developing countries). Finding a way to address agriculture's credit needs was important to agricultural development.

The federal government entered the U.S. agricultural credit market in the early 1900s in response to long-time and widespread complaints from farmers, agribusinesses, and other rural people concerning what they perceived as inadequate, high-cost, and unreliable agricultural credit. In rather quick succession, three separate II acts created the foundation for the Farm Credit System (FCS). The Federal Farm Loan Act of 1916 founded the federal land bank system. The Agricultural Credit Act of 1923 provided the foundation for the intermediate credit banks and later the Production Credit Associations (PCAs). The Agricultural Marketing Act of 1929 provided the basic structure for the banks for cooperatives. Originally, twelve farm credit districts were created, and three types of banks were created under federal sponsorship, with capital stock provided temporarily by the federal government.

The general policy has been for the FCS to provide a full line of credit for agriculture, at competitive rates of interest, in all regions of the United States. The federal land banks (FLB) were the first to offer long-term farm mortgages as a common banking practice. These mortgages relieved much of the worry associated with short-term borrowing and permitted farmers to make more definite long-term plans. In the depression of the 1930s the FLBs played an important role in refinancing millions of farmers who were in financial distress. Emphasis on long-term farm mortgages was critical in helping many family farmers survive the Great Depression. Since then FCS has been a significant provider of credit to agricultural firms and has provided competition for the banks and other private sector lenders.

In recent years about 75 percent of the credit used by American farmers and agribusinesses is provided by private lenders: commercial banks, savings and loan associations, life insurance companies, and agribusiness firms. Federal and state governments regulate all of these. Private individuals also provide a significant amount of credit to farmers.

The remaining 25 percent of the agricultural debt is held by two public agencies, the **Farm Credit System (FCS)** and the **Farmers Home Administration (FmHA).** The larger share of this debt is held by the FCS operating banks that are federally sponsored, but cooperatively owned by their patron borrowers. Thus, the FCS is another example of II. The remainder is held by the Farmers Home Administration (FmHA), an operating agency of the federal government, making the government a MP. It is funded through the U.S. treasury, to which it returns its net earnings. For many years, it served primarily as a direct-lending agency for farmers and others who could not get adequate credit from other sources. Since reorganization in the late 1980s it has served primarily as a guarantor of loans made by commercial banks and others. Also, it is directed to provide services for its borrowers such as limited help with management and book-keeping.

The experience of the FCS over its long history illustrates that the federal government's support of agricultural credit programs can be important in the development and growth of agriculture. The programs are especially important for refinancing the farm sector in times of financial stress or crisis. However, in the long term, it will be important to remember that both the FCS and the FmHA are subject to all the hazards and uncertainties of specialized lenders to agriculture and rural communities. Therefore, although the banks of the FCS, and the loan and guarantee programs of the FmHA, continue to develop a favorable credit experience, it is unlikely that the agencies ever will be completely independent of federal government support. In other words, prudent management and operation must involve a balanced consideration of the unique credit requirements of agriculture, the farm sector especially, and the more general public interest in continued development and growth of agriculture.

Agricultural Business Development Policy

The process of transforming traditional agriculture requires the simultaneous development of an efficient agribusiness sector and substantial government-supported infrastructure. This can be seen in the difference between the contributions of agribusiness to the agricultural sectors to developed and less-developed nations. As discussed in Chapter 9, by far the largest part of the food and agricultural economy in developed countries is not the farms but the firms and industries that supply inputs to the farm sector and that market commodities produced on farms and ranches and sold to consumers.

Governments contribute in important ways to development of agribusiness. They do this by establishing legal and financial infrastructure (II) in which agribusinesses can develop and prosper, by having a favorable tax system, and by providing public infrastructure, such as roads and schools. Also, governments occasionally develop specific policy measures to encourage or regulate agribusinesses. We note four II examples in the United States—the establishment of the Interstate Commerce Commission, the regulation of livestock trading and processing firms, the regulation of the organized grain trading markets, and regulations making it easier for farmers to organize cooperative ventures.

- In 1887, establishment of the **Interstate Commerce Commission (ICC)** provided for controlling railroad rates and services in ways desired by farmers and other shippers. The ICC was created in response to farmer's complaints of high freight rates and poor service, resulting in part from the transport monopoly enjoyed by railroads in the early years of their development.

- The Packers and Stockyards Act of 1921 and subsequent amendments regulate the operation of livestock commission firms, firms that arrange the sale for the livestock owner, in public stockyards. The policy prohibits meat packers engaged in interstate commerce from employing unfair, discriminatory, or deceptive practices, manipulating or controlling prices, or otherwise creating a monopoly and restraining commerce. These laws provide the foundation for operation of U.S. livestock markets today.

- The Grain Futures Act of 1922 and later amendments regulate the nation's grain markets. The act made illegal the manipulation of prices and the attempts to do so. Commission merchants, floor brokers, and the governing boards of trade were made subject to regulation under the broad jurisdiction of the Secretary of Agriculture. Also, the industry has a code requiring ethical practices to maintain public confidence in these markets.

- The Capper-Volstead Act of 1922, sometimes called the "Magna Carta of cooperative marketing," authorized farmers' cooperatives to issue capital stock and engage in interstate trade. It also prohibited the cooperatives from violating antitrust laws by restraining trade. This permitted the growth of larger cooperatives, which soon came to exert market power (Chapter 10). Such cooperatives have continued to grow and flourish, generally free from government intervention, provided they do not abuse their power.

Food Safety Policy

For most developed economies, the goal of producing sufficient basic commodities to provide an adequate diet was achieved long ago. The agribusiness sector now provides year-round access to the full range of food and fiber products. Even perishable products such as fresh fruits, vegetables, and flowers are available as a result of worldwide production and the international marketing system. The quality and variety of highly processed foods continues to grow.

As the quantity of food has become assured, increased attention has focused on the health aspects of the food supply. Broadly, food risks can be divided into two categories: (1) contamination by bacteria and other organisms, which can cause illness or death upon consumption, and (2) contamination by chemicals, which can cause adverse health effects in the long run. There are many more incidents of illness and death caused by bacterial infection than by chemical contamination. Microbial illnesses are generally the result of inappropriate handling of food, often in the home or in a restaurant. Public-health policies require regular inspections of food-processing firms and restaurants to ensure that the food supply is clean and wholesome. Even so, all cases of food contamination or even of food poisoning cannot be prevented. Public pressure for increased resources to address this problem has not been great as might be expected due to the common but scattered nature of the symptoms.

THE SANITARY AND PHYTOSANITARY PROVISIONS IN TRADE AGREEMENT

Problem: How does a country ensure that the imported food being supplied to its consumers is safe to eat by the standards it considers appropriate? And, how can countries be prevented from using overly strict health and safety regulations to protect their domestic producers?

The Agreement on the Application of Sanitary and Phytosanitary Measures, approved in the Uruguay Round of world trade negotiations, sets out the basic rules for food safety (sanitary) and animal and plant health (phytosanitary) standards.

It allows countries to set their own standards. But these regulations must be based on science and be applied only to the extent necessary to protect human, animal, or plant life or health. Also, regulations are to not arbitrarily or unjustifiably discriminate among countries where identical or similar conditions prevail.

Member countries are encouraged to use international standards, guidelines, and recommendations where they exist. However, members may use measures that result in higher standards if there is scientific justification. They can also set higher standards based on appropriate assessment of risks so long as the approach is consistent, not arbitrary. The agreement still allows countries to use different standards and different methods of inspecting products.

FOOD VARIETY

As a result of the increase in demand for food away from home, and the capability of the agribusiness sector to efficiently move food products, moderate-sized metropolitan areas will likely have restaurants offering entrees from many cultures. This development has resulted in greater diversity of choice. It is no longer necessary to travel the world or to visit major metropolitan areas to enjoy the cuisine of other nationalities, as was the case only a few decades ago. There also has been a dramatic proliferation of food chains and franchised "fast food" and full-service restaurants across the nation and around the world. To the extent that food chain franchises replace local, independent restaurant operators, the variety of options diminishes. In some areas, there is concern that the chain's uniform architectural style does not conform to the distinctive regional surroundings.

Although the documented incidence of adverse consequences directly attributable to chemical contamination is small, and some is unreliable, there has been considerable public concern about this issue. Hence, policies have been adopted to control the pesticides used in agricultural production, and some pesticides have been removed from the market. Guidelines have been developed for the rate and timing of use and over the last several decades the testing requirements for approval of new pesticides have been strengthened considerably. The demand for organic products, that is, those grown without the use of chemicals, has shifted out rapidly. Recognizing this increase in demand and to provide better information, congress mandated the USDA to issue standards for organically grown foods.

Important issues remain however. Some pesticides approved several years ago were not required to pass the same stringent tests as pesticides recently developed or currently being registered. Some argue for substantial increases in efforts to accomplish these evaluations. What new controls may be required, and how much control is socially optimal, remain serious points of contention.

In addition, debate continues on what should be the acceptable levels of pesticide residues in food products. The debate is often framed around the trade-off between higher costs of producing food and lower production, perhaps with lower quality if pesticides are restricted or eliminated on the one hand, and the possibly higher probability of consumers contracting major health problems, such as cancer.

Finally, food imports are tested for pesticide residues, but the volume is so great that these tests cannot cover all possibilities of contamination. So, because of the difficulty in testing adequate quantities, the importation of fruits and vegetables is likely to become an increasingly important issue in food safety. International agreements on pesticide and disease control are likely to become ever more important.

SCIENCE AND TECHNOLOGY

Technological advances have been central to American agricultural development. For example, it has been estimated that between 1948 and 1976 growth in productivity accounted for 40 percent of the total growth in agricultural output. One of the important factors determining the direction of effort in science and technology is the concept of **induced innovation.** The idea behind induced innovation is that scientific endeavor responds to economic forces. Researchers, inventors, and firm managers look for new technology that will increase profits, generally by conserving costly inputs. The incentive for innovation is strongest in cases where the innovator can realize the benefits by selling or licensing the innovation. A recent example is the dot com Internet-related companies.

In the early years of the United States, labor was scarce and costly while land was abundant and cheap. Indentured servitude and slavery were early means of reducing labor costs in agricultural production. Innovators were also aware of the large share of labor in production costs and devoted efforts to reduce this share. The result was the invention of a series of mechanical devices that increased labor productivity by replacing labor in agricultural production. The first was the cotton gin, invented by an African American and patented by Eli Whitney in 1791. The gin increased the maximum amount of cotton that could be cleaned by one person from 18 to over 800 pounds per day. While the gin eliminated the labor constraint in removing seeds and made cotton production extremely profitable in the South, manual labor was still necessary to pick cotton. An unfortunate result was that the use of slaves became more profitable, creating an economic incentive to maintain slavery after most other nations abolished it. The mechanical cotton picker was not made commercially profitable until 1944.

Other mechanical implements were developed throughout the 1800s to conserve labor. In 1797 Charles Newbold invented a cast-iron plow. In 1837, John Deere used high-grade steel to make a significantly superior plow and laid the foundation for the company that still bears his name. In 1826, Patrick Bell invented a reaper that enabled a worker with a team of horses to harvest as much grain as it had formerly taken six workers to harvest. Power was also expensive, and innovators provided periodic improvements in the delivery of power to agriculture. In 1855, Obed Hussey developed a self-propelled steam engine for plowing, but it was slow and cumbersome. In 1876, Nikolaus Otto in Germany developed the first practical 4-cycle internal combustion engine. By the 1920s, although many individuals produced tractors, Henry Ford emerged as the dominant manufacturer. Shortly after that, John Deere, McCormick-Deering, and Allis-Chalmers produced tractors that dominated the market.

As the land frontier in the United States came to a close early in the twentieth century and labor demand from the industrial sector pulled workers off the farm, the demand for different kinds of innovation developed and innovators responded. Two complementary land-augmenting innovations increased crop yields. These were nitrogen fertilizer, and plant varieties that effectively utilized higher levels of fertilization. The development of a series of increasingly efficient processes for converting atmospheric nitrogen to fertilizer resulted in a falling price of nitrogenous fertilizer. This price decline, combined with technological improvements in production of other fertilizer ingredients, such as phosphate and potash, contributed to a substantial long-term decline of fertilizer prices relative to other inputs.

Simultaneously, plant breeders developed many new crop varieties. Traditional varieties had been selected by farmers (and nature) for their ability to tolerate low levels of plant nutrients, and many did not respond favorably to higher levels of fertilization. The new seed varieties were developed to effectively use large quantities of fertilizer. In some crops such as corn, the development of hybrid vigor (the "children" of parents from different varieties of seed perform better than their parents, but the grandchildren revert to the original performance) generated the hybrid-seed industry. The "seed-fertilizer" revolution of the mid-twentieth century contributed major increases in productivity of land in the United States.

The effects of induced innovation for corn, wheat, and soybeans in American agriculture can be seen in Figure 16-1. Similar graphs could be constructed for other crops. Although the rise in productivity is by far the most dramatic in the case of corn, the more than doubling of wheat and soybean productivity is also impressive.

The steady decline in the number of farms and farm workers means that a graph of output per worker would also show an increase in labor productivity. From the early 1800s to 1935, U.S. wheat yields were about 15 bushels per acre. During the same period, however, labor productivity on farms rose dramatically. It has been estimated that during the early 1800s, it took 373 days of work to produce 100 bushels of wheat. By 1900, it took only 108 days to produce the same yield, and by 1935 only 67 days. Between 1935 and 1970 labor productivity continued to increase and by 1970 only 9 days of farm work were required to produce 100 bushels. Since then many more sophisticated labor-saving innovations have become commercially profitable, and the on-farm labor requirement has continued to drop.

What are the expected directions for induced innovation in agriculture? In the next two sections we categorize the direction of agricultural research under two headings—biotechnology and information technology.

Biotechnology

Biotechnology refers to the use of living organisms to make or modify products, to improve plants or animals, or to develop microorganisms for specific uses. These techniques are suplanting the modification of species through identification of desirable mutations, cross breeding, and selection. In advanced biotechnology, changes induced

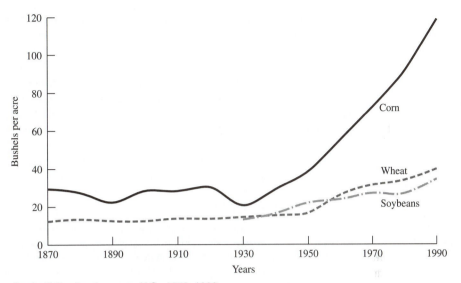

FIGURE 16-1   Productivity of major crops, U.S., 1870–1990.

by genetic engineering in animal reproduction involve three general methods: (1) recombinant DNA (rDNA) techniques, also called "gene splicing," (2) monoclonal antibody production, and (3) embryo transfer. The products of rDNA techniques are commonly termed **genetically modified organisms (GMOs).** The development of these organisms have generated significant levels of public controversy.

Because of the power of rDNA technology to alter life forms of plants and animals (including humans), it is regarded as one of the greatest and most controversial achievements of biological science. At the heart of this technology, DNA can be inserted into a variety of bacteria, yeast, and animal cells, where they replicate and produce useful proteins.

Monoclonal antibodies, extremely pure antibodies, may increase farm productivity by allowing passive immunization of calves against scours; replacing traditional vaccines, antitoxins, and antivenoms; facilitating the identification of the sex of livestock embryos; improving the detection of food poisoning; targeting and killing of cancer cells; and preventing rejection of organ transplants.

GROWTH HORMONES, GENE TRANSFER, AND MILK PRODUCTION

One application of biotechnology that generated controversy is the manufacture of growth hormones for injection into animals to increase their productive efficiency. Injections of bovine growth hormone into dairy cows have resulted in increases of 10 to 40 percent in milk production. Of greater importance for new generations, genes for new traits can be inserted into the reproductive cells of livestock and poultry, thus permanently endowing future generations with the desired traits, such as disease resistance or rapid growth. Some individuals have predicted increases in milk production at nearly 4 percent per year as this technology progresses. Actual increases have been approximately 2 percent per year in the 1980s and 1990s.

Embryo transfer is used for rapidly upgrading the quality and productivity of live-stock, particularly cattle. In the process, a female is artificially inseminated, and the resulting embryos are removed for implantation in surrogate mothers. Before implantation, the embryos can be sexed with a monoclonal antibody, split to make twins, fused with embryos of other animal species, or frozen in liquid nitrogen for storage until the estrous cycle of the surrogate mother is in synchrony with that of the donor. Vigorous pursuit of research in these areas could lead to marketing large numbers of genetically improved embryos containing genes that improve fertility and fecundity. The results might include improved rates of gain among meat animals with lower body fat, improved carcass characteristics, increased milk production, and increased resistance to disease.

BIOTECHNOLOGY PRODUCTS

Other new products will likely be used to diagnose and treat diseases that have reduced the productivity of livestock and poultry on American farms by an estimated 20 percent annually. For instance, a better vaccine has been genetically engineered for hoof-and-mouth disease in cattle, which is still a serious disease throughout Africa, South America, and parts of Asia and recently in Europe. Advances in using genetically engineered interferons that regulate the body's immune response can reduce viral infections. Genetic engineering of the microorganisms that permit ruminants—cattle, sheep, and goats—to digest forages may permit these livestock to digest feedstuffs that were formally virtually useless to them.

The use of genetically modified crops has become a major controversy in the twenty-first century. One major development was varieties of corn, cotton, and potatoes that produce one or more proteins that originated in *Bacillius thurengiensis.* These proteins are toxic to the corn borer (and other Lepidopteran insects), a pest that can do major damage. The other is the development of new varieties of soybean that carry a modified growth-regulating enzyme that is not affected by glyphosate, a herbicide that kills most plants but degrades quickly. With the new soybean variety, glyphosate can be sprayed on the growing crop killing all weeds without harming the soybean plants.

There has been considerable opposition to some of these genetically modified crops. Some opponents argue that it is unethical to manipulate the genetic code of any species, and that if we learn to manipulate the genetic codes of animals we will be more likely to manipulate the human code, which they find unacceptable. Others fear that there might be unintended consequences of developing and releasing new species of plants or animals. Also, some oppose these developments because they fear adverse effects on family farms. For example, there was strenuous opposition to the use of bovine growth hormone to increase the production of dairy cattle when it was first introduced. Some opponents express concern about the effect on the cows and the viability of family dairy farms but the major lingering concern is from those who fear consumption of milk products from treated cows. In the cases of GMO corn and soybeans, concern (most strenuous in Europe) is primarily focused on food safety and has resulted in calls for banning the technology or labeling all food products produced from grain produced using this technology.

The U.S. dairy industry has operated for years under a price-support program that has resulted in large surpluses of dairy products. Clearly if milk production per cow increases by 20 to 40 percent, fewer cows and fewer dairy farms will be needed to supply the nation's dairy needs. Furthermore, it is clear that large farms with high levels of management find it easier to use bovine growth hormone (bGH) than smaller and less well managed operations. Thus the adoption of this production technology has the same effect of increasing farm size, as did many previous technologies adopted over the last several decades. Fewer resources are used to produce an agricultural commodity, meaning fewer but more productive farmers.

Information Technology

For the farmer of the early 1900s, the development of the radio gave unprecedented access to price and other market information and helped to make agricultural markets more competitive. Scientific advances in information technology since then have made major changes in American agriculture and are expected to have even more dramatic effects in the future. New information technologies in various stages of development will greatly improve management and increase productivity of the biotechnologies. For example, on-farm communication and information technologies can include a central computer system, a radio link with operators of tractors, combines, and other implements, all tracked with the global positioning system, and an on-farm weather station. Many processes in plant and animal production are organized and monitored by computers. Some systems are designed to provide more information to the farm operator. Others operate automatically, based on a continuing flow of information; examples include devices to regulate the flow of irrigation water, monitor and control pests, meter the rate of application of fertilizer as equipment moves across the field, and regulate the rate of livestock feeding and the environment for livestock. Other applications include devices to identify individual animal performance for feeding, reproduction, and genetic improvement. Newer systems are providing rapid analysis of pesticide and field management information, reports of new or unknown pests, general pest survey information, and specified field locations with pest severities.

New telecommunication systems enable farmers to have rapid, inexpensive, and reliable access to information on markets, weather, and a wide array of other subjects. This provides information crucial for improved management decisions. Remote-sensing technology will be used increasingly to detect, process, and analyze data on weather and crop prospects, providing an array of information useful for production and marketing decisions.

The effect of new technologies on agricultural development and growth depends on how effective they are in increasing productivity, and on how profitable they are for farmers to use. In other words, the effect they will have in sustaining or increasing agricultural development and growth depends on the growth in demand for commodities they help to produce, and their effect on production costs and supply. This reinforces the importance of international trade for national prosperity and for continuing expansion of agricultural markets.

SUMMARY

The foundation of agricultural development in the United States has been (1) the development of physical and institutional infrastructure, (2) the transfer to individuals of large amounts of free land, and (3) support for agricultural research and education. In contrast with the development of American agriculture in the nineteenth century, in the twentieth century development became almost totally dependent on the successful application of science and technology, and the growth of an efficient agribusiness sector. Lastly, the development of a global market and liberalization of agricultural trade has become more essential for sustaining agricultural development in the future.

LOOKING AHEAD

In Chapter 17, we explore some major issues in U.S. farm price and income policy. This chapter deals more specifically with the economics of price supports, government payments, and other subsidies to farmers, and some additional policy issues in international trade.

IMPORTANT TERMS AND CONCEPTS

Agricultural Credit Act   409
agricultural credit policy   409
Agricultural Marketing Act   409
biotechnology   414
Capper-Volstead Act   411
educational policy   408
Engel's Law   400
Farm Credit System (FCS)   409
Farmers Home Administration
  (FmHA)   410
Federal Farm Loan Act   409
Food and Agriculture
  Organization (FAO)   401
food safety policy   411
genetically modified organisms
  (GMOs)   415
Grain Futures Act   411
Hatch Act   409
homestead policy   406
induced innovation   413

information technology   417
institutional infrastructure   402
Interstate Commerce
  Commission   410
irrigation policy   406
land policy   405
land retirement policy   408
market modification   404
market participant   403
Morrill Act   408
Packers and Stockyards Act   411
physical infrastructure   401
public land policy   407
public policy   399
settlement policy   405
Smith-Hughes Act   409
Smith-Lever Act   409
structural transformation   400
World Bank   401

QUESTIONS AND EXERCISES

Name That Term

Read the following sentences carefully and fill in the missing term or terms.

1 _____ involves the plans and decisions that people make in creating and guiding governments in solving problems that require a public decision.

2 _____ involves the government paying farmers to withhold land from production as part of agricultural production controls and erosion control programs.

3 The _____ established the land-grant college system in the United States.

4 The _____ provided a grant of 160 acres to each family who settled on the land and lived there for five years.

5 The _____ authorized farmers' cooperatives to issue capital stock and engage in interstate trade.

6 _____ of an economy in the process of economic development involves a relative decline in employment in agriculture, in the proportion of economic activity in the agricultural sector, and in expenditure of food relative to nonfood goods and services.

7 _____ says that the share of food in consumption expenditure declines as incomes rise.

8 One of the important factors determining the direction of effort in science and technology is the concept of _____.

9 _____ is a term that refers to any technique that uses living organisms to make or modify products, to improve plants or animals, or to develop microorganisms for specific uses.

10 The products of rDNA technology are commonly termed _____.

True/False

Read the following sentences, then decide whether each statement is true (T) or false (F) and mark it accordingly.

T   F   1 Economic development involves increasing expenditures for food relative to nonfood goods and services.

T   F   2 Agriculture's share of GDP tends to be larger in countries with higher GDP per capita.

T   F   3 In the early 1800s a policy was adopted forcing almost all Native Americans off of land east of the Mississippi River.

T   F   4 The Federal Reclamation Act of 1902 reclaimed for the federal government much of the land given to settlers after the Homestead Act of 1862.

T   F   5 The federal government owns substantial amounts of land, particularly in the western states.

T   F   6 Starting in the 1930s, the federal government subsidized REA cooperatives to bring electricity to farms and rural areas not served by the private sector.

T   F   7 In the United States, individual property owners have full control over what they do with their land.

T   F   8 In several countries in the world, governments are the only active participants in the market for goods and services.

T   F   9 The Ordinance of 1785 provided land in each township that would be used to support elementary and secondary education.

T   F   10 The amount of U.S. foreign assistance per capita or percent of GDP is quite low as compared to most other industrialized nations.

Multiple-Choice Questions

Circle the letter of the response that best answers the question or completes the statement.

1 Which of the following is characteristic of countries with low incomes?
   a low GDP per capita
   b most labor resources are employed in agriculture
   c agriculture's share of GDP is high
   d food expenditures account for a large share of total expenditures
   e all of the above

2 Which of the following are things a government can do to assist in the process of economic development?
   a provide roads, harbors, bridges, and other pieces of physical infrastructure
   b enforce rules and regulations
   c provide utility, transportation, and other services that might be provided by private firms
   d alter market outcomes
   e all of the above

3 Which of the following was not a reason for American farmers after the Civil War to join the Populist Revolt in hopes of gaining support for agriculture?
   a a strong dollar
   b monopolistic structure in the industries that served the farm sector
   c low import tariffs of industrial goods
   d weak demand for farm exports
   e low farm prices

4 Which of the following has been an element of the federal government's land policy through U.S. history?
   a settlement of agricultural land by private farmers
   b control of land use to achieve environmental objectives
   c management of land to achieve income objectives
   d creation of large amounts of public lands
   e all of the above

5 Which of the following is an objective of current U.S. land-management policy?
   a support of farm incomes
   b increased agricultural productivity
   c conservation of land resources and protection of the environment
   d increasing agricultural exports
   e all of the above

6 Which of the following founded the cooperative extension system for agriculture and home economics?
   a the Morrill Act of 1862
   b the Smith-Lever Act of 1914
   c the Smith-Hughes Act of 1917

    d the Capper-Volstead Act

    e none of the above

7 Which of the following is not an example of physical infrastructure?

    a roads

    b legal restrictions

    c harbors

    d railroads

    e none of the above

8 A well-functioning system of credit markets

    a is crucial to the efficient transfer of capital resources from savers to users.

    b provides short-term financing of farm production operations.

    c provides long-term financing of agricultural real estate.

    d is important to agricultural development.

    e all of the above.

9 The Interstate Commerce Commission

    a provided for controlling railroad rates and services in ways desired by farmers and other shippers.

    b was created in response to farmer's complaints of high freight rates and poor service.

    c resulted in part from the transport monopoly enjoyed by railroads in the early years of their development.

    d is an example of market modification by the government.

    e all of the above.

10 The Capper-Volstead Act

    a is sometimes called the "Magna Carta of cooperative marketing."

    b authorized farmers' cooperatives to issue capital stock.

    c authorized farmers' cooperatives to engage in interstate trade.

    d prohibited the cooperatives from violating antitrust laws by combining or conspiring in restraint of trade.

    e all of the above.

Questions for Thought

1 Most of the developed countries have given agriculture considerable help in developing infrastructure, and some have subsidized farm prices and incomes. What is the basic reason or rationale for this combination of policies? Why have the less-developed countries not done the same?

2 List the four major roles of government in development and give at least one example of each that is not mentioned in the text.

3 It has been suggested that rich nations tend to have a farm problem while poor nations have a food problem. Is this true? Why, or why not?

4 How did U.S. agricultural development policy change between the era of land settlement which ended in the early twentieth century, and the era that followed?

5 How does the contemporary setting of agricultural development in America differ from the setting of past policy development?

6 Proponents of biotechnology support it because of its potential to increase productivity. Negative publicity about this technology may cause consumers to doubt wholesomeness of the products produced. Assume biotechnology's proponents are correct in their assertion and opposition does lead to a reduction in demand for a new product. What would happen to the price and the quantity of the product produced? What would happen to the number of farms producing the product?

7 What effect do you think economic development in other countries has on U.S. agriculture? Is it possible that the increased income in the developing country means increased demand for U.S. products, including food?

ANSWERS AND HINTS

Name That Term   **1.** Public policy; **2.** Land retirement; **3.** Morrill Act; **4.** Homestead Act; **5.** Capper-Volstead Act; **6.** Structural transformation; **7.** Engel's law; **8.** induced innovation; **9.** Biotechnology; **10.** genetically modified organisms

True/False   **1.** F; **2.** F; **3.** T; **4.** F; **5.** T; **6.** T; **7.** F; **8.** F; **9.** T; **10.** T

Multiple Choice   **1.** e; **2.** e; **3.** c; **4.** e; **5.** e; **6.** b; **7.** b; **8.** e; **9.** e; **10.** e

17

AGRICULTURAL PRICE AND INCOME POLICY

OVERVIEW

There are few (if any) countries in which market forces alone determine the prices received by farmers and paid by consumers; and market forces alone do not dictate the amount and quality of agricultural products to be produced and sold or stored for the future. Most developed countries use an array of policies and programs to support farm prices and incomes, to stimulate or regulate production and marketing, and to promote or restrict international trade in agricultural commodities. In this chapter, we examine policies used by the United States to influence agricultural prices and incomes. We start with a brief history of American agricultural price and income policy.

LEARNING OBJECTIVES

This chapter will help you learn:
- The reasons for agricultural price and income policy
- The foundations of modern American agricultural price and income policy
- How to determine consequences of agricultural price and income policy
- The major elements of contemporary policy

POLICY FOUNDATIONS

We have agricultural price and income policies because society has decided that without them, some outcomes of a market economy are not desirable. There are many economic, social, and political justifications for implementing agricultural price and income (or any other) policy. Examples of policy goals include:

- Ensuring an adequate supply of food
- Saving family farms
- Stabilizing agricultural prices
- Protecting the resource base and the environment
- Maintaining economic viability of rural businesses and communities
- Minimizing dependence on food imports
- Expanding agricultural exports
- Promoting political stability in agriculture and the rural community

Seldom does any policy receive unanimous support. Policies are chosen to achieve some combination of goals, often by compromise among the proponents of alternate approaches. To appreciate why certain policies are used today one must understand the conditions that existed when these policies were adopted. Since the basic structure of current American policy has evolved throughout the twentieth century, we examine conditions that existed as the various policies were developed.

Economic Setting for Policy Formulation

Supply and demand conditions for agricultural commodities differ from those of many other products. Once incomes have reached a moderate level, demand for food is both price- and income-inelastic. Although there can be quite dramatic shifts among types of food, individuals buy about the same quantity despite price and income changes. Thus, the U.S. demand for food grows slowly as population increases.

The twentieth century saw dramatic outward shifts of the agricultural supply curve, initially as cultivated area grew and later from the technological changes discussed in the previous chapter. Short-run agricultural supply is price-inelastic. It is also subject to shifts caused by economic and technological factors, as well as by changes in weather and other natural phenomena.

Figure 17-1 illustrates the structure of many agricultural markets. In the short run, both demand and supply curves are inelastic. In a closed economy, a shift of either curve

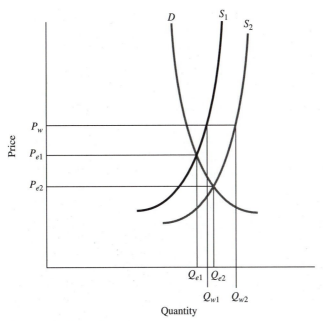

FIGURE 17-1   Agricultural demand and supply.

causes a large change in the equilibrium price and a small change in the equilibrium quantity. For example, if bad weather shifts the supply curve inward, price rises substantially, but the equilibrium quantity falls only slightly. If the economy is open and domestic prices are determined primarily in international markets, a shift in either curve would have little effect on the domestic price. Instead, the changes would alter quantities traded and either domestic production (if the supply curve shifts) or domestic consumption (if the demand curve shifts), or both.

The first twenty years of the twentieth century were exceedingly prosperous for American agriculture (the "golden age"). Domestic demand for food was shifting outward rapidly because of rapid growth of industrial employment, and population growth was at an all-time high, with high birth and immigration rates. Furthermore, foreign demand for American agricultural goods was strong. In the 1910s, farm exports accounted for about 20 percent of farm income. Agricultural output was also growing rapidly as homesteaders cultivated new lands. However, demand growth was more rapid than supply growth, and farm incomes soared.

World War I enhanced the already good times for farmers. Foreign demand for American agricultural exports, partly financed by the U.S. government, skyrocketed during the war. The nominal value of agricultural exports reached a level in 1920 not achieved again until 1951. The high profits in farming resulted in attempts by farmer owners to expand their farms and by others to enter farming. Since land supply was inelastic, land prices were bid up sharply. In the most productive farming areas, the price of land more than doubled between 1914 and 1920.

The end of World War I also marked the end of the good times for American agriculture. Foreign demand for agricultural products fell, as did the prices of most agricultural products. Some major export crops, such as wheat and cotton, were hit especially hard. Most of the farm sector experienced severe financial difficulties. Few farmers made a profit. Many farmers who had borrowed to purchase land expecting continued high prices could not meet their payments and were soon in or near bankruptcy. Agribusinesses that supplied variable farm inputs were losing money, and some of them also faced bankruptcy.

As we noted in Chapter 16, the American government implemented policies in 1922 and 1923 to improve the functioning of livestock and grain markets, to help agricultural cooperatives, and to provide an additional source of short-term credit for farmers and some agribusinesses. However, these policies did not offset the fundamental changes in international markets that caused farm prices and incomes to fall.

From 1924 to early 1928, some influential farm and agribusiness leaders pushed for an export subsidy program for major agricultural crops—primarily grain, cotton, and tobacco—to address the decline in farm income. No action was taken during that period. The Great Depression of the 1930s further worsened an already bleak picture. Incomes around the world fell dramatically, and demand for American exports plummeted. While domestic demand for farm products fell, foreign demand for American agricultural products fell even more. The value of agricultural exports was cut in half. By the beginning of the 1930s, exports accounted for only about 10 percent of farm income.

When foreign demand was strong, it helped to stabilize domestic agricultural markets. However, restrictions on international trade helped to bring on the depression, and the stabilizing effects of foreign markets were reduced. The United States became more like a closed economy. Agricultural prices became more susceptible to price fluctuations caused by inelastic domestic demand and supply. Supply increases without corresponding increases in demand tended to bring greater price declines because export markets were no longer available. In addition, in some major crop areas, unusually severe crop failures because of drought and insects reduced farm output. The drastic decline in agricultural incomes and increased variability in agricultural prices led the U.S. government to become more heavily involved in managing agricultural prices and farm incomes.

A Failed Policy   In 1929, Congress passed the Agricultural Marketing Act. The act created thirteen new banks for cooperatives (discussed in Chapter 16), established the Federal Farm Board, and authorized $500 million to implement a new marketing program under the board's direction. The $500 million was to be loaned to cooperatives through the new cooperative banks to hold farm commodities off the market to raise prices.

The act instructed the Federal Farm Board to use this money to prevent surpluses from causing undue fluctuations or depression in prices. No provision was made to control production, to subsidize domestic demand, or to subsidize exports. The board carried out its broad mandate by lending money to cooperatives to purchase and store commodities, primarily wheat and cotton. The board's activities resulted in a temporary outward shift in the demand curve, as suggested in Figure 17-2. The market price was higher than it would have been without the board's program, which meant some

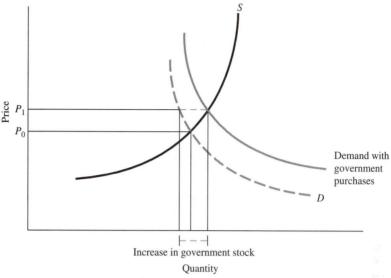

FIGURE 17-2   Effects of government purchases.

increase in production and some decrease in the quantity demanded. However, once the $500 million was expended, the board had no further power to influence markets or to reduce supply, and demand fell back to D. The government did not allocate additional funds to continue the program. Also the question of how to dispose of the government stocks without further depressing the market price became an issue. The board concluded that prices could not be supported for long without production controls.

Foundations of Contemporary Policy 1930–1990

The Agricultural Adjustment Act of 1933 was adopted after intense discussion among farm and agribusiness leaders, members of Congress, and finally the newly elected president, Franklin D. Roosevelt. Production control, in the form of a new acreage allotment program, was at the heart of this act. Additional legislation over the next three years introduced an astounding array of provisions, including most of the components of current price and income policy. However, in 1936 the Supreme Court held that portions of this legislation were unconstitutional, including the processing tax used to fund the program. A new act was adopted that included support from the U.S. Treasury. Compliance with acreage restrictions was generally voluntary, but farmer eligibility for government price support and income payment programs became dependent on meeting the requirements of each program. Furthermore, if two-thirds of the producers approved, the secretary of agriculture was authorized to invoke marketing orders and impose penalty taxes on producers who planted more than their allotted acreage.

The policy initiatives of the 1930s laid the foundation for almost all agricultural price and income policy developed subsequently. Since low farm incomes and unstable prices were seen as the key problems, the central policy goals were to raise farm income and make prices more stable. Two types of policy instruments were used to meet these goals: (1) production control to shift the supply curve inward and raise prices, and (2) price supports to raise and stabilize farm prices.

Since the 1930s, there has been almost continual pressure for change in agricultural price and income policy. During the 1950s and the 1960s, the basic instruments of policy did not change much, but the level of support varied as market conditions changed.

AGRICULTURAL POLICY PROVISIONS, 1933 TO 1936

- Voluntary acreage controls, tariffs on some imports, surplus disposal program for hunger relief, and a processor tax to finance the acreage controls
- A program to pay farmers to plow up cotton
- Mandatory acreage allotments to control tobacco and cotton production
- Use of 30 percent of the customs receipts to promote exports, domestic consumption, and industrial uses
- Establishment of the Commodity Credit Corporation to operate a nonrecourse loan program for major crops
- Mandatory controls based on the support of two-thirds of the farmers in a referendum
- Tobacco acreage allotment program by type of tobacco
- Marketing agreements which implemented controls without the use of a tax
- A rice production control program that operated through the rice millers
- Price supports with mandatory production for milk and some fruits and vegetables
- Tariffs and quotas on sugar imports
- Standards for agricultural labor in sugar production
- Distribution of surplus food to the poor

The conflicts concerning agricultural price and income policy moderated greatly during the 1970s, largely because of three changes outside agriculture. These three factors worked together to improve farm incomes.

1 The U.S. government abandoned the gold standard and allowed the value of the dollar to be determined in world financial markets. As a result, the dollar declined, reducing the cost of American commodities and increasing the quantity demanded for export.

2 The former Soviet Union changed a key agricultural policy. Instead of allowing livestock numbers to fluctuate with domestic feed supplies, it began to supplement domestic feed production with imports. This resulted in substantial growth in Russian grain imports.

3 During the 1960s and 1970s, several developing countries and Japan had rapid income growth, and this increased their demand for meat and for imports of American feed grains.

The events of the 1910s were replayed in the end of the 1970s. Rapid growth in demand for agricultural products translated into higher commodity prices and, in turn, higher land prices. Increased funds available through the farm credit system made some of the purchases possible. As a result, the nation entered the 1980s with an inflated land market, inflated production costs, and high prices for farm products.

The macroeconomic events of the late 1970s and early 1980s described in Chapter 14 caused record-high interest rates, and increased demand for dollars by investors caused the value of the dollar to increase. Recession in many parts of the world reduced foreign demand for agricultural commodities and other exports. High domestic support prices and an overvalued exchange rate exacerbated the effect on agriculture. The result was that American farm commodities were priced out of the import markets of many countries. Surpluses began to accumulate, and market prices fell well below support price levels. Low market prices, combined with continuing high interest rates, resulted in a crash in farmland values and put many farmers in serious financial difficulties.

We turn now to the major policy provisions. American agricultural policy has become extremely complex. There are special provisions for many crops, with differences in details. However, the key instruments of U.S. agricultural policy in the early 1990s were production control through land set aside, target prices, nonrecourse loans, and deficiency payments. We next examine these major farm policy provisions. While it seldom occurs that all of these provisions are in effect at the same time, these are illustrative of the type of policy options typically considered.

Programs to Reduce Production

The use of production controls was a central part of the agricultural legislation of the 1930s, and major portions of that legislation remained until the mid 1990s. A key element is the land set-aside program, used primarily with wheat, corn, and cotton. Participation is voluntary, but a farmer cannot receive federal payments without participating.

Each year, the federal government indicates how much land must be held out of production, or **"set aside,"** to reduce supply enough to raise the predicted domestic price to a level set by Congress or the administration. Each farm has a **base acreage** that is defined on a historical basis. If the farmer participates in the program, the amount of land that can be planted is the farm's base acreage less the percentage to be "set aside." For an individual farmer, the decision to participate depends on how much land must be taken out of production, the expected price, and government payments. Figure 17-3 illustrates what will occur if there is no trade. The domestic supply curve shifts to the left, from S_1 to S_2, and the domestic price rises from P_1 to P_2. With a given demand and no trade, the domestic price increases, and the quantity demanded domestically falls.

The reason S_2 is drawn as a broad line is that **slippage** occurs. If the government decides that 20 percent of the land should be set aside, the supply curve will shift inward by somewhat less than 20 percent. Slippage occurs for several reasons.

- Participating farmers take their least productive land out of production, resulting in a less than 20 percent reduction in output with a 20 percent reduction in planted acreage.
- Some farmers do not participate in the program, and some of these may increase the amount of land they plant.
- With higher expected prices farmers use more variable inputs.

This effect is illustrated for the individual producer in Figure 17-4. In long-run equilibrium, without government intervention, the farmer would operate at Q_e, P_e. The set-aside program shifts the cost curves from ATC_1 to ATC_2 and from MC_1 to MC_2. If the market price increases to P_c, production shifts to Q_c. The firm and the industry are not in a long-run equilibrium because excess profits are being realized. Over time, the attempts

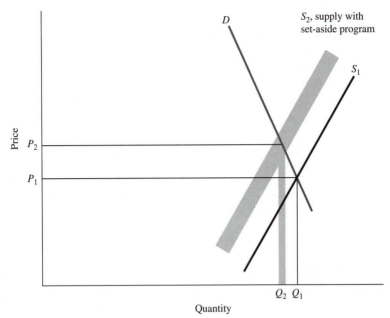

FIGURE 17-3 The effects of area restrictions to reduce supply.

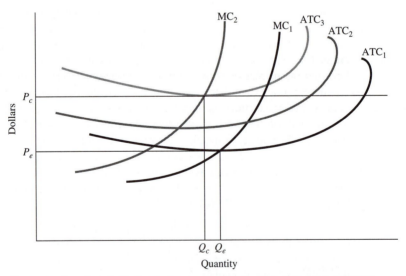

FIGURE 17-4 Cost curve for a hypothetical grain farmer under a program of acreage reduction and price supports.

of individuals to enter the industry or to expand operations will bid up the price of land to ATC_3 where the cost of production equals the target price. Thus the major beneficiaries of the program are the owners of the land when the program is first implemented.

When international trade is not a strong factor in determining domestic prices, the domestic price increases if cultivated area falls, *ceteris paribus*. However, for any one farmer, reducing the area in crops generally results in lower profit. A change in one farmer's production has no effect on price. Therefore, since participation is voluntary, a farmer must be given an incentive to participate. This incentive must be enough to offset the profit lost by keeping some land out of production. Clearly a farmer will not participate in an acreage allotment or set-aside program unless the net effect on expected income is positive. These incentives have taken two basic forms: nonrecourse loans and deficiency payments.

Programs to Encourage Participation

For a production control policy to be effective, a means of ensuring participation is needed. For an individual farmer, the ideal situation would be to produce as much as possible while everyone else reduced production. To induce farmers to participate in production control programs, the government has used several different incentive schemes.

CROSS COMPLIANCE

Recent agricultural acts have increased the **cross compliance** provisions of the legislation with which a farmer must comply to be eligible for price and other support payments. An important example is that farmers operating on erodible land must develop and implement a soil conservation program.

Parity Prices   In the 1920s, farm prices fell more than nonfarm prices, and the goal of parity of farm with nonfarm prices emerged in the farm community. The administrators of the Agricultural Adjustment Act of 1933 interpreted the **parity** goal as establishing agricultural prices that would give farmers the same purchasing power per unit of output as in the period from 1910 to 1914. Since that time there has been a continuing debate on the appropriateness of the parity concept, and it has been criticized on several grounds. To begin with, parity is backward-looking. Parity prices are based on conditions in a base period, not current or anticipated market conditions. As farm size increases over time, the profit per unit necessary to generate an equivalent level of income decreases. Also, as new cost-saving technologies are adopted, the price levels necessary to generate acceptable levels of farm income decreases.

A related approach has been to focus on parity incomes instead of parity prices. However, since income support programs typically operate through support of prices, the difficulties in establishing appropriate prices remain. Because of these difficulties, parity has been discontinued as the basis for most farm price and income goals, although the concept is supported by some.

Nonrecourse Loans   One of the first incentives used to encourage participation in acreage reduction programs was a **nonrecourse loan.** The government sets a floor price called the loan rate—a fixed price per unit of grain, for example. Farmers can borrow from the government using grain at the floor price as collateral. If the market price is below the loan rate, the farmer can default on the loan and the government takes possession of the grain. (The government has "no recourse;" it must take the grain.) Hence, farmers have

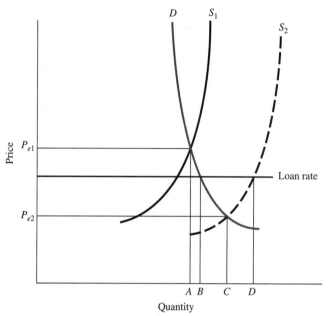

FIGURE 17-5 Effects of the nonrecourse loan program.

the choice of receiving the market price or the loan rate. There are two principal effects of the nonrecourse loan program (see Figure 17-5). First, a floor is put under the price farmers receive and therefore under the market price. For example, if good weather shifts the supply curve outward, the market price in a closed economy would normally fall from P_{e1} to P_{e2}. With the nonrecourse loan program, if the market price is below the loan rate, farmers can default on their loans and receive the loan rate for their grain. Second, government stocks increase by the amount BD. If the supply curve shifts back in a future period, the government can sell its stocks at the higher market price. However, if the loan rate is generally above the equilibrium market price, government stocks will continue to grow.

If the economy is open, the loan rate still acts as a floor price. In this case, it protects against large drops in the world price due to shifts in world supply and demand.

Target Prices and Deficiency Payments

The current policy instruments of **target prices** and **deficiency payments** were introduced on a limited basis in the Wheat-Cotton Act of 1964. Farmers participating in the program received a certificate for the portion of their crop that they could sell in the domestic market. This certificate could be redeemed for a payment equal to the difference between a target price set by the government and the market price. For example, suppose 70 percent of the wheat crop was projected to be sold in the domestic market in a given year and that the U.S. Department of Agriculture (USDA) set a target price 50 cents per bushel above the market price. Then a farmer would receive a payment equal to 50 cents for 70 percent of the quantity produced. Of course, participating farmers may be required to take a specified percentage of land out of production to be eligible.

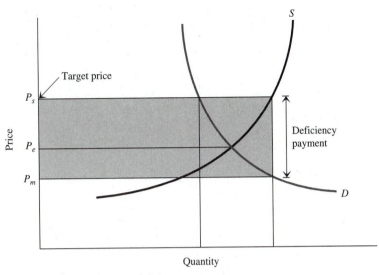

FIGURE 17-6   The effects of a target price-deficiency payment program.

Target prices and deficiency payments have become the mainstay of price and income support programs since the 1970s when the USDA declared that American farm policy would begin using a market approach. In the early 1990s, although the nonrecourse loan provisions were not removed, loan rates were lowered substantially.

To implement a target price program, the government chooses a desired or target price prior to the time crops are planted. If the market price after harvest is above the target price, the government does not become involved in the transaction. However, if the market price is below the target price, the government pays the difference between these two prices as a deficiency payment.

The effects of this policy differ dramatically from those of the nonrecourse loan (Figure 17-6). In both cases, farmers receive a price higher than would have prevailed without the program. But with the deficiency payment, the market price, which might be the world price, can be well below the target price. Consumption increases because of lower market prices. Instead of the costs of dealing with government stocks, there are the costs of deficiency payments.

PAYMENT LIMITS

One of the politically sensitive aspects of the price support program is the size of payments received by large farms. Without limits, some large farm operations may qualify for millions of dollars of payments. As a result, the law specifies maximum payments per farm. One consequence of these limits is to restrict the effectiveness of the supply control aspects of the program, because the largest farms are not eligible to participate to the fullest possible extent. However, this effect is minimized by the fact that many operators of large farms find ways to get around such limits through changes in land ownership or leasing contracts.

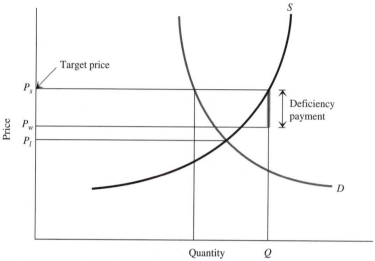

FIGURE 17-7   The effects of a target price-deficiency payment program with world prices.

Some crops have had both deficiency payment and nonrecourse loan programs in effect. The combined effects depend on the locations of the domestic demand and supply curves as well as the levels of the target price, the loan rate, and the world price. For example, in Figure 17-7, the world price is below the target price and above the loan rate. In this case, farmers participating in the program by reducing acreage would receive a deficiency payment equal to the difference between the target price and the world price. However, the payment would be made only on a base yield, established by the USDA and somewhat less than the expected actual yield.

SUPPLEMENTAL INCOME SUPPORT

Early in 2001 the Commission on 21st Century Production Agriculture proposed that congress consider a Supplemental Income Support (SIS) program. Under this program, if gross farm income fell below the five year moving average by more than a set amount, farmers would receive federal support payments. The commission estimated that this approach would be less costly to the government than crop-specific target-price programs.

American farm legislation has included many other programs that have some effect on market prices and farm incomes. Next we examine a few of these programs.

SOIL BANK AND CONSERVATION RESERVE PROGRAMS

Most acreage reduction programs include a goal of maintaining the agricultural resource base and reducing environmental degradation. Both the Soil Bank Program (SBP) of the 1950s and the Conservation Reserve Program (CRP) begun in the 1980s emphasized long-term land retirement and soil conservation. The primary objective of the soil bank was to maintain soil fertility by controlling erosion. The primary objective of the CRP is to reduce sedimentation of lakes and streams, again through erosion control. In each program, the government makes cash payments to farmers who agree to take land out of production.

Under the soil bank program, payment rates were established at the local level, and farmers decided whether to participate by comparing the payment rates with the expected profit level from crop production. In a few areas, many farmers put the entire farm into the program. One result was that agribusiness firms in those areas suffered because their sales were reduced. Subsequently, the program was modified to limit the amount of land taken out of crop production in a given locality. Actually, it was funded at less than half of the targeted goal in the original legislation. Therefore, the SBP did not reach its production targets and was deemphasized in favor of the CRP.

To participate in the CRP, farmers submit bids to the government indicating the annual payment for which they would agree to a 10-year contract to take some of their land out of production and control erosion. The government accepts as many bids as necessary to meet its acreage reduction targets. The bidding process has been carried out several times since the program was first implemented in 1985, allowing the government to change the payment levels to attract land into the program. As the program developed, other objectives were added and the criteria for eligibility modified accordingly, as discussed in Chapter 15.

COMMODITY-SPECIFIC ALLOTMENT PROGRAMS

Commodity-specific allotment and quota programs provide income support to producers of tobacco, peanuts, sugar beets, and sugarcane. The tobacco program is the only government program that has consistently placed a quota on land that can be used for a crop. It has been very effective in stabilizing tobacco prices at levels acceptable to farmers. To control the quantity of tobacco produced, the government determines the acreage that may be planted and it limits imports. Farmers may not produce tobacco for sale without an **acreage quota.** The cost of the quota is either an explicit or an implicit cost of production to the farmer.

Tobacco producers can buy and sell production quotas separate from the land, thereby creating a market for the quotas. The operation of the market for quotas has several important economic implications. First, the quotas can be transferred to farms in areas with a comparative advantage in tobacco production and to farmers who are efficient tobacco producers. Second, the expected profitability of tobacco production determines the equilibrium price of the quota. If the government increases the tobacco support price, the value of the quotas increases. Quota owners at the time of an increase generally earn higher profits, or they can sell the quota at a higher price. Thus, a farmer who wants to start growing tobacco would have to pay a higher quota price. The separation of quota and land ownership makes more clear the fact that the benefit of an increase in the support price of a commodity is realized only by the current owner: here, the owner of the quota; and for other crops, the owner of the land.

The tobacco program and the peanut program are two among a few that are essentially compulsory. For example, both tobacco and peanuts have been placed under more stringent types of acreage allotments or quotas, generally with the strong political support of the growers. (In 1998, 97.5 percent of tobacco growers voted in favor of the quota/price support program.) In each case, the advantage for the grower to comply with the quota is so great, and the output of tobacco and peanuts is controlled so effectively, that the program for each commodity is the equivalent of a producer cartel. The cost to the federal government is minimized, and the higher grower prices are passed forward through the marketing system to consumers. As with more voluntary programs, each year the government must set certain goals and regulate the program accordingly. However, because of the public criticism of government expenditures for a price support program for tobacco (reflected in the No Net Cost Tobacco Act of 1982), producers are required to contribute to a fund that reimburses the government for any losses incurred in the operation of the program.

The U.S. government has intervened in the sugar industry almost since the founding of the nation. Until 1934 the sugar program consisted primarily of tariffs on imported sugar. From 1934 to 1974 the main policy instrument was a quota on imports. The quota system was eliminated in 1974, but changes in the world sugar market led to the reintroduction of a sugar program in 1981. Two principles guide the current program. First, it is structured to operate at no cost to the federal government. Revenues from other parts of the industry must offset any payments to producers or processing firms. Second, it should allow producers a reasonable rate of return. To achieve these goals, the program uses tariff-rate import quotas, acreage allotments, and supplementary deficiency payments to growers.

TARIFF-RATE QUOTA

Under a tariff-rate quota a certain amount of a product may be imported at a low or zero tariff. A prohibitively high tariff is set for any amount of imports greater than the "quota" limit. The effect is the same as a quota, in that imports are effectively limited to a fixed amount.

The sugar program has a substantial effect on the U.S. sugar market. Domestic sugar prices average some three to four times the world price. This high domestic price of sugar has encouraged the development of the high-fructose corn syrup industry, which competes with beet and cane sugar.

The annual net cost of the sugar program to American consumers varies, depending on the price gap between domestic and world markets. Annual estimates have varied from $2.5 to $5.0 billion. Translated to individuals, this would be $10 to $20 per person. In addition, sugar growers are encouraged to use domestic resources in an inefficient way. Instead of producing goods in which the United States has a comparative advantage, the resources are used to produce sugar that could be produced at lower cost elsewhere.

MARKETING AGREEMENTS AND ORDERS

The Agricultural Marketing Agreements Act of 1937 and the Agricultural Act of 1949 provide the policy framework for the operation of several marketing orders (Chapter 9) that control the production or marketing of various agricultural crops. The **Agricultural**

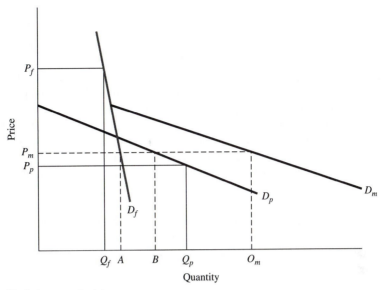

FIGURE 17-8   Market segment pricing.

Marketing Service of the USDA makes the marketing decisions for federal orders after input from producers, processors, and any other interested parties. Milk, tobacco, fruits, vegetables, and specialty crops are marketed under such provisions.

Market segmentation is one of the techniques used in many marketing orders. For example, a market might be divided into two segments: "fresh" and "processed." As suggested in Figure 17-8, charging different prices in the two segments of the market enhances revenues. Because the preferred fresh product is characterized by a less elastic or more inelastic demand curve (there are fewer substitutes), a higher price can be charged, $P_f Q_f$. The portion of the product that cannot be sold in the higher-priced fresh market is sold on the lower-priced processed market, $P_p Q_p$. Note that if all of the product were sold in an aggregate market, the demand curve would be D_m and the market would clear at price P_m. Somewhat more fresh and less processed product would be sold, as indicated by A and B on Figure 17-8. Clearly the total revenue would be less than if the product is sold in a segmented market. Milk is an example in which market segmentation increases revenues. Fluid milk has a very inelastic demand, and so a 10 percent increase in price might increase total revenue by significantly more than 10 percent. For manufactured dairy products (cheeses, etc.), a 10 percent increase in price might decrease total revenue because demand for these products generally is elastic.

Sometimes, the cooperative operating the marketing order can separate the market into three segments. For example, there might be two distinct segments in the domestic market, with different elasticities of demand, and an export market that has a more elastic demand than either of the two domestic segments. If sales in the market segment with the lowest price is below the desired minimum profitable price, producers will be more receptive to production controls. Many cooperatives operating marketing orders conduct aggressive advertising campaigns with the objective of shifting outward the demand curve for the high-priced segment or for both segments. If the cooperative is large enough to have a significant effect on market prices, it can manage its supply to increase total market receipts. It does so by reducing its sales in the inelastic domestic market and selling more abroad in the more elastic export market. Shifting demand such as D_{f1} outward would allow the sale of more of the product at the higher price. Total sales could be increased by producing more or selling less in the low-priced market.

Before turning to export subsidy programs, we note that there have been two domestic programs of aid to the poor that tend to increase the demand for agricultural commodities. First, the **Food Stamp Program**, is discussed next. The second program involves the actual distribution of surplus agricultural commodities to needy individuals. Cheese purchased in the operation of the dairy price support program is a commonly distributed product. Most of this distribution is accomplished through school lunch and similar programs. Millions of individuals participate in these programs, especially during a recession.

FOOD STAMP PROGRAM

The food stamp program is an important component of agricultural and welfare policy. The expenditures for this program are larger than any other component of the agricultural policy. Under this provision eligible low-income families are given free food stamps. To be eligible household net income must be below a level based on the poverty level (gross income less than 130 percent of the poverty level). The family also must meet work requirements and have minimal assets. These stamps are used as money except that they can be used only to purchase food. (Recipients are no longer given stamps that look like money. They are given an Electronic Benefits Transfer (EBT) card that is the equivalent of a bank debit card. This EBT card is used to purchase food.) The primary goal of the food stamp program is to eliminate, or at least reduce, the incidence of hunger in the United States. A secondary goal is to increase farm income by shifting outward the demand curve for food.

In 1975, 17 million individuals were served by the program at a cost of $4.6 billion. In 1995, 26.6 million were served at a cost of $24.6 billion. Due to good performance of the economy the number served fell to 18.2 million, at a cost of $21.2 billion, in 1999.

We can analyze the effects of the food stamp program on recipient's expenditure patterns using the budget line and indifference curve model we developed in Chapter 2. In Figure 17-9 the Jones family has a budget of B_1. Given their indifference curves, they choose point a and thus will purchase F_1 of food. If the Jones family is given food stamps in the amount of FS and their indifference curve is I_2 they will be able to achieve point b and will purchase F_2 of food and O_2 of other goods. In this case the effect is exactly the same as if the Jones had been given the support in cash.

With indifference curves I_3 and I_4 instead of I_2 the situation changes. If the Jones are awarded FS of food stamps their budget line shifts outward to $B_1 + FS$. Thus the Jones family would like to choose the combination of food and other goods indicated by point c. But, they cannot do so because they only have sufficient cash to purchase quantity O_4 of all other goods. In this case they will choose to consume at point d where they will

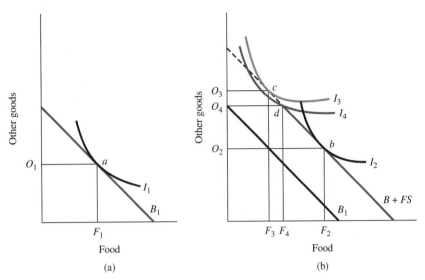

FIGURE 17-9   Effects of food stamps on expenditure patterns.

spend all of their cash to purchase quantity O_4 of all other goods and F_4 of food with the food stamps.

There are several implications to be drawn from this analysis.

• For families that receive a relatively small food stamp supplement to their income from other sources, the effect is the same as giving them a cash award. The consumption of both food and other goods is increased.

• For families for whom food stamps represent a major portion of their income, the effect is different from awarding cash. They are "forced" to purchase more food and less of other goods than they would have purchased under a cash award. In this case the family achieves a lower indifference curve or level of satisfaction.

• The food stamp program does increase the demand for food, especially for families for whom food stamps are a large portion of their disposable income.

This analysis also shows that the net effect of the food stamp program is to increase the consumption of other goods as well as the consumption of food. Some argue that this is inappropriate. They argue a means should be found of assuring that all of a food stamp allocation is spent on food, or at least that the net effect is not to increase the purchases of drugs, alcohol, cigarettes, or other such products.

EXPORT SUBSIDY PROGRAMS

Export subsidy programs for agriculture were rejected by Congress in the 1920s but gradually began to be used in the 1950s. The government had accumulated stocks of the major agricultural commodities through the operation of the price support programs and had no acceptable way of disposing of them. One of the largest subsidy programs was authorized by Public Law 480, the Agricultural Trade Development and Assistance Act

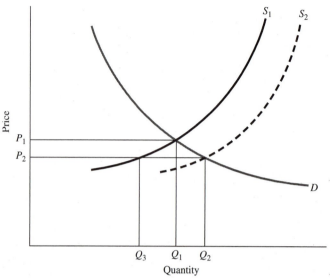

FIGURE 17-10   The effects of food aid on the recipient country.

of 1954. It was designed to finance the export of agricultural commodities to selected poor countries in need of food. This appeared to be the dream answer to a policy maker's problem because it addressed several objectives. American agriculture would benefit. Surpluses of commodities could be reduced. Poor people in other countries could be helped. Economic development could be promoted in the recipient countries, and they, in turn, could become stronger markets for American exports. The federal tax-payer would bear the cost, but reducing surpluses in government-financed storage would ease the cost.

However, there were two major drawbacks to the policy: the costs to the U.S. tax-payers and the negative effects on agricultural development in the recipient countries. The negative effects on the recipient country are suggested in Figure 17-10. The donated food shifts the supply outward from S_1 to S_2, making more food available as suggested by the shift from Q_1 to Q_2. However, prices also fall from P_1 to P_2. The reduction in prices results in less production by and income for domestic producers, which exacerbates the country's food and poverty problems.

EFFECTS OF FOOD AID ON A DEVELOPING COUNTRY

- Provides additional resources to an economy to support development
- Reduces malnutrition and hunger
- Reduces prices and profitability of agriculture
- Increases the potential for the LDC government to ignore the development needs of the agricultural sector

---

By the early 1970s, strong commercial demand and the low value of the dollar resulted in the liquidation of most government-held stocks. The value of foreign food aid financed by the American government fell to about one-tenth of the food aid given to domestic consumers.

In 1982, the United States implemented an **Export Enhancement Program (EEP)** to increase the quantity of products exported. Under the program, the federal government awards certificates redeemable in government-owned commodities. The effect is to allow the exporting firms to sell commodities to specified countries at prices below the U.S. market price. The export credit guarantee program assures repayment of private short-term credit for up to three years. This program stimulates exports by making more credit available to exporters and by reducing the risk of default normally faced by the exporting firm.

EU COMMON AGRICULTURAL POLICY OF VARIABLE LEVIES AND TARIFF-RATE QUOTAS

The European Union (EU) has used trade policy instruments to keep domestic prices for grains much higher than world prices. First, for most of its history a variable levy was placed on imports. This tax changed every day and was equal to the difference between the world price and the support price within the EU. As a result of WTO negotiations the variable levy system was replaced by a tariff-rate quota system. Now a limited amount of imports are charged a low fixed tariff and any additional amounts are taxed at a much higher rate.

MODERN U.S. AGRICULTURAL POLICY—THE 1990s

The title of the 1990 "farm bill"—the Food, Agriculture, Conservation, and Trade Act of 1990—reflects the comprehensive nature of the legislation. This legislation was replaced by the Federal Agriculture Improvement and Reform Act of 1996. It became known as the FAIR or the Freedom to Farm Act. The major provisions of these acts were:

- A food stamp program (both acts)
- Foreign food aid, credit sales, and development aid (both acts)
- A farmer-owned grain reserve program (eliminated in the 1996 act)
- Voluntary production controls for grains with price supports, market discretionary loans, target prices, and deficiency payments (eliminated in the 1996 act but reinstated in subsequent emergency legislation)
- Soybean price supports, marketing loans, an assessment fee, and oilseeds price supports (eliminated in the 1996 act but reinstated in subsequent emergency legislation)
- Cotton and rice price supports, target prices, and marketing loans (eliminated in the 1996 act but reinstated in subsequent emergency legislation)
- Transition payments for the period 1990 to 2002 (1996 act)
- Sugar price supports, import quotas, assessment fees, acreage limits, and no net cost provisions (both acts)

• Dairy price supports, assessment programs, and a mandate to consider a new program (both acts)
• Wool, mohair, peanut, tobacco, and honey price support programs (both acts)
• Payment limits of $50,000 in 1990 act, $75,000 in the 1996 act
• Sodbuster, swampbuster, conservation reserve (40 million acres), insecticide record requirements, water and food safety emphasis (both acts)
• Export credit sales, subsidies, and promotions (both acts)
• Subsidized credit for qualifying farmers (both acts)
• Comprehensive research program with emphasis on sustainable agriculture (both acts)
• Reforestation program and global warming studies (both acts)
• Rural development program (both acts)

The adoption in 1996 of the FAIR or the Freedom to Farm Act represented a major change in direction. The major change was the elimination of the target price/deficiency payment programs for wheat, feed grains, cotton, and rice. Producers are allowed to plant any amount of any crop except tobacco, peanuts, or fruits and vegetables. In place of the price supports the government provided cash "transition" payments for the years 1996–2002. These "transition" payments depend on pre–1996 plantings and yields, but not current plantings, yields, or market prices. The farmer-owned reserve was also eliminated.

Since the passage of this legislation, crop prices have fallen significantly due to the increased crop production allowed under the legislation, favorable weather for crop production, and the financial crisis in Asia that reduced export demand. The federal government responded by providing several billion dollars of emergency cash payments and increasing the cash payment limitation to $75,000.

The directions of future legislation will depend on many factors: the economic performance of the agricultural sector, the health of the American economy, the outcome of international trade negotiations, the philosophy of the administration and Congress, and the health of international economies, particularly the developing countries and their agricultural sectors.

SUMMARY

This chapter describes the basic goals of American agricultural price and income policies. It discusses the economic conditions that prevailed during the 1920s and 1930s, when the basic agricultural goals that have shaped U.S. policy were formulated. The chapter then describes and analyzes the major elements of the price support and income support provisions of these policies. Particular emphasis is placed on the price support, acreage reduction, and international trade provisions of the legislation.

LOOKING AHEAD

Having worked our way through microeconomics and macroeconomics, considered the role of agriculture and agribusiness, analyzed the role of economics in important policy issues, we next show how all of this ties together. In Chapter 18 we also return to several of the questions we asked in Chapter 1 and show how we can use economics to provide answers.

IMPORTANT TERMS AND CONCEPTS

acreage quota   435
Agricultural Act of 1949   436
Agricultural Adjustment Act of 1933   427
Agricultural Marketing Agreements Act
  of 1937   436
Agricultural Marketing Service   437
Agricultural Trade Development
  and Assistance Act of 1954   439
base acreage   429
Conservation Reserve
  Program (CRP)   434
cross compliance   431
deficiency payments   432
Export Enhancement Program   441

export subsidy   439
Federal Agriculture Improvement
  and Reform Act of 1996   441
Food Stamp Program   438
market segmentation   437
No Net Cost Tobacco Act
  of 1982   436
nonrecourse loan   431
parity   431
Public Law 480   439
set-aside   429
slippage   429
Soil Bank Program   434
target prices   432

QUESTIONS AND EXERCISES

Name That Term

Read the following sentences carefully and fill in the missing term or terms.

1 Under _____ pricing, farmers could receive a price for a unit of their commodity equal in purchasing power to a unit of that commodity in 1910–1914.
2 Government payments that reduce the price of a product for foreign buyers are _____.
3 _____ loans are loans to farmers whereby the farm commodity may be delivered to the government for full settlement of the loan.
4 _____ is the requirement that to qualify for price support and similar programs a farmer must also satisfy the provisions of other federal programs.
5 _____ are federal government funds paid to farmers when farm prices are below the target price.
6 To implement certain provisions of price support programs, the USDA uses a _____ acreage, the number of acres on a farm that were in crop production in a base period.
7 _____ acreage is the land a farmer must hold out of production in order to qualify for the federal price-support program and other benefits.
8 The _____ is an agency of the USDA that administers grading, inspection, market order, and promotion programs related to agricultural commodities.
9 The _____ Program in the United States permits low-income people to obtain coupons (for free or at reduced prices) that can be exchanged for food at retail stores.
10 Under the _____, landowners contract for ten years to receive annual payments for taking highly erodible land out of crop production and controlling erosion.

True/False

Read the following sentences, then decide whether each statement is true (T) or false (F) and mark it accordingly.

T   F   1 The end of World War I marked the beginning of prosperous times in U.S. agriculture.

T   F   2 Farmers' incomes are enhanced when federal marketing orders allow markets to be segmented and higher prices charged in those segments with less elastic demand.

T   F   3 From the perspective of an individual farmer, the ideal situation is to produce an economically efficient level of output while other farmers' outputs are restricted.

T   F   4 The Agricultural Marketing Service reports timely information on the supply, demand, and prices of most agricultural commodities.

T   F   5 Other things equal, export demand for U.S. commodities is inversely related to the strength of the dollar on international exchange markets.

T   F   6 Public Law 480 is an example of an export subsidy program that was used by the United States.

T   F   7 In an open economy, an outward shift of the domestic demand curve for an agricultural product would cause a large increase in domestic production of the product.

T   F   8 For an agricultural good in a closed economy, an increase in the quantity supplied leads to a relatively large increase in equilibrium price because the demand for food is price inelastic.

T   F   9 The price support program for tobacco involves a significant cost to the government.

T   F   10 Supporting prices for agricultural commodities tends to benefit the individual who owns land at the time the supports were put in effect.

Multiple-Choice Questions

Circle the letter of the response that best answers the question or completes the statement.

1 Which of the following was not a factor in creating the "golden age" of agricultural prosperity between 1900 and 1920?
  a Between high birth rates and high immigration rates, population was growing at a record rate.
  b Export markets were strong, especially during World War I.
  c Government policies were in place that restricted output in order to ensure a high price for farmers.
  d Growth in industrial employment meant people had more income to purchase food.
  e None of the above.

2 Which of the following is an economic consequence of the U.S. policy of imposing compulsory quotas on tobacco production?
  a increased production of tobacco
  b decreased price of tobacco
  c increased start-up costs for new tobacco producers
  d decreased income for the owners of the quotas
  e all of the above

3 Which of the following is used to increase incomes from products under marketing orders?
  a set-aside requirements
  b market segmentation
  c deficiency payments
  d nonrecourse loans
  e all of the above

4 Which of the following crops are not marketed under federal marketing orders?

  a milk

  b fruits

  c vegetables

  d corn

  e all of the above

5 Which of the following *did not* occur during the 1970s to augment farm incomes and reduce conflicts concerning U.S. agricultural price and income policy?

  a The dollar weakened, leading to increased exports.

  b Income grew rapidly in Japan and several developing countries leading to increased demand for meat and grains.

  c The Soviet Union changed its economic policy in a way that generated the need for increased imports of grain.

  d A low inflation rate kept farm values stable.

  e None of the above.

6 Compared to a competitive market in a closed economy without any government programs, a program that pays deficiency payments based on a target price leads to

  a higher prices for consumers.

  b less spending by the government.

  c higher quantity produced.

  d all of the above.

  e none of the above.

7 Which of the following is *not* a correct statement concerning the cause of farm problems, or the farm problem, in the United States?

  a As the U.S. economy has grown, the domestic demand for farm products has increased relatively less than the demand for all nonfarm products.

  b The rate of increase in farm productivity has generally not kept pace with rates of increase in the rest of the economy.

  c The increases in demand for off-farm use of farm labor and management have generally not kept pace with the decrease in demand for farm labor and management.

  d All of the above.

8 Under a parity based price-support concept

  a the prices of farm commodities should vary inversely with changes in the prices-paid index for farmers.

  b because of productivity increases in farming, prices of farm products should rise by a corresponding amount.

  c the incomes of farmers should remain constant, regardless of increase or decrease in the prices or the products they buy.

  d farm prices or farm incomes should be the same as in a base period, in real terms.

9 What are "set aside," acreage allotment, and soil bank programs all designed to do?

  a reduce the supply of agricultural products

  b make the demand for farm products more price elastic

  c bolster the demand for agricultural commodities

  d accelerate the movement of human resources out of farming

10 Target prices and deficiency payments

  a shift a major part of the costs of farm income supports from the consumer to the federal government (taxpayers).

b are an indirect way of subsidizing agricultural exports.

c are an indirect way of subsidizing consumers as well as farmers.

d all of the above.

11 Marketing agreements and order programs

    a have been used to manage the marketing of commodities that face inelastic demands.

    b cannot use market segmentation in pricing because this is unconstitutional.

    c have been used most often for major commodities such as wheat and cotton, which face inelastic domestic demands.

    d deal primarily with agricultural commodities at the farm level and thus have little effect on retail markets.

Technical Training

Suppose your agency is charged with increasing the price farmers receive for wigwacks. Your job is to analyze the effects of three different policies being considered. You know that, at present, with no government program in place, the wigwack market is perfectly competitive with the following aggregate supply and demand curves:

$$Q = 600 - 20P \text{ (demand)}$$

$$Q = -16 + 24P \text{ (supply)}$$

1 What is the equilibrium price and quantity under the present conditions?

Consider a proposal to impose a strict quota that permits farmers to sell only 220 units in aggregate.

2 Compared to the present situation, by how much does the price received by farmers increase?

3 How does the quantity of wigwacks consumed privately change? By how much?

4 How much would the government pay to wigwack farmers under such a quota policy?

Next consider a program where farmers are offered nonrecourse loans at a support price of $19 per bushel of wigwacks. Compared to the present situation:

5 By how much does the price received by farmers increase?

6 How does aggregate wigwack consumption change? By how much?

7 How does aggregate wigwack production change? By how much?

8 How many surplus wigwacks end up being stored by the government?

9 How much does the government end up paying wigwack farmers if supply and demand conditions remain unchanged?

Now consider a third possible program. Suppose the government sets a target price of $19 per bushel of wigwacks and offers deficiency payments. Compared to the present situation:

10 By how much does the price received by farmers increase?

11 How does aggregate wigwack consumption change? By how much?

12 How does aggregate wigwack production change? By how much?

13 How many surplus wigwacks end up being stored by the government?

14 What price do consumers pay for wigwacks?

15 How much does the government end up paying wigwack farmers?

Questions for Thought

The following questions are designed to help you apply the concepts you have learned to real-life situations. Answering them will help prepare you for discussion questions based on the material in this chapter.

1 Why were the policies developed in the 1920s and 1930s not successful?
2 Contrast the policies of the 1930s with the current agricultural policies.
3 Explain the difference in the operation of a nonrecourse loan and a target-price/deficiency-payment program in their effects on the farmer, consumer, and taxpayer.
4 Why are acreage allotment and set-aside acreage reduction programs implemented uniformly across producing areas instead of selectively by withdrawing the least productive land?
5 Explain the advantages of market segmentation by a market order. Would the same advantages be realized if the elasticity of demand was the same in all market segments?
6 Explain how the export enhancement programs are a form of market segmentation.
7 Will export enhancement programs be effective if they are adopted by all significant exporting nations? Who would benefit from this situation?
8 Discuss the role of positive, normative, and prescriptive economic analysis in the development of agricultural policy.
9 Consider the following solution to the problem of low corn prices: restrict corn production and move up the price inelastic demand curve for corn until prices are suitably high. What do you see as potential problems with this policy? How would you keep farmers from producing too much corn at high prices? What would the policy do to our export markets in corn?
10 Suppose you were operating a farm and needed to take forty acres out of production. What kind of land would you stop working? Would you work the remaining land more or less intensively?
11 Two goals of farm policy are expanding agricultural exports and minimizing dependence of food imports. Can all nations simultaneously achieve both goals? Given that it is unlikely to be able to have both, should a nation with a comparative advantage in producing food favor free trade in food or protection for domestic producers?
12 An often-heard lament is that the federal government is being inconsistent by both trying to discourage smoking and supporting tobacco farmers' income. Is there an inconsistency in federal government behavior toward tobacco?

ANSWERS AND HINTS

Name That Term   **1.** parity; **2.** export subsidies; **3.** Nonrecourse; **4.** Cross compliance; **5.** Deficiency payments; **6.** base; **7.** Set-aside; **8.** Agricultural Marketing Service; **9.** Food Stamp; **10.** Conservation Reserve Program (CRP)

True/False   **1.** F; **2.** T; **3.** T; **4.** T; **5.** T; **6.** T; **7.** F; **8.** F; **9.** F; **10.** T

Multiple Choice   **1.** c; **2.** c; **3.** b; **4.** d; **5.** d; **6.** c; **7.** d; **8.** d; **9.** a; **10.** d; **11.** a

Technical Training   **1.** P = $14; Q = 320; **2.** P increases $5; **3.** Q is reduced by 100; **4.** nothing; **5.** P increases $5; **6.** Q consumed is reduced by 100; **7.** Q produced increases 124; **8.** 220 are stored; **9.** $4180; **10.** P increases $5; **11.** Q consumed increases by 120; **12.** Q produced increases by 120; **13.** nothing is stored; **14.** $8.00; **15.** $4840.00

18

PUTTING IT ALL TOGETHER: THE ECONOMIC SYSTEM

OVERVIEW

This book is about the principles of economics, about the economic system, and about how agriculture—which is defined to include natural resources, production agriculture, agribusiness, and consumers—is governed by these principles. The book indicates that agriculture is an important part of the economic system.

Each chapter describes a set of principles that relate to a component of the economic system and illustrates these principles with examples from agriculture. Although some links between the various components of the economic system are identified in the preceding chapters, an integrated picture is not presented. This chapter brings these components together to provide a comprehensive view of the economy.

In the second section of this final chapter we return to some of the questions asked in Chapter 1 and show how you can answer them using material presented in this book. You will find that you have learned a lot. You should already have a much better understanding of the economic aspects of the world around you. Furthermore, the analytical tools learned from this book provide a solid foundation for further studies of agriculture and economics.

This chapter concludes with several projections of important issues that may face future economists.

LEARNING OBJECTIVES

This chapter will help you:

- Understand more fully how the major components of the economic system fit together
- Understand the answers to some of the economic questions raised in the first chapter
- Think about some important economic issues for the future

THE ECONOMIC SYSTEM

An economic system is incredibly complex. Individuals, firms, and agencies interact to allocate among individuals with unlimited wants the goods and services produced from limited resources. There is order within that complexity, however. Businesses combine natural resources with the efforts of workers and equipment to produce goods and services. Households buy the final goods and services from businesses to meet their demands. Households earn income to pay for these goods and services by providing the services of the natural and human resources they own. This circular flow of goods from businesses to meet household demands and productive services from households to businesses, with the corresponding counterflows of dollars, constitutes the basic economic system. It functions as it does because of attempts by consumers to improve their well-being as much as possible by consuming goods and services, and because of the profit-maximizing goal of businesses. Servicing the flow of goods and services are markets and the intermediary institutions such as the financial system and governments.

The diagram in Figure 18-9 summarizes the major components of the economic system, the microeconomic decision criteria, the interactions, the macroeconomic linkages, and the role played by the government in managing the system. In the following pages we use Figures 18-1 to 18-8 to describe the components of this diagram to review and integrate the material presented in the book.

Consumption

We began our presentation in Chapter 2 by showing that a consumer's purchases can be explained using information about that person's tastes and preferences for goods and services, the prices of the goods and services, and the consumption budget. This is suggested in Figure 18-1, which shows a consumer's indifference map and the demand curve that can be derived from it. The demand curve is derived by finding the combination of goods that maximizes utility given the money the consumer has to spend on these goods at various prices. In Chapter 2 we also showed that a demand curve could be derived from the fact that a consumer strives to maximize utility. Consumers maximize utility by buying goods in quantities such that the ratio of marginal utility to price is the same for all goods and services consumed: $MU_i \div P_i = MU_j \div P_j$. Adding horizontally

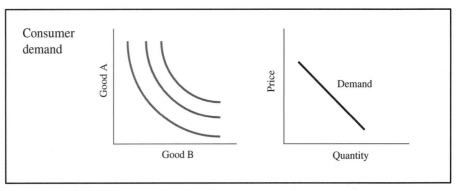

FIGURE 18-1   The consumer demand segment of the economic system.

the demand curves of many individuals for a good generates a market demand curve, as shown in Chapter 5.

Production

The production activities described in Chapters 3 through 5 are represented in Figure 18-2. In any economy there are a myriad of firms that produce goods and services. Many of these firms sell directly to consumers. Other firms sell intermediate goods to firms that sell to consumers. In the agricultural sector, farmers and ranchers produce commodities that processors and retailers—agribusinesses—transform into consumer-ready products.

A basic goal of a firm is to maximize profits. This goal is pursued by hiring workers, borrowing money or selling shares, buying natural resources and intermediate inputs, and using capital equipment to produce an output. A firm determines how much of each input to purchase by operating where the cost of the last unit of the input is equal to the value of additional output produced: $P_i = \text{MPP}_i \times P_o$ (MRP). This is represented by the MRP curve in Figure 18-2.

The firm decides how much output to produce to maximize profits. It does this by increasing output until the marginal cost of producing an additional unit equals the marginal revenue from the sale of that unit: $\text{MR} = \text{MC}$. These relationships are indicated by the graph of the cost curves in Figure 18-2. The marginal cost curve is the firm's supply curve. For an individual good, we add (horizontally) the supply curves from all firms to find the market supply curve (Chapter 5).

As indicated in Chapter 10, one of the major differences between large and small firms is the complexity of the management structure. In small firms, decision making is concentrated in the hands of one or a few individuals. In large firms, individuals and groups, each specializing in a particular area such as personnel and marketing, provide management functions. Teams of individuals make most decisions. Considerable effort must be devoted to coordination in order to generate the most profitable strategy.

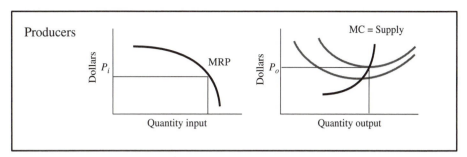

FIGURE 18-2   The producer segment of the economic system.

Markets

Markets, discussed in Chapters 5 through 9, are where buyers and sellers come together to conduct transactions. They provide the linkages among firms as well as between firms and individuals. The several supply-demand diagrams in Figure 18-9 indicate markets. A typical supply-demand diagram is shown in Figure 18-3. In markets, buyers and sellers exchange goods at mutually acceptable prices. This process determines prices, which are the signals that guide adjustments of the levels of production and consumption. If prices are at equilibrium levels the market clears with neither shortages nor surpluses.

The managers of small firms, such as most farms, sell their output in competitive markets where the individual firm cannot influence price. In contrast, some firms, including many large agribusinesses, can influence the prices paid for inputs and the prices received for outputs, as discussed in Chapters 9 and 10. Most large agribusinesses operate in markets that approximate the conditions of monopolistic competition or oligopoly. They have some control over the prices they pay for inputs or the prices they get for the products or services they sell. They can influence prices by changing the quantity of a good or service sold. The firm must anticipate the reaction of consumers and other firms in the market to price changes to determine the profit-maximizing quantity and price. The nature of interactions among firms depends upon the number of firms in the market, the extent of differentiation of products, conditions of entry, and the availability of information to buyers and sellers. A pure monopolist (or monopsonist) has complete control of prices, and the government generally regulates the firm by controlling or influencing the prices charged. Regardless of the market organization, the basic concepts of marginal revenue equals marginal cost ($MR = MC$) and marginal revenue product equals marginal resource cost ($MRP = MRC$ and MRC equals the price of the input in a competitive market) still guide production decisions.

Resources

Resources are essential to produce the goods and services that improve people's welfare. Only a few resources contribute directly to welfare. Combining resources in a

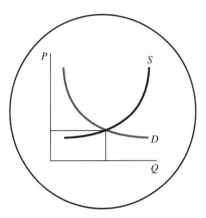

FIGURE 18-3   The market function in the economic system.

production process produces useful goods. Technical change enhances the efficiency of resource use. There are three categories of resources (sometimes called *factors of production*): natural resources, human resources, and capital resources.

Natural Resources   Figure 18-4 represents the natural resource base of an economy, as discussed in Chapter 15. They include both renewable (e.g., water and living things) and nonrenewable (minerals and energy) natural resources. Decisions about their use depend largely on the assignment of property rights. Most natural resources must be combined with human and capital resources to produce desirable goods and services. However, some natural resources are consumed directly—for example, recreational and scenic resources. This relationship is indicated by the overlap of the natural resource and environmental quality elements of Figure 18-9.

The United States has a comparative advantage in many agricultural products because it has spent a great amount in building up its human capital and is richly endowed with land and other natural resources. It is not surprising that the United States is an exporter of food and other agricultural commodities.

The quantity of nonrenewable natural resources available to a nation is essentially fixed (added resources can be acquired by trade, negotiation, or conquest). The rate of use is determined by resource owners or by the government in centrally planned economies. In market economies owners of these nonrenewable resources compare their expectations about future prices and the rate of return on other assets in determining the rate of resource use. The rising price path in the nonrenewable resource diagram in Figure 18-4 suggests this. The result is that the resource is economically exhausted when its price reaches a level at which substitute resources are available. As the price of a resource rises, the incentive to search for substitutes increases.

Renewable resources are available to society indefinitely if managed appropriately (improper management, usually resulting from an inappropriate specification of property rights, can result in exhaustion). Owners, sometimes organized as common property managers, compare the cost of additional effort in harvesting with the added returns in determining the appropriate rate of use. This is suggested by the diagram of the

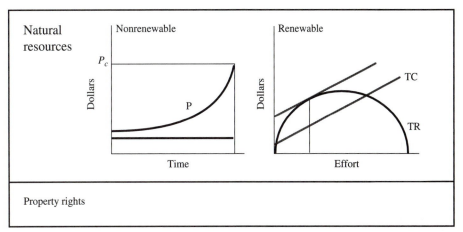

FIGURE 18-4   The natural resource segment of the economic system.

optimal management of a renewable resource, such as a fishery, shown at the right side of Figure 18-4.

The property rights base in Figure 18-4 reflects the importance of the specification of a set of property rights to the management of resources and draws attention to the fact that the property rights system varies significantly among nations. A clearly defined system of property rights, a legal system, policies, and institutional structures are required for any market to function, as discussed in Chapter 17. If these institutions do not exist, the market system fails. One of the great challenges facing the countries of the former Soviet Union is to develop these institutions so that they can develop market economies.

Human Resources   The second category of resources is human resources or human capital. The number of people in an economy is less important than the quality of their education and training in determining their productivity and thus their earnings. Individuals provide labor and management services to the economy, and receive wages and profits in return. Wages are determined in a labor market. (Households also own natural and capital resources, for which they receive rental income when the resources are used in production.)

Capital Resources   The third category is capital resources. These include the buildings, factories, and equipment that have been constructed by individuals, businesses, and governments. Business owners make investment decisions by comparing the expected rate of return with the rate of interest on borrowed capital. Expansion of capital resources is made possible by savings that flow from consumers through the financial institutions to producers.

Environmental Quality

The goods produced by the economy are eventually returned to the environment as wastes (services, the largest part of GDP, do not produce these wastes). The connections

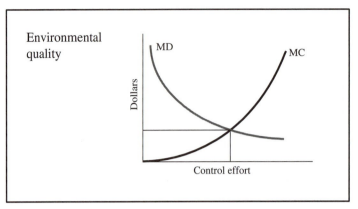

FIGURE 18-5   The environmental component of the economic system.

from consumption to the environment and from producers to the environment reflect this flow. The flow includes used and discarded products, packaging materials, food wastes, and pollutants, such as sediments, plant nutrients, and pesticide residues generated by producers. Some of these wastes reduce aesthetic values more than they degenerate in the natural environment. Other wastes, such as toxic chemicals, remain in the environment and cause health or other damages. Some discharges to the environment adversely affect the productivity of the natural resource base.

The appropriate level of control of a pollutant is suggested in Figure 18-5. As pollution control efforts increase, the marginal costs of pollution control rise and the marginal damages from pollution fall. The appropriate goal is to equate the marginal costs to the marginal damages. If a market for pollution control existed, this level of control would occur. Without a market, the government can increase efficiency by intervening. In many cases, the government sets pollution limits at the estimated level and enforces them with regulations. Another approach, favored by economists when it is feasible, is for the government to establish property rights and impose taxes or subsidies to provide the incentive to individuals or firms to reach the appropriate level of conservation.

The environment also provides benefits that individuals enjoy but do not pay for in a market. For example, country vistas, national parks, hiking, camping, or fishing often provide benefits well above the cost to those who enjoy them. However, there is an opportunity cost of the time, travel, and so on. In some cases, the government provides these resources through agencies, by establishing common property management structures, or by regulating private property owners. In other cases, these benefits are positive externalities that are a by-product of an economic activity.

The segments of the economic system already described in this chapter are combined in Figure 18-6. This figure pictures the components usually identified as the agricultural sector. The farmers and ranchers use natural resources to produce commodities sold to agribusiness firms that process, deliver, and sell to the final consumers. This diagram omits the flow of inputs to the producers and resource owners. These input firms are

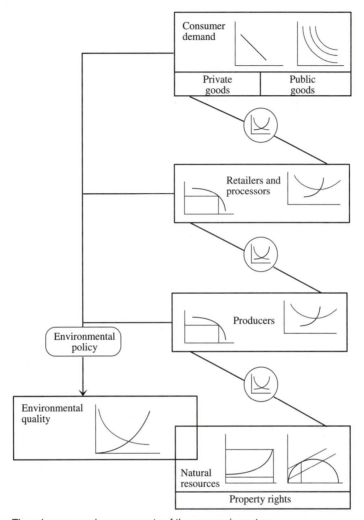

FIGURE 18-6   The microeconomic components of the economic system.

another important part of the agricultural sector. Consumers and firms produce wastes and other pollutants that flow to the environment, in many cases affecting the quality of the natural resource base. Viewing Figure 18-6 in the context of the total economy reveals the many firms that sell to each other and to consumers.

In Figure 18-7 the flows of income to households are shown. Individuals provide labor and management to firms and receive wages and salaries in return, with the transactions occurring in the labor markets. Individuals who own businesses earn profits and individuals who own resources earn income as rent in return for providing the services of these resources.

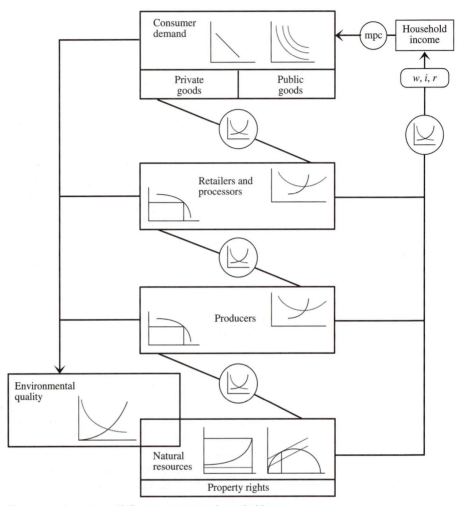

FIGURE 18-7   The economic system with flows to consumer households.

In Figure 18-8, financial institutions and the government are added to Figure 18-7. The figure also reflects the allocation of income by consumers. The government takes some income as taxes. The remainder is divided between consumption and saving, as reflected by the marginal propensities to consume and save. Most consumers spend a majority of household income to buy goods from private sector firms. The marginal propensity to consume (mpc) indicates the percentage of income spent for these goods.

Savings flow to financial institutions, which lend these funds to businesses for investment purposes. The rate of investment depends on the interest rate.

Figure 18-8 also shows the flow of services from the government to consumers, usually providing public goods that are not provided by the private sector.

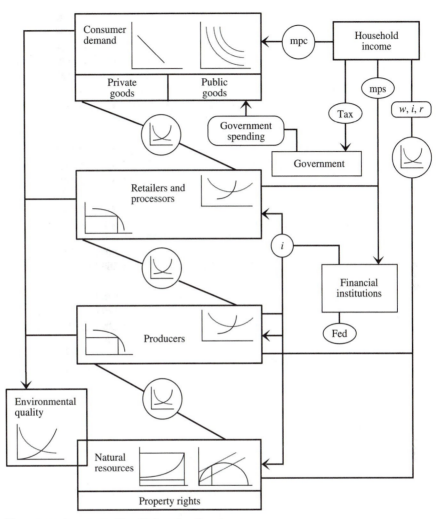

FIGURE 18-8   The economic system without international exchanges.

Macroeconomic Policy

Macroeconomic analysis, in contrast to microeconomic analysis, embraces the whole economy. The goals of an economy are to use the available resources (natural, human, and capital) to produce goods and services for consumption now or later. Two of the potential problems addressed in the management of an economy are inflation and recession. If the demand for goods and services outpaces productive capacity, inflationary price increases result. If demand is sluggish, resources will be unemployed and the economy will be in a recession.

There are two components of macroeconomic policy used to deal with inflation and recession. The first, monetary policy, is determined by the Federal Reserve System (the

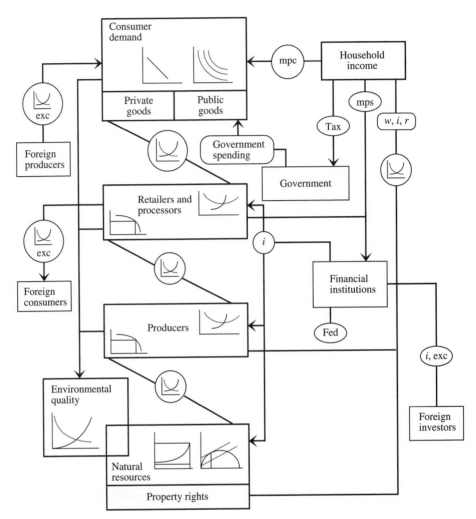

FIGURE 18-9   The economic system.

Fed). It is represented in Figures 18-8 and 18-9 by the link from the Fed to the financial institutions. The Fed can stimulate the economy by increasing the money supply. The increase in money supply reduces the interest rate and encourages more investment. Increased investment stimulates demand through purchases of plants and equipment. The Fed reduces the money supply to control inflation by forcing up interest rates, increasing the cost of capital for investment and thus reducing investment expenditures.

The second component of macroeconomic policy is fiscal policy. The public sector (the government) provides some goods and services (national defense, infrastructure, regulations, and the post office) and collects taxes and fees to support them. The government influences the performance of the economy by changing taxes or government spending. If the government spends more than it collects in taxes, the effect is to stimulate the economy. If tax collections are greater than expenditures, the effect is to slow the economy and reduce inflationary pressures. Fiscal policy decisions are made in the political arena. The appropriate adjustments to monetary and fiscal policy are an ongoing issue in the United States and in most other nations.

International Linkages

Trade among nations occurs as producers in countries with a comparative advantage in a product sell to consumers in another country. Figure 18-9 shows that foreign producers sell into the domestic market and domestic producers sell into foreign markets. For goods with active international markets (for example, most food and other agricultural commodities), international prices, exchange rates, and governmental interventions in international transactions influence domestic prices. Governments often intervene in international markets by establishing tariffs, quotas, or subsidies.

Besides the trade in goods and services, there is an active international financial market represented in Figure 18-9 by foreign investments in the United States and U.S. investments in foreign countries. Interest rates in various countries and the exchange rates determine the flow of funds among countries. As indicated in Chapter 16, there is a relationship between the flows of merchandise and currency. For the last nineteen years of the twentieth century, record U.S. imports were "paid for" with funds borrowed from abroad, and the United States became the world's largest debtor country.

In total, Figure 18-9 shows that the economic system is a complex, interlinked system. The decisions made by individuals, firms, and governments come together to determine our well-being.

IRRATIONAL EXUBERANCE

Alan Greenspan, Chairman of the Federal Reserve, is often quoted for his remark that "irrational exuberance" explained why the stock market was higher than he felt it would be if it were valued in line with traditional economic measures. Many others have conducted experiments and have shown that individuals do not always make decisions consistent with "rational behaviors" under economic theory.

While some economists may argue that all individuals and firms adhere rationally to economic theory, all, or essentially all, the time, we do not make such an assertion. We accept that individuals have many different motivations as they decide what and when to purchase goods. We accept that firms have many goals and that profit maximization is only one of them. We also accept that both producers and consumers make mistakes. They do not always make decisions that are consistent with their objectives, be those objectives based on economics or something else.

For these reasons and others, the economic system as represented in Figure 18-9 is not analogous to a finely tuned machine that is always perfectly allocating scarce resources among competing uses. Rather it is a huge and complex set of relationships among millions of individuals all trying as best they can, day to day, to "do their thing". The point is that economics provides a set of tools to understand a basic set of motivations (utility and profit maximization), relationships (markets and institutions), and constraints (government policy) that are at the core on the market system. The market system has performed so well that it has replaced all competing systems. While it is far from perfect, either in concept or in operation, it is more successful than any of the alternatives tried to date.

TYPICAL ECONOMIC QUESTIONS

In Chapter 1 we asked several "typical" economic questions. Here we return to those questions and see how we can use the material in the text to answer some of them.

Some Everyday Questions

Why isn't the price of a Big Mac twice, or half, its current price?
The price of a good, such as a hamburger, is determined by the interaction of its demand (discussed in Chapter 2) and supply (Chapter 4) in the market for hamburgers (Chapter 5). The supply of hamburgers depends on the cost of the variable inputs used (such as meat, buns, ketchup, and labor) and the nature of competition among sellers in the market. The demand for hamburgers depends on the tastes and preferences of consumers and their budgets. Since hamburgers are easy to make and the inputs needed to make them are inexpensive, it is easy to enter and exit the hamburger business (Chapter 7 discusses categories of market competition). Hence, the hamburger industry approximates pure competition. The price of a hamburger is determined by the cost of the inputs plus a normal return to the owner's investment.

Hamburger marketing firms distinguish their products by advertising brand names (such as Big Mac, Burger King's Whopper, and Wendy's Deluxe), with subtle differences in ingredients that affect taste, and by locating outlets in convenient places. These efforts sometimes sustain small differences in price, but any hamburger firm that charges significantly more than its competitors will lose sales. Ultimately, such a firm must either lower its price or be forced out of business.

Why did the price of hogs fall to record lows in 1998/99 and the retail price of pork not fall as significantly?
The price of hogs fell because the supply curve shifted outward more rapidly than did the demand curve. This was in part because of the entry into the market of large hog production facilities that took advantage of economies of scale as well as lower feed costs. The retail price of pork did not fall in large part because pork processing and marketing industries are not perfectly competitive. Thus the firms have some control over the prices. They were not forced to reduce the retail prices, at least in the short run. Over time we should expect that the supply curve will shift inward and that retail prices will decline. History suggests periods of over and under production will generate cyclical hog prices. However, over time we can expect that the prices will approximate equilibrium levels but with somewhat higher than normal profits in the processing sector due to imperfect competition.

Why can I buy a good lunch for half the price of a good dinner?
Restaurant meals fit the model of monopolistic competition (Chapter 9). The restaurant business is easy to enter and exit, and each restaurant produces a somewhat different product. For example, although a meal at an Italian restaurant might provide the same nutrients as a meal at a Chinese restaurant, they are hardly perfect substitutes! Furthermore, an evening meal is not the same product as lunch. On the supply side, the cost of labor at night is somewhat higher than during the day. However, the major differences arise on the demand side of the market. Most noon restaurant meals are "just" meals;

they are consumed quickly before continuing the day's activities. In comparison, many restaurant meals in the evening provide more than nourishment; they provide a dining occasion. People go to restaurants in the evening to celebrate events, to transact business, or to relax and enjoy attentive service, as well as the meal. If a restaurant provides these services, the cost of the meal can be a small part of the total value of the meal (note that fast-food restaurants do not increase their prices in the evening). Restaurants that specialize in providing amenities are relatively small in number and charge higher prices, as expected in a market that is not purely competitive (Chapter 9).

What explains the drastic changes in the value of the dollar in international currency markets?
The value of the dollar in international currency markets is determined by interaction of the demand for and supply of dollars in these markets (Chapter 14). The demand for dollars comes from foreigners who want to buy U.S. goods, invest in U.S. companies, and buy U.S. financial assets. The supply of dollars comes from U.S. individuals and firms that want to buy foreign currency to purchase foreign-made goods, invest in foreign companies, and buy foreign financial assets. If economic conditions change to shift either the demand for or the supply of dollars in international currency markets, the value of the dollar also changes.

The fluctuations in the value of the dollar were more pronounced at the end of the century because economic conditions changed quickly. The most important change was the rapid growth in the federal budget deficit in the 1980s (see the end of Chapter 14). This growth pushed equilibrium GDP above full-employment GDP. In a closed economy, this condition would have resulted in a general increase in prices (an inflationary gap, see Chapter 12). However, the United States is an open economy. In an open economy, the growth of GDP to near full-employment levels results in an increase in imports, a decrease in exports, and a smaller increase in prices than in a closed economy. Therefore, the growth of GDP resulting from deficit spending produced the trade deficit. Simultaneously, the Federal Reserve reduced the domestic money supply to raise real interest rates in the United States to counteract inflationary pressures (Chapter 13). These high rates of return on U.S. assets attracted foreign investors, and they demanded more U.S. dollars. This outward shift of the demand for dollars in the foreign currency market caused the value of the dollar to rise. Later in the 1990s, the federal deficit declined (and was in surplus at the end of the century), foreign interest rates rose to compete with the returns received in the United States and domestic interest rates fell because of the inflow of funds. As a result, the foreign demand for dollars decreased which reduced the value of the dollar. Late in the 1900s the U.S. economy prospered and the stock market became attractive to domestic investors who kept their funds at home and to foreign investors who purchased U.S. stocks. This created a demand for dollars that increased the exchange rate.

What determines the inflation rate?
As indicated in the macroeconomic section, fiscal policy and monetary policy are major determinants of inflation. Fiscal policy refers to the balance between the level of government spending and tax revenues. Increased spending or reduced taxes tend to stimulate the economy and may result in inflation. One of the primary goals of the Federal Reserve System (the Fed) is the control of inflation which it does by controlling

the supply of money in the economy. Fiscal and monetary policy operate among a myriad of other economic factors that influence the performance of the economy such as imports, exports, weather, producer expectations, and consumer confidence.

Why did the same farmland in the U.S. Corn Belt cost $4000 per acre in 1980, $1500 in 1985, $4000 in 1997, and $2500 in 2000?

The value of farmland, like any other good or resource, is determined by the value of the goods that can be produced from it and the opportunity cost of capital (see Chapters 6 and 13). In the 1970s, world demand for corn and soybeans, the major crops produced in the Midwest, was strong and exports boomed. The prices of corn and soybeans were high. The cost of capital borrowed to buy land was low because the inflation rate was high (much of the land was purchased with variable interest rate loans). Thus, the opportunity cost of buying land was low as compared with buying an interest-bearing asset. Also, the several decade history of continuously increasing land prices led many buyers to expect continued increases in the future. This combination of factors led to record land prices in the U.S. Corn Belt.

In the early 1980s, a strong dollar caused real interest rates to rise sharply as the rate of inflation moderated and the quantity of U.S. corn and soybeans demanded for export shifted inward causing grain prices to fall. These factors also resulted in lower land prices. The fall in land prices and increase in interest rates resulted in some purchasers being unable to service their loans. Many were forced to default (lenders foreclosed). These failures placed more land on the market, which pushed land prices rapidly downward to even lower levels.

As the 1980s progressed, the dollar weakened and there was economic growth in Asia, shifting foreign demand out for corn and soybeans. Also, the real interest rate declined from the record levels of the early 1980s. The rate of farm failure fell, and land values rose again to levels that reflected the value of the crops produced on the land.

At the end of the century a change in federal policy reduced the restrictions on the amount of acreage that could be planted resulting in an increase in supply of commodities. At the same time there was a significant reduction in exports due in part to an economic downturn in Asia. These resulted in a fall in commodity prices that was reflected in the price for land. Thus, while there may be temporary fluctuations, the price of crops that can be grown on the land and the earning potential from investments in other assets determine the price of land.

How important is rainfall to the price of agricultural commodities such as wheat?

Rainfall, and other weather conditions, can have a major effect on crop prices when the weather conditions are widespread. For example, a drought that affects major crop producing areas can shift the supply curve inward significantly and thus increase prices. The price fluctuations due to weather are much less than would be the case without the international market.

Why are some farmers reluctant to adopt soil and water conservation practices?

The conservation issue is discussed more fully in Chapter 15. To summarize, like any other economic activity, soil and water conservation practices use scarce resources. A farmer has limited resources and wants to maximize the return from them. Resources allocated to conservation practices have an opportunity cost equal to what they could earn in other uses. Landowners do realize some of the benefits of conservation practices, such as making some land more productive by improving crop-growing conditions, or maintaining productivity by reducing erosion. Farmers can be expected to invest in profitable conservation activities.

The second type of return from conservation is to other members of society. The off-farm costs of soil erosion and water pollution are externalities that are reduced by practices such as soil conservation. Conservation can reduce siltation of drainage systems, streams, rivers, and lakes. Water conservation can preserve wildlife habitat, reduce pollutants, and reduce seasonal variability in stream and river flows. But landowners do not realize these returns and therefore have no direct economic incentive to implement the conservation practices.

As discussed in Chapter 15, there are several ways to address this problem. One option is to allow producers to use production practices that generate runoff. This implies that farmers have the right to pollute the streams and that the off-farm costs must be borne by other members of society. Another approach is to consider unpolluted water a right of other members of society and force farmers to incur the cost of reducing the runoff or pay for any damages created. In most of the policies adopted, the farmer pays part of the cost and others in society incur the remaining costs. These policies include educational programs to inform producers of the problem and the available control measures, technical assistance and cost sharing for installation of conservation measures, incentives to remove erosive land from production, and regulations that require farmers to adopt conservation measures to qualify for other programs. The government also subsidizes the dredging of silt from watercourses.

Why are most U.S. farms larger than most farms in less-developed countries (LDCs)?
In addition to non-economic factors such as tradition and land ownership policies, many developing countries have a comparative advantage in labor-intensive activities because they have relatively abundant labor resources (Chapter 9 discusses comparative advantage). Developed countries have relatively abundant capital resources, and so they have a comparative advantage in capital-intensive activities. Capital-intensive production activities are more likely to have economies of scale than labor-intensive activities. In addition, the United States has abundant land resources well suited for agricultural production. So farming in the United States is capital-intensive, and farms are large to exploit economies of scale in the use of capital equipment, such as tractors and harvesters, and highly qualified workers (human capital).

Why are crop yields on similar land higher in Europe and Japan than in the United States and Australia?
In Japan and Europe farm level crop prices are supported by the government and are higher than in the United States or Australia. The higher prices result in farmers using more variable inputs per acre.

Will consumers abandon shopping malls in favor of purchases via the Internet?
While one cannot predict the distant future, it is clear that the web has the potential to be a major factor in retail markets. Malls and other such outlets will likely continue to exist

because they can provide the direct "hands-on" shopping without waiting for delivery. As web-based marketers continue to develop their services such as a reliable return policy, web sales will continue to grow. As this growth occurs it will have significant effects on the market if it provides price competition that approximates competitive market conditions. Policy discussions about whether or not to, and if so how, Internet sales should be taxed have been occuring for sometime. Such policy issues will continue to grow in importance and complexity. For example, as international Internet sales begin to grow, as one must expect they will, questions of import and export policy will arise and the taxation of Internet sales will become an even more difficult issue.

Fundamental Economic Questions

The questions in this category are at the heart of what economics is all about. Economists continue to search for more acceptable answers to these questions. Obviously, it is not possible to answer them in a few paragraphs. Instead, we try to examine some elements that go into an answer.

What determines value? Why does a pair of designer jeans cost more than ordinary jeans? Why is the salary of a steelworker four times the salary of a worker in a fast-food restaurant?

The value of goods, such as jeans, depends on how the goods satisfy wants and desires (see Chapter 2 on utility). The value of the service of a resource, such as a worker, depends on the productivity of the resource (Chapter 3) and the desirability of the output produced from the resource. But it is not enough for a good to be desired; the potential consumer must have the resources to buy it (see Chapter 2 on budget constraint). Also, the characteristics of the market in terms of the number of firms and their conduct influence the prices of resources and products (see Chapter 10). Unions also influence the price of labor. So value is determined by a combination of factors including availability of resources, productivity of the resources, desirability of the product, and ability of potential consumers to pay for the product.

What determines the distribution of economic goods? Why are many Americans rich and many people in Somalia hungry?

The availability of goods depends first on the availability of resources and second on the infrastructure development (Chapter 17). The United States is blessed with abundant resources and a significant investment has been made to improve their productivity. Somalia's resources are much more limited, and less development of them has occurred. Another important determinant of the availability of goods is the technology to convert resources efficiently into goods (Chapter 3). Many years of research by both the private and the public sector in the United States have produced advanced technologies to use resources efficiently. In Somalia, on the other hand, much less technology is available to take advantage of the Somalian resource endowment.

However, perhaps the most important determinant of the availability of goods is the institutional environment. A well-functioning government implements laws and regulations that foster strong economic performance. Public goods such as transportation and communications infrastructure, educational and public research systems, and market information and product standards are also crucial to economic productivity (Chapters 11, 12, and 17).

What determines growth in the availability of economic goods? In the 1980s, why did the South Korean economy grow at 7 percent per year and the U.S. economy at only 2.5 percent per year?
For an economy to grow and make more goods available, it must increase resource availability in two ways (Chapter 13). First, by saving part of today's output and investing it in productive assets, more output can be produced tomorrow. In 1989, gross Korean savings were 34 percent of consumer income, while the comparable figure for the United States was only 6 percent.

Second, technical and institutional change can make existing resources more efficient. South Korea made major changes to the structure of its economy after the end of the Korean War. These changes allowed its resources to be used much more efficiently. Also, the South Korean economy was adopting already proven technologies, while the U.S. economy was developing and adopting new technologies.

LOOKING AHEAD

Of course this book and the summary in this chapter do not cover nearly all that can be said about the topics. It is, after all, an introductory text. However, as the twenty-first century unfolds, much of what has been presented in this book will have relevance to understanding the changes that will take place. No person can predict exactly what these changes will be, but we can be certain that economics will continue to play a major role in determining the course of human events. Advantages will accrue to those who have made and continue to make prudent investments in their own human capital, and to those for whom they are responsible. Therefore, continuing investment in research and education is wise for the individual regardless of career plans, and it is wise for a country regardless of its status among nations.

Projections

An interesting challenge is to predict the economic issues that will be important in the future. We do not pretend to have a crystal ball that can foretell the future. Instead, we rely on "educated" guesses about where the events of the day are likely to lead.

The Food Supply   As this is being written, the world commodity prices are low because the growth in production has exceeded the growth in demand and stocks are at high levels. This is in part because the United States has all but eliminated policies that restrict production. But this picture may change, even dramatically. In the short run, only a year or two of bad weather or disease in the major producing areas could greatly reduce stocks, increase prices, and cause hardships in the poorer importing countries.
The long run presents a more challenging problem for world agriculture. Continued growth of world population and income, even if birthrates fall, means continued rapid

growth in demand for food. The higher population levels would mean a larger domestic market for American agriculture and higher demand for U.S. farm exports.

In the United States, as well as in many other places around the world, more land can be brought into production but at higher economic or environmental cost than the land that is currently in use. Much of the land that could be added to the productive base is susceptible to erosion, requires costly irrigation, or has other characteristics that have kept it out of production to date. Also, in some cases, the conversion of this land to agricultural use conflicts with the goal of maintaining habitat for wildlife. Some land currently in production is becoming less productive because of erosion, salinity problems, or waterlogging. Increased irrigation, more widespread use of fertilizers and higher yielding seed, and the increased use of capital inputs can increase production beyond current levels. These changes will mean abundance for U.S. consumers, but they are unlikely to generate the doubling or tripling of food needed to feed the projected population of the earth. If food supplies are less than adequate, the populations of the less developed countries will be affected the most.

The continued development of improved crop yields is required to meet the long-term growth in demand from population and incomes. As indicated in Chapter 16, we expect that there will continue to be important developments in biotechnology and information technology, and these will change production technology throughout the agricultural sector. However, large-scale payoffs to biotechnological research are unlikely to come soon unless agricultural research receives substantially more resources. Opposition, especially in Europe, to genetically modified organisms (GMOs) is a significant problem for those developing these products. It is possible that a crisis may focus attention on the food supply issue. Given the tendency of political systems to respond to a crisis, such a situation would focus political attention on the food supply issue and lead to increased receptivity to GMOs and other new developments.

Trade   Trade will continue to be an important issue. It will grow in importance and complexity as nations seek to move toward freer world trade while simultaneously seeking to gain an advantage in international markets. Complicating the picture will be the continued growth in importance of the regional trading blocks, such as the European Community, the North American Free Trade Agreement, the Asia-Pacific Economic Cooperation, the proposed Free Trade Agreement of the Americas (FTAA), and cartels such as the Organization of Petroleum Exporting Countries. Also, various special interest groups are pressuring trade authorities to make changes in trade policy to achieve their goals—be they working conditions, product qualities, environmental goals, and so on. Some of these groups likely will continue to oppose freer trade, others are likely to use the potential economic gains from trade as a lever to achieve their goals.

Economic Structure   After more than a decade of downplaying the role of government, there will be a renewed interest in seeking ways for the government to stimulate growth. Nations will experiment with various relationships between industry and governments (corporate regulations, tax laws, development support, and even different educational models). A common model will involve a significant governmental presence in a market system that attempts to capture the incentives present in pure competition. An interesting arena for experimentation of this type will be in the nations emerging from the former Soviet Union and its satellites. In these countries, whole new economic systems must be created. Some countries will succeed; some will fail. In others, political changes have the potential to bring rapid economic development or chaos. Here are several potential economic "hot spots" that may command public attention over the next decade or so.

1 The countries of the former Soviet Union are the focus of considerable attention as they struggle with the basic challenge of developing the economic institutions of a market economy while holding off the negative political consequences of declining incomes and cultural and ethnic differences.

2 China is making substantial progress in agriculture, and younger individuals have replaced aging leaders. The new leaders are adopting economic policies that move toward increasing reliance on markets. Their success in implementing these policies is important beyond China's borders. A successful transition and continued economic reforms are likely to lead to continued rapid growth of the Chinese economy as inefficiencies are reduced. In this case, China could become a major force in the world economy. Failure of these policies could result in political turmoil and economic stagnation, or worse.

3 Indonesia, the fifth largest country in the world, has undergone a change in political leadership and is dealing with internal turmoil related to economic problems and attempt of some areas to secede. While the potential to continue economic growth is great, these political challenges are formidable.

4 Many of the African countries face significant economic problems due to the lack of physical and institutional infrastructure. The spread of AIDS and other diseases is placing an additional enormous burden on many of these countries as they provide for the ill. Due in part to AIDS many children are not receiving the care and education they need to develop into productive members of society.

Environmental Standards   Barring a severe worldwide food shortage that diverts the public's attention, pressure on production agriculture to conform to higher environmental standards will continue. This pressure will come from inside the countries as urban residents demand safer food, a cleaner environment, and a more aesthetically pleasing rural environment. Pressure will continue to come from other countries, as they demand adoption of higher environmental standards for products they buy in international markets. These standards are likely to include increased regulation of toxic chemicals, plant nutrients, and sediments.

In developing countries, the pressure to keep population and food supply in balance is likely to hinder their efforts to reduce environmental degradation. Without substantial transfers from the developed countries, tropical deforestation will continue and result in loss of biodiversity and continued concerns about possible climatic impacts.

As the costs of regulation for achieving environmental goals become more evident, the use of pollution taxes, markets for pollution rights, and other economic incentives will become more common.

Finally, the possibility of oil shortages may be the darkest cloud on the horizon. The search for oil reserves, the pressure for environmentally responsible use of these resources, and the search for alternatives to oil will all be important issues.

Questions for Thought

The following questions are designed to help you apply the concepts you have learned to real-life situations. Answering them will help prepare you for discussion questions based on material throughout the book.

1 Why do some children in Africa starve while the U.S. government spends millions of tax dollars to keep agricultural land out of production?
2 Does participation of the United States in the world economy make U.S. macroeconomic policy more effective, less effective, or is the effect the same?
3 Why do presidents of many large companies make as much money as they do?
4 Currently the federal budget is in surplus. Would you prefer to reduce taxes, increase spending, or pay down the debt? Why? How would your decision affect the performance of the economy?
5 Many believe that humans have dominion over all things on this earth. Does this mean that the value of everything on this planet is determined by our ability to make use of it?

APPENDIX A: MATHEMATICAL AND GRAPHING TOOLS

A Note on Graphs

Economists often use graphs to convey information. You will see them in this text, in lectures, and in news media stories about the economy. You will also see them on your tests. Thus, understanding graphs and how to use them is important to you.

Consider the following hypothetical graph of the relationship between the average number of calories consumed per day by a person and that person's weight. The graph conveys a large amount of information.

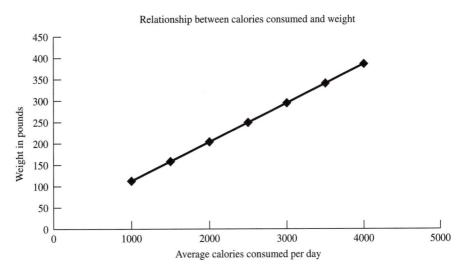

Relationship between calories consumed and weight

Here are some of the important things for you to appreciate or know about the graph:

1 It is easy to appreciate that as a person eats more calories per day, his or her weight increases. This is called a direct or positive relationship. In other graphs the line will slope downward, and in those cases the relationship is called indirect or negative. (Many of us would like a negative relationship between calories and weight, at least for a month or so.)

2 In this hypothetical case, the line on the graph is straight. This is called a linear relationship.

3 You can find the weight associated with any number of calories consumed between 1,000 and 4,000. To do this, you draw a line from the number of calories given on the horizontal (often called the X) axis up to the line on the graph. Then you draw a line from that point left to the vertical (often called the Y) axis where you can "read off" the weight.

4 It is often important to understand the cause and effect relationship. In this case, it is obvious that the number of calories is the cause and weight is the effect. Calories consumed determines weight, weight does not cause the consumption of calories. In economics, we often refer to the causal variable as the independent variable and the affected variable as the dependent variable. In most graphs the independent or causal variable is plotted on the X axis and the dependent variable is plotted on the Y axis. However, economists sometimes reverse and place the independent and dependent variables on the Y axes. This means you must pay attention to the description of the graph.

5 It is obvious that many things other than calories consumed affect a person's weight. The person's metabolism and the amount of exercise are obvious examples. But in this graph, we are holding everything but calories constant or unchanging. For graphs of economics, we always presume that everything else is unchanging. This is called the *ceterius paribus* assumption. It is discussed in the text.

6 The data to support the creation of the graph are given in the following table. In some cases you will take data from a table and create a graph by plotting the several values. That is, find the point in the graph associated with 1,000 calories on the X axis (by reading up from the horizontal axis) and with 115 pounds on the Y axis (by reading across from the horizontal axis). Then find the point associated with 1,500 calories and 160 pounds, etc. In other cases, you will be given a graph and be asked to find a particular value. For example, if you were ask how much a person would weigh if they consumed 2,500 calories per day, you would draw a line from 2,500 calories up to the line and then go to the right and read the value on the Y axis, in this case 250 pounds.

7 It is also possible to present the same information in an algebraic formula. In this case $W = 25 + 0.09C$, or weight equals 25 pounds + 0.09 times the number of calories.

RELATIONSHIP BETWEEN THE NUMBER OF CALORIES CONSUMED
AND THE PERSON'S WEIGHT

Calories	Weight
1,000	115
1,500	160
2,000	205
2,500	250
3,000	295
3,500	340
4,000	385

8 The slope of a line is useful in understanding the relationship. In this example, the slope is change in weight divided by the change in calories between any two points on the line. In this case every 45-pound change in weight is associated with a 500 calorie change. In economics we often use Δ, the standard notation for a change. Thus the slope is 45 ÷ 500 or 0.09, or $\Delta X \div \Delta Y = 0.09$. This tells us that for each additional calorie this person consumes per day, his or her weight will increase by nine hundredths of a pound. (Oh, how those hundredths can add up!) If the line slopes downward, the slope will be a negative number.

9 The intercept of a line is also useful in understanding a relationship. In our example, if you extend the line to the left, it will intercept the vertical axis at 25. This suggests that if a person eats zero calories, he or she can expect to weigh 25 pounds. (Of course this is silly, and that is the reason the curve wasn't drawn all the way to zero.)

10 Now that you have a slope and an intercept, you can write an equation for our line. If you take the intercept and add to it the slope times a value from the x axis, you will get the associated observation on the y axis. A general equation for a straight line is X = intercept + slope times X or $Y = a + b(X)$. For example, if you want to know how much a person will weigh if they consume 2,100 calories, you substitute 2,100 in the equation for X and multiply 2,100 by the slope coefficient of 0.09. This gives you 189. Then you add the value of the intercept, 25 pounds, and get 214 pounds. It is also possible to solve the equation the "other way". That is, if you want to know how many calories are associated with a given weight you can calculate it from the equation. Take 175 pounds $175 = 25 + 0.09(X)$, or $150 = 0.09(X)$ or $X = 150/0.09$, or $X = 1666.67$ calories.

11 What if you are "thrown a curve"? Suppose that the actual relationship between calories and weight is more like the curve shown in the following graph. In this case, the slope of the curved line changes as you move along it. The slope of a straight line that just touches—is tangent to—the curve has the same slope as the curve at the point where the two lines touch. To get the slope at any point, you draw a straight-line tangent at that point and then calculate its slope. (Or if you know calculus, you can take the first derivative on the curve at that point—but if you have studied calculus, you are well aware of this fact!)

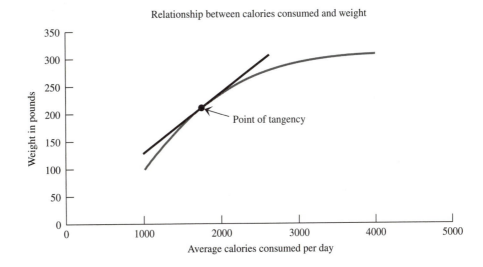

Relationship between calories consumed and weight

APPENDIX B: CONNECTIONS

Industry and Academic Resources

American Agricultural Economics Association   *http://www.aaea.org/*

American Meat Science Association   *http://www.meatscience.org/*

American Society of Animal Science   *http://www.asas.org*

The Burlington Northern Santa Fe-transportation information   *http://www.bnsf.com/*

Illinois Corn Page   *http://www.ilcorn.org/*

Illinois Soybean Association   *http://www.ilsoy.org/*

Institute of Food Technologies   *http://www.ift.org/*

PorkNet   *http://porknet.outreach.uiuc.edu/*

Roman L. Hruska U.S. Meat Animal Research Center   *http://sol.marc.usda.gov/*

National and International Organizations

Bloomberg Financial Information on Markets   *http://www.bloomberg.com*

The CIA World Fact Book
   http://www.cia.gov/cia/publications/factbook/indexgeo.html

The Congressional Budget Office   *http://www.cbo.gov/*

The Economic Research Service of the USDA   *http://www.ers.usda.gov/*

FDA Center for Food Safety and Applied Nutrition   *http://vm.cfsan.fda.gov/list.html*

Federal Reserve Bank   *http://www.frb.org/home/subeconomic.cfm*

Food and Agricultural Organization (FAO)   *http://www.fao.org/*

Food and Drug Administration (FDA)   *http://www.fda.gov/*

International Food Policy Research Institute   *http://www.ifpri.cgiar.org/*

International Monetary Fund (IMF)   *http://www.imf.org/*

National Soybean Research Laboratory   *http://www.nsrl.uiuc.edu/nsrlhome.html*

The U.S. Census   *http://www.census.gov/*

United States Department of Agricultural (USDA)   *http://www.usda.gov/*

USDA Food Safety and Inspection Service   *http://www.usda.gov/agency/fsis/html*

World Bank   *http://www.worldbank.org/*

World Health Organization   *http://www.who.ch//*

World Trade Organization (WTO)   *http://www.wto.org/*

Data Sources

AGCENTRAL   *http://www.agcentral.com/*

AgLinks Links for the Agriculture Industry   *http://www.gennis.com/aglinks.html*

Agricultural Data and Tools.   *http://www.farmdoc.uiuc.edu/*

Agriculture Network Information Center (AgNIC)   *http://www.agnic.org/*

agriculture Online   *http://www.agriculture.com*

The Bureau of Economic Analysis   *http://www.bea.doc.gov/*

Bureau of Labor Statistics   *http://stats.bls.gov/*

FAO Statistical Database   *http://apps.fao.org/*

Federal Statistics Related to Agriculture   *http://www.fedstats.gov/programs/agriculture.html*

National Agricultural Library   *http://www.nalusda.gov/*

National Agricultural Statistic Service   *http://www.usda.gov/nass/*

Virtual Library: Agriculture   *http://vlib.org/Agriculture.html*

Yahoo! Agriculture   *http://www.yahoo.com/Science/Agriculture/*

Yahoo! Economics   *http://dir.yahoo.com/Social Science/Economics/*

GLOSSARY

absolute advantage   The ability of one country to produce a good more efficiently than another country. Absolute advantage in producing a good does not necessarily mean that it will be exported under free trade. A country must have a comparative advantage in a commodity to gain from exporting it under free trade.

acquisition   In an economic context, the term refers to the purchase of one firm by another firm. More generally, it means the purchase of a good or service.

acre   a measure of the amount of land. One acre equals 0.405 hectare. There are 640 acres in a square mile.

acreage-allotment program   The program that determines the total number of acres that are to be used to produce various agricultural products and allocates these acres among individual farmers. The farmers are required to limit their plantings to the number of acres allotted to them to receive the support price for their crops and other federal program benefits.

acreage quota   The number of acres a farmer may plant to a specified crop. In some cases such as with the tobacco quotas, the farmer may sell the quota separate from the land.

acreage reduction program (ARP)   A program that requires farmers to reduce the amount of a crop planted to less than the base acreage to qualify for price supports and target prices for that crop.

adjustment costs   The costs that are incurred when a firm or a nation shifts from one equilibrium situation, or policy, to another.

aggregate demand   A schedule or curve that shows the total quantity of goods and services that will be demanded (purchased) at different income levels. It is the sum of all the individual demands for the output of the economy.

aggregate demand-aggregate supply model   The macroeconomic model that uses aggregate demand and aggregate supply to determine and explain the price level and real domestic output.

aggregate expenditures   The total amount spent for final goods and services in the economy.

agribusiness   Firms engaged in producing and distributing agricultural inputs or in marketing, processing, or distributing agricultural commodities.

agricultural conservation program (ACP)   A program in which producers agree to carry out specified conservation practices on their farms and receive payments to help pay part of the cost.

Agricultural Marketing Service (AMS)   An agency of the USDA that administers grading, inspection, market order, and promotion programs related to agricultural commodities. The agency also reports timely information on the supply, demand, and prices of most agricultural commodities.

Agricultural Research Service (ARS)   An agency of the USDA that conducts research on all aspects of agriculture and related concerns, both domestic and international in scope.

alternative uses   A good may be used for different purposes, such as for production or consumption. The decision is generally based on benefits and costs.

arc elasticity   The price elasticity of demand (supply) calculated from the observations at two points on a demand (supply) curve. (See also *point elasticity* and *elasticity coefficient).*

asset   Anything of monetary value owned by a firm or an individual.

autonomous consumption   That portion of consumption expenditures unaffected by the level of disposable income. The level of consumption that a consumer would make out of savings or from borrowed funds if income fell to zero.

average fixed costs (AFC)   The total fixed cost of a firm divided by its output (the quantity of product produced).

average physical product (APP)   The total physical product (TPP) for a given quantity of input divided by the quantity of that input.

average revenue (AR)   The total revenue divided by the quantity of output of a product sold.

average total costs (ATC)   The sum of average fixed costs and average variable costs. It is the total costs divided by the amount of output produced.

average variable costs (AVC)   The total cost of the inputs that are used in varying quantities in the short run divided by the amount of production.

balance of payments   A statement showing all of a nation's international transactions, including trade in goods and services as well as international capital flows.

balance of trade   A statement showing a nation's international purchases and sales of goods. When services (or "invisibles") are included, total balance is called the *current-account balance.* The balance is positive if exports exceed imports, or negative when imports exceed exports.

balance sheet   A statement of a business' financial condition at a given point in time.

balanced budget multiplier   The effect of equal increases (decreases) in government spending for goods and services and in taxes. The effect is to increase (decrease) the equilibrium gross domestic product by the amount of the equal increases (decreases).

bank reserves   Bank reserves held at a Federal Reserve Bank plus bank-vault cash.

bargaining association   An association of firms (for example, in agriculture, a group of farmers) that bargain for higher prices or other favorable economic conditions.

barrier to entry   Anything that prevents the entry of firms into an industry, such as patents, high start-up costs, or loyalty to the brands of other producers.

barrier to trade   Restrictions, such as quotas, tariffs, and product standards, prohibiting an individual or a firm from making an international purchase or sale.

barter The exchange of one good or service for another good or service. No money is used in the exchange.

base acreage The number of acres on a farm that were in crop production in a base period. It is used by the USDA in implementing certain provisions of price support programs.

base year The year with which other years are compared when a price index, or any index, is calculated.

benefit-cost analysis A procedure for comparing the benefits and costs of a project or program (for the production of a good or service) often used in determining whether a government should spend funds on a proposed activity.

benefits-received principle The belief that those who receive the benefits of goods and services provided by government should pay the taxes required to finance them.

bilateral monopoly A market in which there is a single seller (monopoly) and a single buyer (monopsony).

biotechnology A technology that modifies the natural evolution of living systems to develop processes and products for commercial or other purposes.

blended credit A financing plan for export sales in which government credit guarantees, or government credit at lower interest rates, is blended with regular commercial credit to provide lower interest rates and more favorable terms for foreign buyers.

Board of Governors The seven-member group that supervises and controls the money and banking system of the United States. It is formally known as the Board of Governors of the Federal Reserve System, or the Federal Reserve Board.

break-even point The output at which a (competitive) firm's total cost and total revenue are equal. The firm has neither an economic profit nor a loss.

budget constraint The limit an individual consumer's income (and the prices that must be paid for the goods and services) imposes on the ability of that consumer to buy goods and services.

budget deficit (surplus) The amount by which the expenditures of the federal government exceed its revenues in any year. A budget surplus is the amount by which revenues exceed expenditures.

budget line A line that shows the different combinations of two products a consumer can purchase with a given money income. It is the dividing line between affordable and unaffordable combinations of goods.

buffer stocks Supplies of a product stored on farms or in commercial elevators or warehouses to moderate extreme price fluctuations by ensuring a more stable supply. Buffer stocks are usually controlled by the government, while total stocks include both government and privately held supplies.

bushel One bushel equals 0.352 hectoliter.

business cycle Recurrent ups and downs over a period of years in the level of economic activity.

businesses Businesses produce goods and services using factor services from households and other businesses. They sell those goods and services to each other, households, the government, and foreign buyers.

capital Human-made resources such as buildings and machines used to produce goods and services. Capital goods do not directly satisfy human wants.

capital account The section in a nation's international balance of payments in which are recorded the capital inflows and the capital outflows of that nation. A negative balance on the capital account is called a *capital-account deficit;* a positive balance is called a *capital-account surplus.*

capital-account transaction   A transaction recorded in the capital inflows and capital outflows section of a nation's international balance of payments.

capital gain   The increase in value realized when securities or properties are sold for a price greater than the price paid for them. The difference between the purchase price and the sale price, after all expenses are paid, is called the capital gain.

capitalism   An economic system in which most of the economy's resources are privately owned and managed.

cargo preference   A policy requiring that a certain portion of goods or commodities exported from the United States be shipped in American ships.

carryover   The stock of a commodity that has not been used before the harvest of the succeeding year's crop begins.

cartel   A formal written or oral agreement among firms to set the price of the product and the outputs of the individual firms, or to divide the market for the product geographically. A cartel usually seeks to control production or the amount marketed to raise prices and maximize profits.

cash flow   The total funds generated by a firm. This flow of funds must be sufficient to cover the cost of operating the business.

central bank   The government agency responsible for managing a nation's money supply and credit conditions. In the United States, the Federal Reserve System has this function.

central planning   The management of a national economy by a bureaucracy that controls the allocation of resources among productive enterprises.

ceteris paribus   An assumption that other things are equal.

checkoff   A program by which a small amount of money per unit of product is deducted from the proceeds of a farm commodity when it is sold, usually by the first buyer, for the purpose of supporting research or promoting sales of that product.

choke price   The price at which a resource, as it nears exhaustion, becomes too costly to be used in productive activities.

circular flow   The flow of goods and services from businesses to households, the corresponding flow of factor services from households to business, and the equivalent flows of payments.

Clayton Act   The federal antitrust act of 1914 that strengthened the Sherman Act by making it illegal for business firms to engage in certain specified practices.

Clean Air Act of 1990   Legislation embodying a variety of measures to deal with air pollution, urban smog, motor vehicle emissions, ozone depletion, and acid rain.

close-down case   The circumstance in which a firm would experience a loss greater than its total fixed cost if it were to produce any output greater than zero; alternatively, a situation in which a firm would cease to operate when the price at which it can sell its product is less than its average variable cost.

closed economy   An economy that neither exports nor imports goods and services.

cobweb theorem   In some markets, the price and production decisions of producers generate a pattern of adjustments of prices and quantities that looks like a cobweb.

collective   A cooperative managed by the government.

collusion   A situation in which firms agree (collude) to set the price of the product, the amount each firm will produce, the geographic area in which each firm will sell, or to otherwise constrain competition.

command economy   An economic system in which property resources are publicly owned and central economic planning is used to direct and coordinate economic activities.

commercial banking system   All commercial banks and thrift institutions as a group.

commodities   Broadly defined, any goods exchanged in trade. However, the term is usually used to refer to widely traded raw materials and agricultural products such as wheat, corn, and rice.

Commodities Futures Trading Commission (CFTC)   The agency of the federal government responsible for regulating and overseeing the operations of all futures contract markets.

Commodity Credit Corporation (CCC)   A government-owned and government-run corporation authorized to borrow funds from the U.S. Treasury to operate the USDA's price and income support programs, to manage government-owned stocks of agricultural commodities, and to administer their disposal through domestic and export programs. Most activities are carried out by NRCA personnel, although certain programs are administered and implemented through the Agricultural Marketing Service, the Foreign Agricultural Service, and the Food and Nutrition Service.

commodity money   A commodity that is accepted as money. Examples of commodities that have been used as money include gold, cowrie shells, bullet casings, and bullocks.

common agricultural policy (CAP)   The agricultural policy of the European Community.

common property   Property that is managed by a group of individuals or firms for the benefit of the members of the group. In some cases, management occurs under formal government-sanctioned administrative structures. In other cases, the management occurs under informal traditional practices.

communism   As an economic system, communism prohibits the private ownership of most, if not all, productive resources. Communist countries are also characterized by extensive central planning, with the government setting prices and output levels.

comparative advantage   A firm or country that has a comparative advantage in the production of a good or service that has the lowest cost relative to the cost of other goods and services produced. Thus it is advantageous to trade some of this production for goods or services for which others have a comparative advantage.

competition   The presence in a market of a large number of independent buyers and sellers, the freedom of buyers and sellers to enter and leave the market, homogenous products, and perfect information. (See also *pure competition.*)

complementary goods   Goods or services for which there is an inverse relationship between the price of one and the demand for the other. When the price of one falls (rises), the demand for the other increases (decreases).

concentration ratio   The percentage of the total sales of an industry made by the four (or some other number) largest sellers (firms) in the industry.

conduct   The joint activities of firms in an industry with regard to the establishment of prices, quantities, advertising, and other economic decisions.

conglomerate merger   The merger of a firm in one industry with a firm in another industry (that is neither a supplier, customer, nor competitor).

conservation reserve program (CRP)   A program under which highly erodible land is retired from crop production and planted to grass or trees for a period of ten years. For doing this, the owner receives an annual payment from the government.

constant-cost industry   An industry in which expansion by the entry of new firms has no effect on the prices the firms in the industry pay for resources and no effect, therefore, on their cost schedules (curves).

consumer goods   Goods and services that satisfy human wants directly.

consumer price index (CPI)   A price index of the cost of a market basket of goods consumed by a typical American family. The CPI is the most widely used price index.

consumption function   The relationship between income and consumption.

consumption possibilities frontier   A line or curve that shows the consumption tradeoffs facing a nation with a given resource endowment. In a closed economy, the consumption possibilities frontier is the same as the production possibilities frontier. In an open economy, international trade allows the consumption possibilities frontier to be higher than the production possibilities frontier as a nation exploits its comparative advantage.

controlling   One of the four functions of management. It assesses the extent to which the goals of a firm or other organization are being met.

cooperative   A form of business owned by the customers. There are many types of farmer-owned cooperatives that provide supplies and services, or buy and sell agricultural commodities.

cooperative bargaining association   A bargaining association that is operated as a cooperative or by a cooperative.

corporation   A legal entity (person, firm, specified government agency) chartered by a state or the federal government. It is distinct and separate from the individuals who own it.

correlation   Systematic and dependable association between two variables (two kinds of events).

cost-of-living adjustment   An increase in the incomes (wages) of workers that is automatically received by them when there is inflation in the economy. It is guaranteed by a clause in their labor contracts with their employer.

cost-plus pricing   A procedure used by (oligopolistic) firms to determine the price they will charge for a product. A percentage markup is added to the estimated average cost of producing the product.

cost-push inflation   Inflation caused by increases in the prices of inputs, such as sharp increases in commodity prices due to crop failure or increasing oil prices due to supply interruptions.

cost sharing   In some farm programs, the government will share the cost of certain soil conservation and water quality enhancing practices with the farm owner or operator.

Council of Economic Advisors   A group of three persons who advise the President of the United States on economic policy.

credit   An accounting notation that the value of an asset (such as the foreign money owned by the residents of a nation) has increased.

credit union   An association of persons who have a common tie (such as employees of the same firm or members of the same labor union) that sells shares to (accepts deposits from) its members and makes loans to them.

cross compliance   The requirement that to qualify for price support and similar programs a farmer must also satisfy the provisions of other federal programs. For example, on farms with highly erosive soils, the operator must develop and implement a soil conservation program in order to qualify for price supports.

cross-price elasticity of demand   The ratio of the percentage change in quantity demanded of one good to the percentage change in the price of some other good. A negative coefficient for the cross-price elasticity of demand indicates that the two goods are substitutes. A positive coefficient indicates that they are complementary.

cross-price elasticity of supply   The ratio of the percentage change in quantity supplied of one good to the percentage change in the price of some other good. A negative coefficient indicates that the two goods are competitive in production (i.e., some of the resources used are the same). A positive coefficient indicates that the two goods are complementary in production.

currency appreciation   An increase in the number of foreign currency units that can be purchased with the domestic country's currency. For example, when the dollar appreciates relative to the yen, a dollar buys more yen.

currency in circulation   The coins and paper money in the hands of the public.

currency depreciation   The opposite of currency appreciation.

current account   In a nation's international balance of payments, the net value of a nation's exports and imports of goods (merchandise) and services, its net investment income, and its net transfers.

current-account transaction   The purchase or sale of merchandise (good or service) in international trade. (See also *current account.*)

cyclical unemployment   Unemployment caused by insufficient aggregate expenditures; it occurs in periods of decreased economic activity.

debit   An accounting notation signifying an increase in assets or decrease in liabilities. In balance-of-payments accounts it is an import.

debt/asset ratio   A measure used to determine the financial soundness of a firm. Farmers whose debts are equal to 70 percent or more of their assets are considered to be in financial difficulty.

decision-making skills   One of the characteristics of an effective manager. Generally, good skills involve a systematic approach to problem solving that includes setting goals, specifying alternatives, evaluating, choosing, implementing, reevaluating, and revising.

decrease in demand   A decrease in the quantity demanded of a good or service at every price; a shift of the demand curve to the left. (See also *shift in supply or demand.*)

decrease in supply   A decrease in the quantity supplied of a good or service at every price; a shift of the supply curve to the left. (See also *shift in supply or demand.*)

deficiency payment   Federal government funds paid to farmers when farm prices are below the target price. The payment rate is determined by taking the difference between the target price and an average price received by farmers, or the loan rate (whichever is higher). The total payment is determined by multiplying the payment rate times the acreage base times the program payment yield.

deflation   A fall in the general (average) level of prices in the economy.

demand curve   A curve that shows the amounts of a good or service buyers purchase at various prices during some period of time. Generally, the demand curve is downward-sloping because as the price of a good decreases, the amount of the good purchased increases.

demand deposit   A deposit in a commercial bank against which checks may be written; a checking account.

demand-pull inflation   Inflation that is the result of an increase in aggregate demand.

demand schedule   A schedule that shows the amounts of a good or service buyers purchase at various prices during some period of time.

depreciation of the dollar   A decrease in the value of the dollar relative to another currency so that a dollar buys a smaller amount of the foreign currency. For example, if the dollar price of a British pound changes from $1.75 to $2, the dollar has depreciated against the pound.

derived demand   The demand for a good or service that is dependent on or related to the demand for some other good or service; the demand for a resource that depends on the demand for the products it can be used to produce.

descriptive economics   The collection and reporting of relevant economic facts (data).

determinants of aggregate demand   Factors such as consumption, investment, government, and net export spending that, if they change, will shift the aggregate demand curve.

determinants of aggregate supply   Factors such as input prices, productivity, and the legal-institutional environment that, if they change, will shift the aggregate supply curve.

determinants of demand   Factors other than its price that determine the quantities demanded of a good or service.

determinants of price elasticity of demand   Factors that influence and change the price elasticity of demand. Such factors include the closeness of substitutes (or complements), level of income, tastes and preferences, time, variety of goods on the market, quality of goods, purchasing and saving habits.

determinants of price elasticity of supply   Factors that influence the price elasticity of supply. The chief factor is *time,* which is also associated with such factors as ease of substitution among products and among inputs.

determinants of supply   Factors other than its price that determine the quantities supplied of a good or service.

devaluation   A decrease by a nation in the value of its currency.

differentiated product   A product that differs physically or in some other way from the products produced by competing firms; buyers prefer the product of one seller over that of another.

diminishing marginal physical product (MPP)   The declining increase in output resulting from successive increases in one input.

diminishing marginal utility   As an individual consumes more of a good, less additional utility is derived from the last unit consumed.

direct market   A market in which the producers and final consumers meet to exchange goods. Also called a farmers' market or a bazaar.

direct relationship   In mathematics or economics, designating a relation between two variables in which one increases or decreases with the other (the opposite of a direct relationship is an *inverse* relationship).

directing   To give directions, to supervise, to manage, to conduct.

discount rate   The interest rate that the Federal Reserve Banks charge on the loans they make to depository institutions.

discretionary fiscal policy   Changes in taxes (tax rates) and government spending (spending for goods and services and transfer payment programs) by a nation, usually to move toward a full-employment, noninflationary gross domestic product and economic growth.

diseconomies of scale   Forces that increase the average cost of producing a product as the firm expands the size of its plant (its output) in the long run.

disposable income   Personal income less personal taxes; income available for personal consumption expenditures and personal saving.

dissaving   Spending for consumer goods and services in excess of disposable income; the amount by which personal consumption expenditures exceed disposable income.

distribution   The process by which goods move from production to consumption. The distribution process includes storing, selling, shipping, and advertising.

diversion payment   Payments made to farm owners and operators for diverting land from the production of specified crops into conservation uses.

diverted acres   Acres that are taken out of production and diverted to some conserving use under acreage reduction programs.

diversified firm   A firm that produces several different outputs.

divestiture   The sale of a division of a corporation to another firm.

division of labor   Dividing the work required to produce a product into a number of different tasks that are performed by different workers; specialization of workers.

dollarization   The decision by a country to use the U.S. dollar as its official currency, as has occurred in Panama and Ecuador.

domestic capital formation   Adding to a nation's stock of capital by saving part of its domestic output to, for example, add to plant and equipment.

double coincidence of wants   A situation in which an item (good or service) that one trader wishes to obtain is the same item that another trader desires to give up, and the item that the second trader wishes to acquire is the same item that the first trader offers.

double counting   Overestimating the value of intermediate goods in the gross domestic product, due to counting the same good or service more than once.

dumping   The sale of products below cost in a foreign country.

durable good   A consumer good with an expected life (use) of one year or more.

earnings   The money income received by a worker. It is equal to the wage (rate) multiplied by the quantity of labor supplied (the amount of time worked) by the worker.

EC   European Community.

economic analysis   Deriving economic principles from relevant economic facts.

economic efficiency   The production of the maximum possible value of goods and services from the nation's resources. For a firm, operation on or near the production function (technical efficiency) and with the cost minimizing combination of inputs (economic efficiency).

economic exhaustion   The point at which the increase in the price of a resource reaches a choke price that ends the use of this resource.

economic growth   (1) An outward shift of the production possibilities schedule or curve that results from an increase in the supply of inputs or from technological advances; (2) An increase in real output (or gross domestic product) or in real output per capita.

economic integration   Cooperation among and the complete or partial unification of the economies of different nations; the elimination of the barriers to trade among these nations; the bringing together of markets in each of the separate economies to form one large (common) market.

economic model   A simplified picture of reality; an abstract generalization.

economic perspective   A viewpoint that envisions individuals and institutions making rational or purposeful decisions based on a consideration of the benefits and costs associated with their actions.

economic policy   Course of action intended to correct or avoid a perceived economic problem or achieve an economic goal.

economic principle   Generalization of the economic behavior of individuals and institutions.

economic profit   The total revenue of a firm less all its economic costs. It is also called *pure profit* and *above-normal profit.*

economic rent   The payment received for the use of a natural resource above the opportunity cost of the use of the resource.

Economic Research Service (ERS)   An agency of the USDA that conducts research on economic problems concerning agricultural commodities and related issues.

economic theory   Deriving economic principles from relevant economic facts; an economic principle.

economics   A social science concerned with the allocation of scarce resources to obtain the maximum satisfaction of the unlimited wants of society now and in the future.

economies of scale   The forces that reduce the average cost of producing a product as the firm expands the size of its plant (its output) in the long run; the economies of mass production. Also applies to different sizes of plants in the short run.

efficiency   In the context of a firm, producing the largest possible amount of output from the inputs used by the firm.

efficient allocation of resources   The allocation of the scarce resources of an economy among the production of different products and services that leads to the maximum satisfaction of the wants of consumers.

elastic demand   A market situation in which a percentage change in price results in a greater proportional change in the amount purchased. Thus total receipts will be greater with a lower price.

elastic supply   The elasticity coefficient is greater than one. The percentage change in quantity supplied is greater than the percentage change in price.

elasticity coefficient   A number that compares the percentage change in one variable with the percentage change in another variable. In particular, it is the number obtained when the percentage change in quantity demanded (or supplied) is divided by the percentage change in the price of the commodity, another commodity, or income.

eminent domain   The right reserved by government to take land or other assets for public use, generally with compensation at fair market value.

emission fees   Special fees levied against those discharging pollutants into the environment.

employment rate   The percentage of the labor force employed at any time.

Engel's law   As income rises, the share of total expenditure for food declines.

equality versus efficiency tradeoff   The presumption that in order to achieve economic efficiency a society must accept an increase in income inequality.

equilibrium domestic output   The real domestic output at which the aggregate demand curve intersects the aggregate supply curve.

equilibrium gross domestic product   The gross domestic product at which the total quantity of final goods and services produced (the domestic output) is equal to the total quantity of final goods and services purchased (aggregate expenditures).

equilibrium price   The price in a competitive market at which the quantity demanded and the quantity supplied are equal; the price at which there is neither a shortage nor a surplus, and at which there is no tendency for price to rise or fall.

equilibrium price level   The price level at which the aggregate demand curve intersects the aggregate supply curve.

equilibrium quantity   The quantity demanded and quantity supplied at the equilibrium price in a competitive market.

equilibrium real domestic output   The real domestic output that is determined by the equality (intersection) of aggregate demand and aggregate supply.

equity   The net worth of a firm; the net value of property after all debts are deducted.

erosion   The loss of soil due to the action of water or wind. The soil may be moved to a new location on the site (e.g., farm, ranch, construction or off-site), where it may have unwanted effects on the environment.

European Union (EU)   Is the result of a process of cooperation and integration which began in 1951. After nearly fifty years, with four waves of accessions the EU today has fifteen Member States and is preparing for its fifth enlargement, this time towards Eastern and Southern Europe.

excess reserves   The amount by which a member bank's actual reserves exceed its required reserves. It is equal to actual reserves minus required reserves.

exchange rate   The amount of a foreign currency that can be purchased with a unit of domestic currency.

exchange rate appreciation   An increase in the value of a nation's money in foreign exchange markets; an increase in the rates of exchange for foreign moneys.

exchange rate depreciation   A decrease in the value of a nation's money in foreign exchange markets; a decrease in the rates of exchange for foreign moneys.

excise tax   A tax levied on the expenditure for a specific product or on the quantity of the product purchased.

expanding economy An economy in which net private domestic investment is greater than zero (gross private domestic investment is greater than depreciation).

expanding industry An industry in which firms realize economic profits and output is increased by existing firms and the entry of new firms.

expansion path The optimal combinations of two inputs that can be used to produce increasing quantities of a product. The cost-minimizing combinations are given by the tangency of isoquants and isocost curves.

expansionary fiscal policy An increase in aggregate demand brought about by an increase in government expenditures for goods and services, a decrease in net taxes, or some combination of the two.

expectations What consumers, business firms, and others believe will happen or what conditions will be in the future.

expected price The price that a buyer expects to pay or a seller expects to receive in the future for a product.

expected production The expected yield times the acreage planted.

expected yield The yield a producer expects to get given the amount of fertilizer applied and the other production inputs used. Weather and pests usually cause the actual yield to differ from the expected yield.

export certificates A discretionary provision under which the Secretary of Agriculture may make export certificates available to producers participating in the wheat and feed grain programs. The certificates are redeemable for cash when the certificate holder shows that the specified amount of grain has been exported.

export enhancement programs (EEPs) Various programs in which the government pays subsidies in money or in kind to increase the volume of agricultural exports.

export payment in kind (PIK) An export subsidy program under which the government provides exporters with special bonuses in the form of commodities so that they can compete in the international market. Export PIK commodities are given only to exporters selling to countries in which competing exporting countries are also subsidizing exports.

export subsidies Government payments that reduce the price of a product to foreign buyers. In Europe, they are often referred to as restitutions.

export tax A tax on exports, either ad valorem (a percentage of the value) or specific (so much per unit).

exports Goods sold in the market of a foreign country.

externality An uncompensated positive or negative effect of a production or consumption activity on another producer or consumer. Generally a discharge to the environment that adversely affects other individuals and for which there is not a payment based on a market price.

extinction The elimination of a species from life on earth.

factors of production Economic resources whose services can be used in the production of economically useful commodities: natural resources, capital, and human capital.

farm Since 1978, the U.S. Bureau of the Census has defined a farm as an entity that has or would have had $1000 in gross sales of farm products.

Farm Credit Administration An independent agency of the federal government that supervises the Farm Credit System.

Farm Credit System (FCS) The credit institutions established by authority of Congress which are now farmer-owned: the federal land banks, the federal intermediate credit banks, production credit associations, and banks for cooperatives.

farm problem Historically, the relatively low income of many farmers (compared with incomes outside the farm sector) and the tendency for farm income to fluctuate sharply from year to year.

farm-retail price spread The difference between the price received at the farm level and the price charged at the retail level. The difference is accounted for by the costs of services provided by agribusinesses and is determined by the interactions of supply and demand in both farm and retail markets.

farm sector Includes all firms involved in producing crops and raising livestock, usually for sale (limited by definition in the United States to firms that produce $1000 or more of crop or livestock products annually).

Farmers Home Administration (FmHA) An agency of the USDA that provides credit to farmers and some other rural Americans who are unable to borrow from other sources. The agency also provides loan guarantees to other commercial lenders making loans to farmers and rural communities.

FDIC (See *Federal Deposit Insurance Corporation.*)

Federal Communications Commission (FCC) The U.S. agency that licenses radio and television broadcasting stations.

Federal Deposit Insurance Corporation (FDIC) The federally chartered corporation that insures the deposit liabilities of commercial banks and thrift institutions.

federal funds rate The interest rate that lending depository institutions charge borrowing institutions for the use of excess reserves.

Federal Reserve Bank Any one of the twelve banks chartered by the U.S. government to control the money supply and perform other functions.

Federal Reserve Note Paper money issued by the Federal Reserve Banks.

Federal Reserve System The central bank of the United States and member institutions.

Federal Trade Commission (FTC) The commission of five members established by the Federal Trade Commission Act of 1914 to investigate unfair competitive practices of business firms, to hold hearings of the complaints of such practices, and to issue cease and desist orders when firms are found to engage in such practices.

feudalism The system of political and economic control that prevailed in Europe during the Middle Ages, based on landlords providing protection to vassals (or serfs) in exchange for a share of their production.

final-expenditure approach A method for measuring the gross domestic product (GDP), based on aggregate national expenditures on final goods and services bought by consumers.

financial analysis Analysis of the performance of economic units such as individuals, firms, and governments in terms of profit or loss, balance between assets and liabilities, net worth, appreciation and depreciation, and so forth.

financial intermediaries Financial intermediaries serve to transfer money or credit from savers to investors. Examples include banks, savings and loan associations, mutual funds, and insurance companies.

firm An organization that employs resources to produce a good or service for profit.

firm short-run supply curve A curve that shows the quantities of a product a firm will offer to sell at various prices in the short run; in pure competition, the portion of the firm's short-run marginal cost curve that lies above its average variable cost curve.

fiscal analysis The analysis of government taxation and spending. It can also be used in the context of the analysis of the financial performance of a firm or other entities.

fiscal federalism The system of transfers (grants) by which the federal government shares its revenues with state and local governments.

fiscal policy   Changes in government spending and tax collections for the purpose of moving toward a full-employment and noninflationary domestic output.

fixed costs   The costs that do not vary with production levels. Examples are property taxes, land and machinery payments, and other costs that are often referred to as sunk costs. There are no fixed costs in the long run because all inputs are variable.

fixed exchange rate   A rate of exchange that is prevented by national policy from rising or falling.

fixed inputs   Those inputs the firm owns or is committed to pay for regardless of the amount of output produced. Changing the amount of fixed inputs implies changing the size of the firm.

flexible exchange rate   A rate of exchange that is determined by the demand for and supply of foreign money. It is free to rise or fall.

flow   An amount measured over a specific period of time; a rate. Examples of flow are the interest earned on a financial asset and the amount mined each year from a reserve of coal.

food for peace program   The program that permits less developed nations to buy surplus U.S. agricultural products and pay for them with their own moneys (instead of dollars).

food stamp program   A program in the United States that permits low-income persons to purchase for less than their retail value, or to obtain without cost, coupons that can be exchanged for food items at retail stores.

foreign demand   The demand by other nations for the goods produced by the subject nation.

foreign exchange market   A market in which the money (currency) used by one nation purchases (is exchanged for) the money used by another nation.

foreign investment   Expenditures by a government, firm, or individual to purchase assets (resources, firms, goods, or services) in another country.

foreign loan   Loans by a government, firm, or individual to a government, firm, or individual in another country.

form utility   The utility created by a manufacturing process that converts raw materials and intermediate products into a form with more value to buyers.

fractional reserve   A reserve ratio, usually 15 to 20 percent, of the deposit liabilities of a commercial bank that must be held in the bank.

free rider   An individual or firm enjoying the benefits of a group's efforts without sharing in the cost of their provision.

free-rider problem   The situation when there are so many free riders that the goods or services cannot be provided. One solution is to find a means of charging free riders.

free trade   The absence of government-imposed barriers to trade among individuals and firms in different nations.

Free Trade Agreement of the Americas   A proposed agreement among all the democratic countries in North, Central, and South America to reach an agreement to significantly reduce barriers to trade among all these countries.

freedom of entry and exit   The ability of firms to enter and leave an industry without facing legal barriers (patents), raising substantial capital, or other requirements.

frictional unemployment   Unemployment that results from workers voluntarily changing jobs and from temporary layoffs.

FTAA   (See *Free Trade Agreement of the Americas*.)

full employment   Using all available resources to produce goods and services. There is no cyclical unemployment; however, there will be some structural unemployment.

full-employment budget   What government expenditures and revenues and its surplus or deficit would be if the economy were to operate at full employment.

full production   The maximum amount of goods and services that can be produced from the employed resources of an economy; the absence of underemployment.

futures contract   A contract to buy (or sell) a set amount of a commodity for delivery at a future time and place.

game theory   A theory that analyzes the behavior of participants in games of strategy, such as poker and chess, with that of a small group of mutually interdependent firms (an oligopoly).

General Accounting Office (GAO)   An agency of Congress that investigates the operations of various programs and the expenditure of appropriated funds.

General Agreement on Tariffs and Trade (GATT)   A multilateral agreement on trade now called the World Trade Organization.

general equilibrium analysis   A study of the market system as a whole and of the interrelations among equilibrium prices, outputs, and employments in all markets of the economy.

genetic engineering   (See *recombinant DNA technology*).

genetically modified organisms (GMOs)   Organisms created through the use of rDNA technologies.

goods   In economic terms, a good is anything that helps to satisfy human wants or can be used to produce some other things that can satisfy human wants.

government purchases   Government expenditures for currently produced goods or services. The expenditures of all governments in the economy for final goods and services.

government transfer payment   The disbursement of money (or goods and services) by government for which government receives no currently produced good or service in return.

gross domestic product (GDP)   The total market value of all final goods and services produced annually within the boundaries of the United States, whether by U.S. or foreign-supplied resources.

gross domestic product (GDP) per capita   The GDP divided by the total population of the country.

gross national product (GNP)   The total market value of all final goods and services produced annually by land, labor, and capital supplied by U.S. residents, whether these resources are located in the United States or abroad.

guaranteed loan   A government guarantee to repay a loan made by a private lender. The Farmers Home Administration guarantees some loans made by private banks to farmers and other qualified borrowers.

head tax   A tax levied "per head." Because it is collected regardless of the level of income, it is a regressive tax.

hectare   An amount of land. One hectare equals 2.471 acres.

homogenous product   A product such that buyers are indifferent to the seller from whom they purchase if the price charged by all sellers is the same; products of competing firms are perfect substitutes.

horizontal integration   A group of plants in the same stage of a production process that are owned or managed by a single firm.

horizontal merger   The merger of one or more firms producing the same product into a single firm.

household   Economic units (of one or more persons) that own all factors of production; sell services of factors of production to businesses and other households; and buy goods and services for final consumption from businesses, other households, and governments.

hyperinflation   A very rapid rise in the price level of a nation.

IMF   (See *International Monetary Fund.*)

imperfect competition   All market structures except pure competition. Among sellers it includes monopoly, duopoly, oligopoly, and monopolistic competition. Among buyers it includes monopsony, duopsony, oligopsony, and monopsonistic competition.

implicit cost   The income given up by an individual or firm when a resource is not supplied on the market. It is the opportunity cost, what the resource could have earned in the best-paying alternative employment.

import competition   Competition that domestic firms encounter from the products and services of foreign suppliers.

import quota   A limit imposed by a nation on the quantity of a good that may be imported during some period of time.

import tariff   A tax on imports. It may be either ad valorem (a percentage of the import price) or specific (a fixed dollar amount regardless of the import price).

imports   Goods and services purchased by businesses, individuals, or governments from a foreign country.

incentive function of price   The inducement that an increase (a decrease) in the price of a commodity offers to sellers of the commodity to make more (less) of it available, and the inducement that an increase (decrease) in price offers to buyers to purchase smaller (larger) quantities. It is the guiding function of prices.

incentive pay plan   Compensation scheme that ties worker pay directly to performance. Such plans include piece rates, bonuses, commissions, and profit sharing.

income approach   The measurement of the gross domestic product by adding all the incomes generated by the production of final goods and services.

income effect   The effect of a change in the price of a product on the real income (purchasing power) of a consumer, and the resulting effect on the quantity of that product the consumer would purchase after the consequences of the substitution effect have been taken into account.

income elasticity of demand   The ratio of the percentage change in the quantity demanded of a good to the percentage change in income. It measures the responsiveness of consumer purchases to income changes.

income statement   A statement that shows the financial performance of a business over a period of time such as a quarter or a year.

increasing-cost industry   An industry in which expansion through the entry of new firms increases the prices the firms in the industry must pay for resources and, therefore, increases their cost schedules (shifts their cost curves upward).

increasing returns   An increase in the marginal product of a resource as successive units of the resource are employed.

indemnity programs   Programs under which payments are made to producers who sustain losses as a result of pesticides, nuclear radiation fallout, residues, toxic substances, or other causes.

independent goods   Goods or services for which there is no relationship between the price of one and the demand for the other. When the price of one rises or falls, the demand for the other remains constant.

independent variable   The variable that causes a change in another (dependent) variable.

indifference curve   A curve connecting all the combinations of goods among which the consumer is indifferent (that give an individual equal utility).

indifference map   A series of indifference curves, each of which represents a different level of utility and that together show the preferences of the consumer for two goods.

indirect business taxes   Such taxes as sales, excise, and business property taxes, license fees, and tariffs that firms treat as costs of producing a product and pass on (in whole or in part) to buyers of the product by charging them higher prices.

individual demand   The demand schedule or demand curve of a single buyer of a good or service.

individual supply   The supply schedule or supply curve of a single seller of a good or service.

induced innovation   The tendency of science to invent new products in response to the needs of society.

industry   The firms that produce identical or similar products.

industry short-run supply curve   The horizontal summation of the short-run supply curves of the firms in that industry.

inelastic demand   The elasticity coefficient is less than one. The percentage change in price is greater than the percentage change in quantity demanded. A seller facing an inelastic demand will increase total revenue by increasing price.

inelastic supply   The elasticity coefficient is less than one. The percentage change in price is greater than the percentage change in quantity supplied.

inferior good   A good or service of which consumers purchase less (more) at a given price when their incomes increase (decrease).

inflation   A rise in the general (average) level of prices in the economy.

inflationary gap   The amount by which the aggregate expenditures schedule (curve) must decrease (shift downward) to decrease the nominal GDP to the full-employment noninflationary level.

infrastructure   For the economy, the capital goods usually provided by the public sector for the use of its citizens and firms (e.g., highways, bridges, transit systems, waste-water treatment facilities, education facilities). For the firm, the services and facilities that it must have to produce its products, that would be too costly for it to provide for itself, and that are provided by governments or other firms (e.g., water, electricity, waste treatment, education, transportation, research).

innovation   The introduction of a new product, the use of a new method of production, or the employment of a new form of business organization.

input   Items used by a firm in the production of a good or service. For example, seed, fertilizer, chemicals, feed, machinery, fuel, labor, and land are farm inputs.

input sector   The agribusiness firms that produce and sell to farmers and ranchers the inputs used in production.

input-output analysis   Using an input-output table to examine interdependence among different parts (sectors and industries) of the economy and to make economic forecasts and plans.

institutional infrastructure   The system of property rights, the judicial system, rules of incorporation, market rules and regulations, commodity grades and standards, and other arrangements that make it possible for an economic system to function.

integrated pest management (IPM)   The use of a combination of techniques to control pests on crops, usually as an alternative to the exclusive reliance on chemical pesticides. Techniques include tillage, scouting to determine level of infestation, crop rotations, and low rate of pesticides use.

integration   Combining of various steps in the production and marketing of a product under the management or control of a single firm.

interest   The payment made for the use of money (of borrowed funds).

interest income   Income of those who supply the economy with capital.

intermediate inputs   Goods that are purchased for resale or for further processing or manufacturing.

intermediate range   The upward-sloping segment of the aggregate supply curve that lies between the Keynesian range and the classical range.

internally held public debt   Public debt owed to (government securities owned by) citizens, firms, and institutions of a nation.

internalization   A change in property rights resulting in the benefits and adverse effects of an externality being realized by the same individual or firm.

international balance of payments   Sum of the transactions that took place between the individuals, firms, and governments of one nation and those in all other nations during the year.

International Monetary Fund (IMF)   The international association of nations that was formed after World War II to make loans of foreign moneys to nations.

international trade   The sale and purchase of goods for shipment across national boundaries.

international trade barriers   Regulations used by governments to restrict imports from, and exports to, other countries. Examples are tariffs, embargoes, import quotas, and unnecessary sanitary restrictions.

international value of the dollar   The price that must be paid in foreign currency (money) to obtain one U.S. dollar.

Interstate Commerce Commission (ICC)   The commission established in 1887 to regulate the rates and monitor the services of the railroads in the United States. Now regulates and monitors all forms of transport.

inverse relationship   The relationship between two variables that change in opposite directions, for example, product price and quantity demanded.

investment   In macroeconomics, investment is the addition of new capital equipment to the stock of existing capital equipment. For example, construction of a new building is investment, while purchase of an existing building is not. In microeconomics, investment is an addition to the firm's fixed assets by building new, or purchasing existing, facilities.

investment curve   A curve that shows the amounts firms plan to invest at different income levels of GDP.

investment-demand schedule   A schedule that shows rates of interest and the amount of investment at each rate.

investment schedule   A schedule that shows the amounts firms plan to invest at different levels of GDP.

invisible hand   The phenomenon that as individuals, firms, and resource suppliers further their self-interests in competitive markets, they also further the best interests of society.

invisibles   International trade in nontangible services such as airline flights and consulting services.

isocost line   The combinations of two inputs that can be purchased with a given amount of money. The isocost line is used with an isoquant to determine the cost-minimizing combinations of inputs to produce a given quantity of output.

isoquant   The alternative combinations of two inputs that can be used to produce a given quantity of output.

isoquant map   The several isoquants which taken together map the combinations of inputs that can be used to various quantities of output.

isorevenue line   The revenue the firm can earn from various combinations of two products.

Keynesian economics The macroeconomic generalizations accepted today by most (but not all) economists. According to Keynesian economics, a capitalistic economy does not always employ its resources fully; thus government fiscal and monetary policies are needed to promote full employment.

Keynesian region The horizontal segment of the aggregate supply curve along which the price level is constant as real domestic output changes.

kinked demand curve The demand curve that a noncollusive oligopolist sees for its output and that is based on the assumption that rivals will follow a price decrease but not a price increase.

labor The physical and mental talents (efforts) of people that can be used to produce goods and services.

labor force Persons 16 years of age and older who are not in institutions and who are employed or are unemployed and seeking work.

labor-intensive commodity A product that requires a large amount of labor, relative to the amount of capital, to produce.

labor productivity Total output divided by the quantity of labor employed to produce the output; the average product of labor or output per worker per hour.

labor union A group of workers organized to advance the interests of the group (to increase wages, shorten the hours worked, improve working conditions, etc.).

land-grant university A university established under the Morrill Act of 1862.

land management Refers to the choices made by a land owner as to the crops grown, livestock produced, and conservation measures employed on the land.

land ownership Refers to the rights of individuals, groups, and other entities to control the use of land and to receive the income generated by its use.

land retirement A provision used in several agricultural production control and erosion control programs whereby the government pays farmers to withhold land from production.

law of demand The inverse relationship between the price and the quantity demanded of a good or service during some period of time.

law of diminishing returns As more of a variable resource is used in combination with a fixed resource, beyond some level of employment the marginal product of the variable resource will decrease.

law of supply The direct relationship between the price and the quantity supplied of a good or service during some period of time.

least-cost combination rule (of resources) The quantity of each resource a firm must employ to produce an output at the lowest total cost occurs where the ratio of the marginal product to its marginal resource cost (price in a competitive market) is the same for all resources employed.

legal reserves (deposit) The minimum amount that a depository institution must keep on deposit with the Federal Reserve Bank, or in vault cash.

legal tender Anything that government decrees must be accepted in payment of a debt.

lending potential of the banking system The amount by which the commercial banking system can increase the money supply by making new loans to (or buying securities from) the public.

less developed countries (LDCs) Countries characterized by a lack of capital goods, primitive production technologies, low literacy rates, high unemployment, rapid population growth, and generally with most of the labor force committed to agriculture.

liability An amount owed by a firm or an individual.

life cycle Refers to the tendency of most products (and industries) to increase slowly during an initial stage, to rise more rapidly in a growth phase, to level off in a mature stage, and to eventually go into a phase of decline.

limited liability   Restriction of the maximum that may be lost to a predetermined amount.

liquid assets   Money or things that can be quickly and easily converted into money with little or no loss of purchasing power.

loan forfeiture   The forfeiting of commodities placed under loan instead of repaying the loan in cash.

loan rate   The price per unit (bushel, bale, pound) at which the government will provide loans to farmers to enable them to hold their crops for later sale.

long run   A period of time long enough for producers to change the quantities of all resources. All resources and costs are variable.

long-run aggregate supply curve   The aggregate supply curve associated with a time period in which input prices (especially nominal wages) are fully responsive to changes in the price level.

long-run average cost curve   A curve that envelopes the average total cost curves of firms operating at different sizes as they employ various levels of the fixed and variable factors of production.

long-run competitive equilibrium   The production and price levels at which firms in pure competition neither obtain economic profit nor suffer losses in the long run, and the total quantities demanded and supplied are equal. At equilibrium the price is equal to the minimum long-run average cost of producing the product.

long-run equilibrium in imperfect competition   The production and price levels at which the total quantities demanded and supplied at that price are equal. Because of product differentiation and downward-sloping demand curves, firms do not produce at the minimum long-run average cost and may realize economic profits.

long-run supply   A schedule or curve that shows the prices at which a purely competitive industry will make various quantities of the product available in the long run.

Lorenz curve   A curve that shows the distribution of income in an economy relating the cumulated percentage of families by income receivers is measured to the cumulated percentage of income.

loss-minimizing case   Occurs when the firm chooses to operate with MC = MR at a level of output where the revenues are greater than AVC but less than ATC rather than to shut down and pay the full amount of fixed costs.

luxury goods   Goods with an income elasticity greater than one. Given a certain percentage increase in incomes, consumption increases by a larger percentage.

M1   The currency in circulation and checkable deposits (not owned by the federal government, Federal Reserve Banks, or depository institutions).

M2   M1 plus noncheckable savings deposits, small time deposits (deposits of less than $100,000), money market deposit accounts, and individual money market mutual fund balances.

M3   M2 plus large time deposits (deposits of $100,000 or more).

macroeconomic policies   Monetary and fiscal policies which affect the general economic environment.

macroeconomics   The part of economics concerned with a nation's economy as a whole; with such aggregates as the household, business, and governmental sectors; and with totals for the economy.

marginal cost (MC)   The change in total cost with one additional unit of output. If the amount of output changes by more than one unit, marginal cost is the change in total cost per unit change in output.

marginal physical product (MPP)   The change in output resulting from one additional unit of a variable input. If the amount of output changes by more than one unit, marginal physical product is the change in total output per unit change in input (MPP = ΔTPP $\div$ ΔQ_{input}).

marginal propensity to consume   The fraction of any change in disposable income that is spent for consumer goods.

marginal propensity to save   The fraction of any change in disposable income that households save.

marginal rate of substitution   The rate at which a consumer will substitute one good or service for another and remain equally satisfied (have the same total utility). It is equal to the slope of an indifference curve.

marginal resource cost   The amount by which the total cost of employing a resource increases when a firm employs one additional unit of the resource.

marginal revenue (MR)   The change in revenue from a one-unit increase in output. If the amount of output changes by more than one unit, marginal revenue is the change in total revenue per unit change in output ($MR = \Delta TR \div \Delta Q$). For a small firm with no control of its market price, $MR = P = AR$.

marginal revenue-marginal cost approach   A firm will maximize its economic profit (or minimize its losses) by producing the output at which marginal revenue and marginal cost are equal, provided the price at which it can sell its products is equal to or greater than average variable cost.

marginal revenue product   The change in the total revenue of the firm when it employs one additional unit of a resource.

marginal utility   The extra utility a consumer obtains from the consumption of one additional unit of a good or service.

marginal value product (MVP)   The marginal physical product (MPP) times the price of output.

market   Any institution or mechanism that brings together the buyers (demanders) and sellers (suppliers).

market basket   A selection of items that is representative of the purchases made by consumers. The changes in prices of these items are used to determine whether consumer prices are increasing or decreasing in the economy. The rate of change is measured as a price index and is called the inflation or deflation rate.

market contract   A formal agreement between a seller and a buyer to deliver a certain good in a certain amount or quality at a specified price. For example, some agricultural producers have contracts with sellers of farm inputs and buyers of farm products, such as broiler producers, hog feeders, cattle feeders, and so on.

market demand   A demand curve that applies to a market. The summation of quantities demanded by all individuals at different prices.

market demand curve   The horizontal summation of all individual demand curves for a good.

market economy   An economy in which only the private decisions of consumers, resource suppliers, and business firms determine the allocation of resources.

market equilibrium   The situation when sellers can sell as much as they desire and buyers can buy as much as they want at the market price.

market failure   The failure of a market to develop or to operate to bring about the allocation of resources that best satisfies the wants of society. For example, the over- or underallocation of resources to produce a good or service (because of externalities or informational problems).

market period   A period of time so short that producers of a product are unable to change the quantity produced in response to a change in its price, in which there is perfect inelasticity of supply, and in which all resources are fixed.

market price   The price at which buyers and sellers willingly exchange goods or services.

market segmentation   Allocating portions of a product for sale in segments of the market with differing elasticities of demand, such as fresh, frozen, and export markets for fruit.

market share   The percentage of the total sales in a market made by a firm.

market supply   A supply curve that applies to a market. The summation of quantities offered by all firms at different prices.

market supply curve   The horizontal summation of all individual producer supply curves for that good.

market system   All the product and resource markets of the economy and the relationships among them. It is a method that allows the prices determined in these markets to allocate the economy's scarce resources and to communicate and coordinate the decisions made by consumers, business firms, and resource suppliers.

marketing certificates   Certificates issued by the USDA as part of a price support program that may be redeemed for cash or commodities.

marketing contract   A contract between a producer and a processor (or other seller and buyer) that assures a market for a commodity at harvest at a set price.

marketing loan program   A U.S. price support mechanism for farm products in which farmers are offered the option of selling a product or placing it in government-approved storage and borrowing, using the commodity as collateral for the loan. It is sometimes called a nonrecourse loan program because the government must accept the commodity in payment of the loan when the market price is lower than the loan on the commodity.

marketing margin   The difference between the price the producer receives for a commodity and the price paid by the consumer for an equivalent amount of the same product.

marketing orders and agreements   Federal government regulatory programs that permit agricultural producers to collectively market a good. It permits producers to regulate marketing of the commodity by means of regulatory restrictions on all handlers of the commodity. Restrictions may involve packing standards, grades, size, price, and limitations on quantities shipped or marketed.

marketing quota   The quantity of a crop determined by the USDA to provide adequate and normal market supplies at the national level. This quantity is translated into acreage or individual farm marketing quotas based on a farm's previous production of that commodity. May apply to livestock products such as fluid milk.

maximum sustainable yield (MSY)   The maximum amount of renewable resources such as fish that can be harvested on a continuous basis.

medium of exchange   Money or legal tender; a convenient means of exchanging goods and services without engaging in barter; what sellers generally accept and buyers generally use to pay for a good or service.

merchandise trade balance   A term used in international trade to measure the net trade of tangible goods (rather than to services or fund transfers).

merger   Combining two or more firms into one operating unit.

microeconomics   The part of economics concerned with individual units within the economy—such as industries, firms, and households—and with individual markets, particular prices, and specific goods and services.

minimum wage   The lowest wage (rate) employers may legally pay for an hour of labor.

monetary multiplier   The multiple of its excess reserves by which the commercial banking system can expand the money supply and demand deposits by making new loans (or buying securities). It is equal to one divided by the required reserve ratio.

monetary policy   Changing the money supply to assist the economy in seeking a full-employment, noninflationary level of total output.

money   Any item that is generally acceptable to sellers in exchange for goods and services.

money market   The market in which the demand for and the supply of money determine the interest rate (or the level of interest rates) in the economy.

money supply   (See *M*1, *M*2, and *M*3.)

monoclonal antibodies   Exceptionally pure antibodies that can be produced by *cloning*. These antibodies can be used in many ways, such as for passive immunization; substitutes for vaccines; sexing of livestock embryos; detection of food poisoning; imaging, targeting, and killing of cancer cells.

monopolistic competition   A market in which many firms sell a differentiated product. Entry is relatively easy, the firm has some control over the price at which the product it produces is sold, and there is considerable nonprice competition.

monopoly   A market in which there is only one seller of a good, service, or resource.

monopsonistic competition   A market in which many firms purchase a differentiated product. Entry is relatively easy, the firm has some control over the price paid for the product, and there is considerable nonprice competition.

monopsony   A market in which there is only one buyer of a good, service, or resource.

moral hazard   The possibility that individuals or institutions will change their behavior in unanticipated ways as the result of a contract or agreement. For example, a bank whose deposits are insured against loss may make riskier loans and investments.

most-favored-nation clause   A clause in a trade agreement between the United States and another nation which provides that the other nation's imports will be subjected to the lowest tariff levied then or later on any other nation's imports into the United States.

MR = MC rule   (See *marginal revenue = marginal cost approach.*)

MRP = MRC rule   To maximize economic profit (or minimize losses) a firm should employ the quantity of a resource at which its marginal revenue product is equal to its marginal resource cost.

multiplier   The ratio of the change in the equilibrium gross domestic product (GDP) to the change in investment, to the change in any other component of the aggregate expenditures schedule, or to the net change in taxes. It is the number by which a change in any component in the aggregate expenditures schedule or in net taxes may be multiplied to find the resulting change in the equilibrium GDP.

multiplier effect   The effect on equilibrium gross domestic product of a change in the aggregate expenditures schedule (caused by a change in the consumption schedule, investment, net taxes, government expenditures for goods and services, or net exports).

mutual interdependence   A situation in which a change in price or some other policy by a firm (or firms) will affect the sales and performances of another firm (or firms).

NAFTA   (See *North American Free Trade Agreement.*)

national income   The total income earned by resource suppliers for their contributions to the production of the gross domestic product.

national income accounting   The techniques employed to measure (estimate) the overall production of the economy and other related economic measures for the nation as a whole.

national-income approach   A method used to measure annual national income based on the total income earned by suppliers of factor services for the production of the gross domestic product (GDP). The sources of national income equal the sum of wages, rent, interest, depreciation, and profits. The uses of national income equals the sum of consumption, savings, and taxes.

natural monopoly   An industry in which the economies of scale are so great that the product can be produced by one firm at an average cost that is lower than it would be if it were produced by more than one firm.

Natural Resource Conservation Agency (NRCA)   An agency of the federal government that provides technical and other assistance to land owners to control erosion and maintain the quality of land and water resources. Previously named the Soil Conservation Service (SCS).

natural resources   Factors of production that are from natural sources such as minerals, energy (coal and oil), water, land, forests, and so on.

negative externality   A pollutant or other agent generated by individuals or firms that has adverse affects on other individuals or agents and for which those adversely affected are not compensated by payments determined in a market.

negative income tax   The proposal to subsidize families and individuals with money payments when their incomes fall below a guaranteed level. The negative tax would decrease as earned income increases.

net domestic product   Gross domestic product minus that part of the output needed to replace the capital goods worn out in producing the output.

net exports   Exports minus imports.

net private domestic investment   Gross private domestic investment less consumption of fixed capital; the addition to the nation's stock of capital.

net worth   The total assets less the total liabilities of a firm or an individual.

New International Economic Order   A series of proposals made by the less developed countries (LDCs) for basic changes in their relationships with the advanced industrialized nations that would accelerate the growth of and redistribute world income to the LDCs.

no net cost programs   Price support programs in which producers or processors are assessed to finance the cost. The tobacco program is an example.

nominal gross domestic output (GDP)   The GDP measured in terms of the level of prices at the time of measurement (unadjusted for changes in the price level).

nominal income   The number of dollars received by an individual or group during some period of time.

nominal interest rate   The rate of interest expressed in dollars of current value (not adjusted for inflation).

nominal wage   The money wage (not adjusted for inflation).

nondurable good   A consumer good with a limited expected life (use), less than one year.

nonmarket transactions   The production of goods and services not included in the measurement of the gross domestic product because the goods and services are not bought and sold or are exchanged in the black market.

nonmerchandise trade balance   A comparison of the value of imports and exports of intangible goods such as services and fund transfers.

nonpoint pollutant   A pollutant that is generated at diffuse sources rather than from sources such as smokestacks or discharge pipes. Plant nutrients and pesticides that are washed from the soil surface or leached through the soil and sediment are examples.

nonprice competition   The means other than decreasing the prices of their products that firms employ to attempt to increase sales. This includes product differentiation, advertising, and sales promotion activities.

nonrecourse loans   Price support loans to farmers to enable them to hold their crops for later sale. The loans are nonrecourse because the commodity may be delivered to the government in full settlement of the loan.

nonrenewable resources   Resources such as minerals that exist in fixed quantities.

nontariff barriers (NTBs)   Any restraint on imports or exports other than a tariff, such as import quotas, licensing requirements, unreasonable product-quality standards, and unnecessary red tape in customs procedures.

normal good   A good or service of which consumers will purchase more (less) at every price when their incomes increase (decrease).

normal profit   Payment that must be made by a firm to obtain and retain managerial ability. It is the minimum payment (income) managerial ability must (expects to) receive to induce it to perform the managerial functions for a firm (an implicit cost).

normative economics   That part of economics that pertains to value judgments about what the economy should be like. It is concerned with economic goals and policies.

North American Free Trade Agreement   An agreement from Canada, Mexico, and the United States to significantly reduce barriers to trade among themselves.

NTBs   (See *nontariff barriers.*)

numeraire   The commodity in a commodity money that serves as a measure of value or unit of account.

official settlements   A term generally applied to international trade in which the outstanding financial accounts are settled by government-approved transactions.

oligopoly   A market in which a few firms sell either a standardized or differentiated product. Entry is difficult, the firm's control over the price at which it sells its product is limited by mutual interdependence (except when there is collusion among firms), and there is typically a great deal of nonprice competition.

oligopsony   A market in which there are a few buyers.

open access resources   Resources such as the ocean that can be used by anyone without restrictions and without regard to the effects on other users or potential users.

open economy   An economy that allows international trade, in contrast to a closed economy, that prohibits the purchase of goods from foreign markets or the sale of goods to these markets.

Open Market Committee   The twelve-member group that determines the purchase-and-sale policies of the Federal Reserve Banks in the market for U.S. government securities.

open-market operations   The market for government securities (the debt of the federal government). When the Federal Reserve Banks buy or sell government securities, they are conducting open-market operations. When the Fed buys securities, it increases the money supply; when it sells securities, it reduces the money supply.

opportunity cost   An implicit cost of using a resource to produce a given product that is equal to the payment that could be received if the resource were used in the production of another product.

optimal amount of externality reduction   That reduction of pollution or other negative externality where society's marginal benefit of damage reduction and marginal cost of reducing the externality are equal.

organic farming   Farming methods that use only organic fertilizers, natural fertilizer such as manure, and no (or minimal) inorganic agricultural chemicals and herbicides.

organizing   In management, organizing refers to the process of transforming plans into actions that are efficient and effective.

own-price elasticity of demand   Same as price elasticity of demand.

paid diversion program   A program that provides direct payments to farmers in return for diverting a specified amount of acreage of certain crops into conserving use.

paper money   Pieces of paper used as a medium of exchange; in the United States, Federal Reserve Notes.

parity A relationship that defines a level of price, income or purchasing power for farmers equal to an earlier base period.

partial equilibrium analysis The study of equilibrium prices and equilibrium outputs or employments in a particular market that assumes prices, outputs, and employments in the other markets of the economy remain unchanged.

partnership An unincorporated business firm owned and operated by two or more persons.

patent laws The federal laws that grant to inventors and innovators the exclusive right to produce and sell a new product or machine for seventeen years.

payment in kind (PIK) A program that provides payment to farmers in the form of commodities for reducing acreage of certain crops and placing that acreage in conserving uses. The term may also apply to export enhancement programs or other programs where payments are made in the form of commodities.

payment limitation A limit set by law on the amount of money any individual farmer may receive annually in farm program payments under the commodity price support programs.

pecuniary economies The ability of a firm to realize savings by purchasing at a lower price or selling at a premium, usually resulting from the size of the firm.

perfect information All buyers and all sellers having information on all products and their prices in a perfect market.

perfectly elastic demand A change in the quantity supplied requires no change in the price of the commodity. Buyers will purchase as much of a commodity as is available at a constant price.

perfectly elastic supply A change in the quantity demanded requires no change in the price of the commodity. Sellers will make available as much of the commodity as buyers will purchase at a constant price.

perfectly inelastic demand A change in price results in no change in the quantity demanded of a commodity. The quantity demanded is the same at all prices.

perfectly inelastic supply A change in price results in no change in the quantity supplied of a commodity. The quantity supplied is the same at all prices.

performance In the context of industrial organization, performance measures the extent to which the product price, product quality, profit, and sales expenses are similar to what could be expected in a competitive industry.

permanent legislation The laws upon which many agricultural programs are based. For the major commodities, the permanent legislation is the Agricultural Adjustment Act of 1938 and the Agricultural Act of 1949. These laws have been frequently amended for a given number of years but would again be in effect if current amendments are not enacted.

personal consumption expenditures The expenditures of households for durable and nondurable consumer goods and services.

personal distribution of income The manner in which the economy's personal or disposable income is divided among different income classes or different households.

personal income The pretax income, earned or unearned of households.

personal income tax A tax levied on the taxable income of individuals (households and unincorporated firms).

personal saving The personal income of households less personal taxes and personal consumption expenditures. It is the disposable income not spent for consumer goods.

pesticides Chemicals that are used to control insects (insecticides), weeds (herbicides), or fungi (fungicides) that damage or compete with crops.

Phillips Curve A curve that shows the relationship between the unemployment rate and the annual rate of increase in the price (inflation).

physical infrastructure   (See *infrastructure.*)

planned economy   An economy in which resources are allocated by a government bureaucracy.

planning   Charting a course for the future. In large firms, this often occurs in many subunits for several time periods. The several plans that must be integrated into an consistent plan for the entire organization.

plant   A physical establishment (land and capital) that performs one or more of the functions in the production (fabrication and distribution) of goods and services.

plant nutrients   Minerals and other substances required by plants for growth. The major nutrients required for crops are nitrogen, phosphorus, and potash.

point elasticity   The price elasticity of demand (supply) at a specific point on a demand (supply) curve.

policy   A course of action. Public policy refers to actions taken by a government body.

policy economics   The formulation of courses of action to bring about desired results or to prevent undesired occurrences (to control economic events).

positive economics   The analysis of facts or data to establish scientific generalizations about economic behavior.

positive externality   An agent generated by individuals or firms that has positive effects on other individuals or agents and for which those benefitted do not make payments determined in a market.

pound   One pound equals 0.454 kilograms or 16 ounces.

poverty   An existence in which the basic needs of an individual or family exceed the economic means to satisfy them at socially accepted levels.

prescriptive economics   An analysis of the economic consequences of alternatives or alternative actions or policies, often for the purpose of recommending which alternative is expected to generate the highest net returns or the best benefit-to-cost ratio.

price   The quantity of money (or of other goods and services) paid and received for a unit of a good or service.

price discrimination   The selling of a product to different buyers at different prices when the price differences are not justified by differences in the cost of producing the product for the different buyers. This practice is illegal when it reduces competition.

price elasticity of demand   The elasticity that expresses the relationship between a 1 percent change in a price and the corresponding percentage change in quantity demanded of a commodity.

price elasticity of supply   The ratio of the percentage change in quantity supplied of a commodity to the percentage change in its price; the responsiveness or sensitivity of the quantity of a commodity supplied to a change in the price.

price index   An index number that shows how the average price of a "market basket" of goods changes through time. A price index is used to change nominal output (income) into real output (income).

prices   (1) spot price, the current price at which a commodity can be purchased or sold; (2) forward price, the price at which a contract may be made with a company to buy or sell a specified quantity of a commodity at some specified time in the future; and (3) futures price, the price at which traders in a futures market (such as the Chicago Board of Trade) can buy or sell a specified quantity and grade (quality) of a commodity at some specified time in the future.

price leadership   An informal method that the firms in an oligopoly may employ to set the price of the product they produce: one firm (the leader) is the first to announce a change in price, and the other firms (the followers) quickly announce identical (or similar) changes in price.

price level   The weighted average of the prices paid for the final goods and services produced in the economy.

price support   The minimum price that government allows sellers participating in a program to receive for a good or service.

price taker   A seller (or buyer) of a commodity that is unable to affect the price at which a commodity sells by changing the amount it sells (or buys).

prime interest rate   The interest rate banks charge the most creditworthy borrowers, for example, large corporations with impeccable financing credentials.

private property   The right of private persons and firms to obtain, own, control, employ, dispose of, and bequeath land, capital, and other assets.

private sector   The households and business firms of the economy.

processing and marketing subsector   Firms and industries that buy farm products and transform them into products for consumption. The subsector includes functions of buying and selling, storing, manufacturing, transporting, packaging, freezing or preserving, advertising, wholesaling, and retailing.

producer subsidy equivalents   The level of subsidy that would be necessary to compensate producers (in terms of income) for the removal of government programs affecting a particular commodity.

product differentiation   Physical or other differences between the products produced by different firms that result in individual buyers preferring (so long as the price charged by all sellers is the same) the product of one firm to the products of the other firms.

product market   A market in which households buy and firms sell the products they have produced.

production control programs   A government program intended to limit production. At various times these programs have been referred to as reduced acreage, set-aside acreage, diverted acreage, acreage allotments, marketing quotas, payment in kind, soil banks, and conservation reserves.

production function   The production function describes the relationship between inputs and output. It is also called the *total physical product curve* (TPP).

production possibilities curve   A curve that shows the different combinations of two goods or services that can be produced in a full-employment, full-production economy in which the available supplies of resources and technology are constant.

production possibilities frontier (PPF)   (See *production possibilities curve.*)

production possibilities table   A table that shows the different combinations of two or more goods or services that can be produced in a full-employment, full-production economy in which the available supplies of resources and technology are constant.

productive efficiency   The production of a good in the least costly way. This occurs when production takes place on the production function.

productivity   A measure of average output or real output per unit of input. For example, the productivity of labor may be determined by dividing hours of work into real output.

profit   Economic profit and normal profit. Without an adjective preceding it, the term refers to the income of those who supply the economy with managerial ability.

profit-maximizing case   The circumstances that result in an economic profit for a firm when it produces the output at which economic profit is a maximum; when the price at which the firm can sell its product is greater than the average (total) cost of producing it.

profit-maximizing rule (combination of resources)   The quantity of each resource a firm must employ if its economic profit is to be a maximum or its losses a minimum. It is the combination at which the marginal revenue product of each resource is equal to its marginal resource cost (to its price if the resource market is competitive).

program yield   The yield for a crop on a given farm used to calculate deficiency payments. Program yields are based on history of past yields and on records of crop sales provided to the local ASCS office by the individual producer.

progressive tax   A tax structure under which the rate of taxation increases as income levels increase.

property rights   The rights of an individual or other owner to control the use of land and other property. The rights of the owner can be limited by government, that is, by eminent domain or by zoning restrictions.

property tax   A tax on the value of property (capital, land, stocks and bonds, and other assets) owned by firms and households.

public assistance programs   Programs that pay benefits to those who are unable to earn income (individuals with disabilities and dependent children). These programs are financed by general tax revenues and are viewed as public charity (rather than earned rights).

public debt   The total amount owed by the federal government (to the owners of government securities). It is equal to the sum of past budget deficits less budget surpluses.

public lands   Land owned by the federal, state, or local government and available for use of some or all individuals for governmentally approved activities.

public good   A good or service provided by government available to all without payment.

public policy   A policy is defined as a deliberate course of action followed by a public body, private firm, family, or individual. A public policy is made by individuals in their roles as citizens; in groups and organizations with a political objective or goal; and in participatory governments at local, state, or national levels.

public sector   The part of the economy that contains all its governments.

public utility   A firm has obtained from a government the right to be the sole supplier of the good or service in the area, and is regulated by that government to prevent the abuse of its monopoly power.

purchasing power parity   The idea that exchange rates between nations equate the purchasing power of various currencies. Exchange rates between any two nations adjust to reflect the price level differences between the countries, and the demand and supply of each currency.

pure competition   A market in which a very large number of buyers and sellers exchange a homogenous product. Entry is very easy, the individual seller has no control over the price at which the product sells, and there is perfect information.

pure monopoly   A market in which one firm sells a unique product (one for which there are no close substitutes), entry is blocked, the firm has considerable control over the price at which the product sells, and nonprice competition may or may not be found.

quantity demanded   The amount of a good or service buyers purchase at a particular price during some period of time.

quantity supplied   The amount of a good or service sellers offer (or a seller offers) to sell at a particular price during some period of time.

quota rent   The value of an import or a production quota, such as a sugar import quota or a production quota for tobacco or peanuts, assigned to a specific firm or producer.

ratchet effect   The tendency for the price level to rise when aggregate demand increases but not fall when aggregate demand declines.

rate of exchange   The price paid in one's own money to acquire one unit of a foreign money; the rate at which the money of one nation is exchanged for the money of another nation.

rate of interest   Payment solely for the use of money over an extended period of time (excluding any charges made for the riskiness of the loan and its administrative costs).

rational The behavior of individuals who make decisions to achieve the declared objective. It describes the behavior of a consumer who uses money income to buy the collection of goods and services that yields the maximum amount of utility and a producer who's decisions result in the maximum profits.

rational expectations theory The hypothesis that business firms and households expect monetary and fiscal policies to have certain effects on the economy. In pursuit of their own self-interests, they take actions that make these policies ineffective.

rationing function of price The ability of a price in a competitive market, by rising or falling, to equalize quantity demanded and quantity supplied and to eliminate shortages and surpluses.

real gross domestic product Gross domestic product adjusted for changes in the price level.

real income Nominal income adjusted for changes in the price level.

real interest rate The rate of interest expressed in dollars of constant value (adjusted for inflation or deflation). It equals the nominal interest rate less the rate of inflation or deflation.

real wage The nominal wage adjusted for changes in the price level.

recessionary gap The amount by which the aggregate expenditures schedule (curve) must increase (shift upward) to increase the real GDP to the full-employment, noninflationary level.

recombinant DNA technology A process by which genetic material from one species is transferred to another species, where it replicates and produces useful proteins and amino acids. The foreign DNA confers desirable new traits such as disease resistance or accelerated growth.

regressive A policy or program that has a greater effect on, or works to the disadvantage of, lower-income persons relative to higher-income persons. A food tax is regressive because poor people spend a higher portion of their income on food than do the rich.

regulatory agency An agency (commission or board) established by the federal or a state or local government to control for the benefit of the public. The regulations control the prices charged and the services offered (output produced) by natural monopolies, workplace conditions, pollutant discharges, food quality, and so forth.

renewable resource A resource that, with proper management, is regenerated. Examples include agricultural crops, forests, and fisheries, as well as naturally occurring resources such as the sun and rainwater.

rent-seeking behavior The pursuit through government of a transfer of income or wealth to a resource supplier, business, or consumer at someone else's or society's expense.

reserve ratio The specified minimum percentage of its deposit liabilities that a member bank must keep on deposit at the Federal Reserve Bank in its district, or in-vault cash.

reserve requirement The legal mandate that a bank must keep a minimum amount on reserve in a Federal Reserve Bank or as in-vault cash. The amount is determined by the Federal Reserve Board.

resource The capital, labor, and land resource used in productive activities. Capital resources include buildings and equipment, labor resources include management and production workers, and land includes renewable and nonrenewable natural resources.

resource market A market in which households sell and firms buy the services of resources.

salinization The buildup of salt in soils, usually due to the evaporation of irrigation water.

saving schedule Schedule that shows the amounts households plan to save or plan not to spend for consumer goods, at different levels of disposable income.

savings The part of income that households do not spend for consumption or use to pay taxes.

savings account   A deposit in a depository institution that is interest-earning and can normally be withdrawn by the depositor at any time.

savings and loan association   A firm that accepts deposits, primarily from small individual savers, and lends primarily to individuals to finance purchases of residences.

scarce productive resources   The fixed (limited) quantities of land, capital, labor, and managerial ability that are never sufficient to satisfy the unlimited wants of humans.

sedimentation   The deposition of eroded soil (in ditches, waterways, lakes, harbors, etc.).

seigniorage   The difference between the value of money and the cost of producing it.

service   That which is intangible (invisible) and for which a consumer, firm, or government is willing to exchange something of value.

set-aside acreage   The acreage a farmer must hold out of production in order to qualify for the federal price-support program, compensatory payments, or other benefits.

shift in supply or demand   A movement of the supply or demand curve as a result of changes in production technology or consumer tastes and preferences. It is sometimes referred to as a change in supply or demand. A shift (or change) is to be distinguished from a change in the quantity supplied or quantity demanded as the price of a product changes, resulting in movement along the curve.

short run   The period of time in which certain costs, called *fixed costs,* cannot be avoided even by shutting down the firm. In the long run, all costs are variable.

short-run aggregate supply   The aggregate supply curve relevant to a time period in which input prices (particularly nominal wages) remain constant when the price level changes.

short-run competitive equilibrium   The price at which the total quantity of a product supplied in the short run by a purely competitive industry and the total quantity of the product demanded are equal. It is equal to or greater than the average variable cost of producing the product.

short-run supply curve   For the firm the supply curve is the MR curve above the AVC curve. For an industry the short-run supply curve is the horizontal summation of the firm supply curves.

shortage   The amount by which the quantity demanded of a product exceeds the quantity supplied at a given (below-equilibrium) price.

slippage   A term used to describe the operation of economic programs, such as a farm production-control program, in which the program fails to fully achieve its goal because of substitutions of inputs or products or some other adjustment by program participants.

slope of a line   The ratio of the vertical change (the rise or fall) to the horizontal change (the run) in moving between two points on a line. The slope of an upward-sloping line is positive, reflecting a direct relationship between two variables; the slope of a downward-sloping line is negative, reflecting an inverse relationship between two variables.

small-country assumption   The assumption that a country's international transactions of a good are so small that changes have no effect on the world price.

social costs of imperfect competition   The payment by consumers of higher prices for a smaller quantity of production. It arises out of the downward-sloping (or kinked) demand curve faced by sellers in imperfect competition.

social indifference curve   An indifference curve that includes all the combinations of two goods toward which a society is indifferent.

socialism   An economic system in which most of the means of production other than labor are publicly owned.

socially optimal price   The price of a product that results in the most efficient allocation of an economy's resources and that is equal to the marginal cost of the last unit of the product produced.

sodbuster programs   A provision designed to prevent the plowing up of highly erodible range or pasture land for the planting of grain crops. If highly erodible land is tilled without appropriate conservation measures, producers may lose eligibility for many federal agricultural support programs.

soil bank program (SBP)   A program operated in the 1950s to achieve objectives of both soil conservation and production control. Under the program, farmers signed contracts for varying periods of time to place part of their acreage into conserving uses.

soil conservation   The practices adopted by farmers, and others, to reduce the loss of soil due to erosion by water or wind.

Soil Conservation Service (SCS)   An agency of the federal government that provides technical and other assistance to land owners to control erosion and maintain the quality of land and water resources. This agency has been renamed the Natural Resource Conservation Agency.

sole proprietorship   An unincorporated business firm owned and managed by one person.

solid wastes   The materials that are disposed of by households and firms including packaging materials, used goods, yard wastes, and materials from razed buildings.

space utility   The utility created by firms that transport raw materials, intermediate products, and final goods to locations where they are desired by buyers.

specialization   The use of the resources of an individual, a firm, a region, or a nation to produce one or a few goods and services.

stages of the production function   Classification of the production function into three stages: in Stage I, marginal physical product is above average physical product; in Stage II, marginal physical product is below average physical product but positive; and in Stage III, marginal physical product is negative. The rational producer will operate in Stage II.

stagflation   Inflation accompanied by stagnation in the rate of growth of output and a high unemployment rate in the economy; simultaneous increases in both the price level and the unemployment rate.

standard metropolitan statistical area (SMSA)   A geographic area defined by the U.S. Bureau of the Census in terms of total population. A certain minimum population is required for an area to qualify as an SMSA.

state farm   A farm in which all productive resources are owned by the government and the workers are employees.

stocks   A stock of a resource refers to the total quantity of that resource. Savings is a stock that earns a *flow* of income in the form of interest payments. A nation uses a *stock* of resources—the number of productive workers, the amount of capital equipment, the acres of cropland—to generate a *flow* of income, called the *gross domestic product* or the *national income.*

strategic goods   A good that is essential for a nation to maintain its position in a military hierarchy of nations; any good that is essential for health, prosperity, and the like.

structural deficit   The difference between national tax revenues and expenditures when the economy is at full employment with expenditures exceeding revenues.

structural transformation   The changes in economic activities in an economy as it moves from a predominantly agrarian society to a predominantly industrial and service society.

structural unemployment   Unemployment caused by changes in the structure of demand for consumer goods and in technology.

structure   The structure of an industry refers to the number, size, and thus market share of the firms in an industry.

subsidy   A payment of funds (or goods and services) by a government, business firm, or household for which it receives no good or service in return.

substitute goods   Goods or services such that there is a direct relationship between the price of one and the demand for the other. When the price of one falls (rises), the demand for the other decreases (increases).

subsitutes   Goods or production inputs for which an increase in the price of one good results in an increase in demand for the other good.

substitution effect   (1) The effect of a change in the price of a consumer good on the relative expensiveness of that good and the resulting effect on the quantity of the good a consumer purchased if the consumer's real income remained constant; (2) the effect of a change in the price of a resource on the quantity of the resource employed by a firm if the firm did not change its output.

superior goods   Those goods that experience increases in demand when income increases that have a income elasticity of demand greater than one.

supplementary   In the context of a firm producing several products, supplementary products are produced using different resources owned by the firm or using the same resources at different times of the year.

supply curve   Connects all the points showing a price and the quantity supplied at that price by an individual or group. The supply curve is upward-sloping because as the price of a good increases the amount of the good supplied increases.

supply elasticity   The percentage change in the quantity supplied associated with a 1 percent change in price.

supply management   A term employed to describe government programs used to influence and control the supply of a commodity to maintain a desired price.

supply schedule   A schedule that shows the amounts of a good or service sellers (or a seller) will offer at various prices during some period of time.

surplus   The amount by which the quantity supplied of a product exceeds the quantity demanded at a given (above-equilibrium) price.

swampbuster bills   Legislation that places restrictions on the draining of natural wetlands for crop production.

tangent   A line that touches, but does not intersect, a curve.

target price   A price for certain crops established by law. If the average market price does not equal the target price, qualifying farmers receive a deficiency payment to make up part or all of the difference. Generally, deficiency (or target price) payments are made if average market prices are below the target price.

targeted export assistance   Subsidy programs designed to increase exports to specific countries.

tariff   A tax imposed on an imported good. *Ad valorem* tariffs are imposed as a percentage of the value of the product. *Specific* tariffs are imposed as a fixed amount per unit.

tariff-rate quota   A tariff system under which a specified amount of a product may be imported at a low or zero tariff. A prohibitively high tariff is set for any amount of imports greater than the "quota" limit.

tax incidence   The income or purchasing power that different persons and groups lose as a result of the imposition of a tax.

taxes   Taxes are collected from businesses and households by various levels of government. Taxes can be either "direct"—that is, taken from the income flow—or "indirect"—taken from transactions on goods and services.

technical efficiency   The extent to which a firm achieves the maximum possible amount of output from the inputs used in the production process.

technological change   Changes in the techniques of production of goods and services which result from experimentation, research, and development. Changes that improve productivity shift the production function upward and the production possibilities frontier outward.

technology   The body of knowledge used to produce goods and services from economic resources.

tenure   The relationship or type of control that a farmer has on the land: owner, part owner, or tenant.

terms of trade   The rate at which units of one product can be exchanged for units of another product; the price of a good or service; the amount of one good or service that must be given up to obtain one unit of another good or service.

tiering   A method of directing benefits of federal price support programs toward smaller or medium-sized farms. Under tiering, deficiency payments per unit of production would be higher for a limited number of bushels and lower for quantities beyond that amount.

time deposit   An interest-earning deposit in a depository institution that may be withdrawn by the depositor without a loss of interest on or after a specific date or at the end of a specific period of time.

time horizon   The planning period used by a firm or household when making economic decisions.

time utility   The utility created by firms that store products for future use.

ton   One ton (metric) equals 1.102 tons U.S.; One ton (U.S.) equals 0.907 ton metric or 2,000 pounds.

total costs (TC)   The sum of total variable costs and total fixed costs.

total demand   The demand schedule or the demand curve of all buyers of a good or service.

total demand for money   The sum of the transactions demand for money and asset demand for money; the relationship between the total amount of money demanded, nominal GDP, and the rate of interest.

total fixed costs (TFC)   The costs that do not change in the short run when the amount of output produced changes.

total physical product (TPP)   The total output of a particular good or service produced by a firm (or a group of firms or the entire economy).

total revenue   The total receipts received by a firm (or firms) from the sale of a product and that equals the quantity sold (demanded) multiplied by the price at which it is sold.

total revenue-total cost approach   The method that finds the output at which economic profit is a maximum or losses a minimum by comparing the total revenue and the total costs of a firm at different outputs.

total variable cost (TVC)   The cost of variable inputs or resources.

trade balance   The export of merchandise (goods) of a nation less its imports.

trade deficit   The amount by which a nation's imports of merchandise (goods) exceed its exports.

trade policies   Usually refers to rules, postures, and laws adopted by nations to influence international trade or to regulate imports and exports of certain goods. Trade policies are subject to negotiation between and among nations or in international organizations. (See World Trade Organization.)

trade surplus   The amount by which the value of a nation's merchandise exports exceeds the value of its imports.

trading possibilities line   A line that shows the different combinations of two products an economy is able to obtain (consume) when it specializes in the production of one product and trades (exports) this product to obtain the other product.

traditional economy An economic system (method of organization) in which traditions and customs determine how the economy will use its scarce resources.

tragedy of the commons The excessive use of an open-access (not common property) resource to the point that the resource is damaged or destroyed.

transactions demand for money The amount of money people want to hold to use as a medium of exchange (to make payments).

transfer payment The disbursement of money (or goods and services) usually by government for which no currently produced good or service is received in return.

twin deficits This term became popular in the 1980s when both the federal budget deficit and the trade deficit grew at roughly the same pace and then shrank together. The two deficits diverged in the late 1980s when the trade deficit continued to shrink while the budget deficit expanded again.

two-price plan A plan that involves supporting that part of production used in the domestic market at one price and selling the remainder for export at world prices.

underemployment Failure to produce the maximum amount of goods and services that can be produced from the resources employed; that is, failure to achieve full production.

unemployment Failure to use all available economic resources to produce goods and services; that is, failure of the economy to fully employ its labor force or other resources.

unemployment rate The percentage of the labor force that is unemployed.

unitary elasticity The elasticity coefficient is equal to one; the percentage change in the quantity (demanded or supplied) is equal to the percentage change in price.

United States-Canadian Free-Trade Agreement An accord signed in 1988 to eliminate all trade barriers between the two nations over a ten-year period.

unlimited wants The insatiable desire of consumers (people) for goods and services that will give them pleasure or satisfaction.

utility The want-satisfying power of a good or service, that is, the satisfaction or pleasure a consumer obtains from the consumption of a good or service (or from the consumption of a collection of goods and services).

utility-maximizing rule To obtain the greatest utility, the consumer should allocate money income so that the last dollar spent on each good or service yields the same marginal utility. Thus the marginal utility of each good or service divided by its price is the same for all goods and services.

utils A hypothetical quantitative measure of the utility provided by a good or service. (It is impossible to actually measure utils.)

value added The value of the product sold by a firm less the value of the goods (materials) purchased and used by the firm to produce the product. It is equal to the revenue that can be used for wages, rent, interest, and profits.

value-added approach to measure GDP A method for measuring GDP based on the value of all products sold by firms less the value of all intermediate inputs used by firms to produce goods or services.

value-added tax A tax imposed on the difference between the value of the goods sold by a firm and the value of the goods purchased by the firm from other firms.

value judgment Opinion of what is desirable or undesirable. It is a belief regarding what ought or ought not to be.

value of money The quantity of goods and services for which a unit of money (a dollar) can be exchanged; the purchasing power of a unit of money; the reciprocal of the price level.

variable costs Costs that vary with the level of production of the firm, applicable to both the short and the long run.

variable factors of production   Productive inputs that are changed to adjust the level of output produced in the short run.

variable input (resource)   An input employed by a firm the quantity of which can be increased or decreased (varied).

vault cash   The currency a bank has in its safe (vault) and cash drawers.

velocity of money   The number of times per year the average dollar in the money supply is spent for final goods.

vertical integration   A group of plants engaged in different stages of the production of a final product and managed by a single firm.

vertical intercept   The point at which a line meets the vertical axis of a graph.

vertical merger   The merger into a single firm of two or more firms engaged in different stages of the production of a final product.

visibles   Tangible goods that are included in the merchandise trade balance.

voluntary export restrictions   The limitations by firms of their exports to particular foreign nations to avoid the erection of other trade barriers by the foreign nations.

wage   The price paid for labor (for the use or services of labor) per unit of time (per hour, per day, etc.).

wants   A concept of need or desire generally without content in economics, unless related to value such as value in exchange.

wastes of monopolistic competition   The waste of economic resources resulting from producing an output at which price is more than marginal cost and average cost is more than the minimum average cost.

wealth effect   The tendency for increases (decreases) in the price level to lower (raise) the real value (or purchasing power) of fixed incomes or assets (*ceteris paribus*) and thus to reduce (expand) total spending in the economy (same as real-balances effect).

World Bank   A bank that lends (and guarantees loans) to less developed nations to assist their growth. It is formally known as the International Bank for Reconstruction and Development.

world market   The world supply and demand of a good or service and the resulting world price. The market in which international trade occurs.

world price   The price at which commodities are purchased and sold in markets that move commodities among nations; the cost, insurance, and freight price of an imported commodity at a principal port.

World Trade Organization (WTO)   A multilateral agreement originally signed by ninety-two countries then called the General Agreement on Tariffs and Trade (GATT) which establishes rules and guidelines for regulating world trade among members and a forum for countries to discuss and resolve trade disputes. An underlying principle is that trade should be restricted only through the use of uniformly applied tariffs.

zoning   The establishment of rules restricting the rights to use land for specified purposes. In metropolitan areas, zones are designated for industrial, commercial, residential, and other uses.

INDEX